High-T_c Superconductivity

High-T_c Superconductivity

The Excitonic Coulomb Perspective

Nie Luo
George H. Miley

CRC Press
Taylor & Francis Group
Boca Raton London New York

CRC Press is an imprint of the
Taylor & Francis Group, an **informa** business

High-T_c Superconductivity: The Excitonic Coulomb Perspective

First published 2021 by Jenny Stanford Publishing Pte. Ltd.

Published 2023 by CRC Press
Taylor & Francis Group
6000 Broken Sound Parkway NW, Suite 300
Boca Raton, FL 33487-2742

ISBN 978-981-4800-57-0 (hbk)
ISBN 978-0-429-27870-9 (ebk)

Visit the Taylor & Francis Web site at
http://www.taylorandfrancis.com

and the CRC Press Web site at
http://www.crcpress.com

British Library Cataloguing-in-Publication Data
A catalogue record for this book is available from the British Library.

Dedicated to

J. Bardeen, V. L. Ginzburg, L. P. Gor'kov, and P. W. Anderson

Contents

Preface

The exciton mechanism of high-critical-temperature superconductivity in copper oxides was initially proposed by Professor John Bardeen. His insight is largely shared by another luminary in superconductivity, Prof. V. L. Ginzburg. The primary author of this book, Dr. Nie Luo, was motivated by their insights to give a geometrical explanation to the excitonic Coulomb interaction. Following on with the new approach, Dr. Luo has developed a unique formalism to understand and predict physical properties of high-T_c superconductors. This work is supported by an increasing number of strong evidences for electron-hole interactions in p-type cuprates. The presence of electrons in hole-doped cuprates has been revealed from the works of the authors and many others, including the late Professor L. P. Gor'kov. The book also tries to understand the interlayer Coulomb (ILC) pairing model by the excitonic Coulomb interaction. Developed by Prof. A. J. Leggett, the ILC theory shares many views with Ginzburg's approach. The other author of the book, Professor G. H. Miley, shares with us his personal interactions in the late 1980s with Prof. Bardeen on the exciton's role in a variety of physics problems, including high-T_c cuprate superconductivity.

The results and predictions of this excitonic Coulomb mechanism have been verified by an increasing number of experiments. This book serves to summarize the current status and fathom the directions to come.

Nie Luo
George H. Miley
Urbana, Illinois
October 2020

Chapter 1

Excitonic Coulomb Pairing in a Variational Approach

1.1 Introduction

This book is about a pairing mechanism for the high critical temperature (T_c) superconductivity in cuprates, with focus on the p-types. Specifically we have been developing a theoretical framework in which the pairing boson is a Coulomb bound state of an electron and a hole. This energy-bound electron–hole (e-h) pair in many ways is similar to the exciton often encountered in semiconductor physics, but with some significant differences. The most notable difference is that the e-h pair in cuprates is thermodynamically stable versus a finite lifetime for excitons in semiconductor structures.

Historically, after the establishment of the Bardeen–Cooper–Schrieffer (BCS) theory (Bardeen et al., 1957) in the early 1960s, there had been investigations as to how high the T_c could reach with the phonon-mediated superconductivity: A widely circulated and well-regarded estimate was an upper limit of $\approx$30 K (McMillan, 1968). As a result, attempts were made to explore intermediate bosons of lighter constituency and higher energy scale than phonons. Hence came the exciton on the stage of high-T_c

High-T_c Superconductivity: The Excitonic Coulomb Perspective
Nie Luo and George H. Miley

ISBN 978-981-4800-57-0 (Hardcover), 978-0-429-27870-9 (eBook)
www.jennystanford.com

superconductivity, as in models proposed by Little, Ginzburg and Bardeen (Allender et al., 1973; Ginzburg, 1970; Little, 1964). In Little's version (Little, 1965), the pairing 'glue' is the attractive Coulomb force associated with electronic excitation in a conjugated organic dye. Since the fluorescence (an electronic de-excitation) in organic dyestuffs is intimately associated with the recombination of optical excitons (*e*-*h* pairs), Little's pairing scheme is legitimate to be referred to as an exciton superconductivity. In the Little model, the exciton is more or less local, typically confined to the size of average fluorophore molecules (e.g., diethylcyanine iodide). Later, Ginzburg, Bardeen and their collaborators generalized this idea to more practical multilayer structures constructed of alternating metal and dielectric (semiconductor) layers. The initial experimental verification was largely indecisive in the 1970s (Ketterson, 2000). However, the discovery of high T_c reveals cuprates that bear striking similarity to Ginzburg and Bardeen's proposition in their quasi-2D (two-dimensional) layered structures, though with thinner layers to the atomic angstrom scale.

With Bardeen's passing away in the early 1990s, the exciton model received less attention than it deserves. This is the model that Bardeen utilized to explain the high T_c of cuprates, echoed by similar approaches of the Ginzburg school.

Then starting from the mid 1990s, the author of the current Chapter, Dr. Nie Luo, came up with a geometrical interpretation of the exciton pairing in terms of 3 electrons and 1 hole ($3e+1h$ or vice versa $3h+1e$) that puts identical particles on an equal footing (Luo, 2001b). The positive identification of attractive interaction indicated the potential and rich physics in the exciton models that were ushered by Bardeen and Ginzburg. Various theoretical approaches were attempted to overcome the limitations of the early exciton models (Luo, 2001a; Luo and Miley, 2002, 2009).

1.2 The Excitonic Coulomb Mechanism

We call this pairing mechanism *excitonic Coulomb* or ECM due to three considerations: (1) to emphasize the importance of the Coulomb force, particularly the long range part (over a few lattices' distance) that is important in cuprates; (2) to emphasize that the attractive part of the Coulomb is from the electron–hole interaction like that of an exciton; (3) to emphasize that the repulsive Coulomb force also plays a role for superconductivity, avoiding condensation into an excitonic solid or excitonic insulator. The balance of the attractive and repulsive Coulomb force is known to be important for exciton complexes, such as trions and biexcitons, to name a few.

The three aforementioned considerations are based on experiments and theoretical considerations. Leggett has shown (Leggett, 1998, 1999a,b) that the cuprate pairing should include inter-particle Coulomb interactions, particularly in its long range part. This is for one example revealed from the coherence length: The on-site interaction by itself hardly explains the in-plane size of the Cooper pair, that spans more than 4 lattice units in cuprates. While pairing by the reduction in kinetic energy has not been able to explain the experiments, repulsion-only models have not shown a conclusive evidence of the off-diagonal long range order (ODLRO) either. In comparison, for models with proper attractive interaction, off-diagonal long range order is an expected result (Yang, 1989; Yang and Zhang, 1990). Properly formulated, the excitonic Coulomb model shares important common features with some exciton mechanisms (Leggett, 2006). The follow-on sections demonstrate that the delicate balance between the repulsive and attractive Coulomb produces the well-known doping-temperature phase diagram for high-T_c cuprates.

Nevertheless, "excitonic" here has a distinction from prior models (Allender et al., 1973; Ginzburg, 1970; Little, 1964). Their excitons are virtual or short-lived excitations that are penalized by the additional energy needed to form in semiconductor-like structures, hence not stable against recombination, leading to a much weakened pairing. In ECM in contrast, excitonic complexes are formed from semimetallic band-edges (or Fermi surface portions) or between spatially separated layers. Therefore, the excitonic complexes are thermodynamically stable. Semimetallic band edges are well known in cuprates of multi-CuO_2-plane, simplified here as multiplane or multilayer, meaning that each primitive unit cell has two or more closely coupled CuO_2 planes, at a separation $\sim$3.1 Å. The semimetallic band edge occurs near the $(0, \pi)$ saddle point, due to interlayer coupling from the close proximity of CuO_2 planes, for multilayer cuprates. This coupling can be understood in part from the traditional band theory.

For single-layer cuprate, a semimetallic band edge still persists near $(0, \pi)$. However, this time it is due to strong Coulomb correlation across the unit cell, at a distance as large as 6 Å or more. It is not from a traditional band-splitting picture. We will show that the semimetallic band-edges are due to the charge-transfer effect from a Coulomb instability, which in turn relates to the excitonic Coulomb interaction and pairing, in Chapter 3 and particularly in Section 3.8.

We described some preliminary considerations of this model some 10 years ago (Luo and Miley, 2009) and termed it exciton-like. Now we can use a more appropriate term in the form of excitonic Coulomb mechanism (ECM). Despite the terminology difference, the basic logic is still the same.

A strong correlation model typically treats interaction/correlation in different orders. The electron and hole in this pairing models can be considered initially from a nearly free electron band-structure. They are more appropriately band electrons and band holes, without correlation effects. Then, the correlation is tackled in increasing strength. In practice this transpires to the following steps. First, we neglect the weaker inter-electron interactions, but consider only the stronger electron–ion Coulomb interaction to get a rough picture of energy states: This is essentially an independent-electron band structure calculation. Second, we then turn on the inter-electron interactions but neglect strong correlation effects: This corresponds to a density functional theory (DFT) self-consistent field (SCF) iteration of the band structure, using correlation functionals like LDA/GGA. This will be discussed in Chapter 4. Third, the strong Coulomb correlation is then turned on near the Fermi energy, with many-body techniques (Chapter 3) or few-body correlation functions (Section 1.14) to see if it brings in qualitative changed to the energy spectrum (e.g., spectral density, self energy and renormalization effects). Of course, other correlation effects may still be left out at the third stage and we then tackle these in the next order.

We should make a subtle but often qualitative difference between electron/hole in this excitonic Coulomb picture and those holelike or electronlike quasiparticle excitations from the BCS superconducting ground state. The electron/hole excited from the BCS ground state do not interact, at least in the first order. They are collective excitations *after* the pairing interaction. In the ECM, the electron/hole are quasiparticles from the (independent or LDA/GGA) band, *before* any strong correlation. In retrospect, such a semantic discrepancy might have caused confusion from the alleged double counting problem against a typical exciton model (Inkson and Anderson, 1973). Here we need to clarify it first before delving into mathematical details: Make a *difference* between the band electron/hole and the electron/hole (excitations) from the SC condensate, or other correlated collective ground-states.

High-T_c mechanisms based on excitons are aesthetically appealing: An exciton is in general more "mobile," involving energy scale higher than that of a phonon and hence should naturally lead to higher T_c. Also notable is that it was the first theory to show the possibility of 100 K plus T_c as a viable electronic mechanism. The theory predicted that transition temperatures of some 100 K or more was achievable, and moreover that this is likely to occur in a strongly layered (quasi-two-dimensional) system. As is well-known now, both predictions were verified in the late 1980s, in the perovskite type cuprates whose most notable structural feature is their semimetallic CuO_2 planes intervened by much less conducting layers such as BaO, LaO, TlO, BiO, etc.

More than 30 years after the initial discovery of high-T_c superconductivity, although there is still a lack of consensus on its mechanism, experimentally we are at a much clarified front: (1) The close proximity of superconductivity to recently established charged density order put the long range Coulomb interaction in a commanding position; (2) high-resolution ARPES and Shubnikov–de Haas oscillations routinely found electron pockets (perhaps better termed as segments) in (predominantly) hole-type cuprates and vice versa; (3) transport experiments clarified the early confusion due to the alleged discrepancies in transverse/longitudinal rates (in turn due to the alleged charge-spin separation); (4) The pseudogap phenomenology and preformed pairs are well established. All these can be readily understood from the excitonic Coulomb interaction.

Besides cuprate superconductors, there are two notable categories discovered in the 21st century, namely the diborides (Nagamatsu et al., 2001) and the iron pnictides (Kamihara et al., 2008). More recently there are reports of high T_c in alleged H_2S and the older chalcogenide superconductors may also be mentioned. In Chapter 5 we show that the ECM theory helps, at least in part, to understand these categories of SC.

All of these *unconventional* SC categories are known to have both electronlike and holelike bands near the Fermi level and therefore the ECM could play a role. To the least degree, the ECM should help reduce the Coulomb repulsion between like charges. Therefore, it at least helps to achieve superconductivity together with other mechanisms. This will be exposed in our attempts to predict high-T_c in new materials systems, in Chapter 5.

The exciton mechanism, cast in the approach of canonical BCS (Bardeen–Cooper–Schrieffer), or the Allender–Bray–Bardeen (ABB) scheme, is difficult to tackle experimental phenomena such as preformed pairs, the pseudogap, peculiar normal state properties and many others. Besides, the virtual semiconductor-type exciton envisioned in these approaches is at a conceptual dilemma: The energy penalty involved in forming such an intermediate boson makes it thermodynamically an unlikely "glue" of superconductivity.

The purpose of this chapter is to study the excitonic Coulomb mechanism in a different approach, in a hope to avoid the problems involved in the BCS. For the purpose of convenience and brevity, we may just talk about the pairing of two electrons (or two holes), although high-T_c pairing in cuprates happens in both ways. The generalization is simple, by a simple flip of terms: $e \leftrightarrow h$.

Of course, there have been many theoretical attempts on HTSC besides the exciton-based approach. Many of them starts from the spin degrees of freedom, for example, the spin bag model (Schrieffer et al., 1988)of Schrieffer, Wen and Zhang, the nearly antiferromagnetic Fermi liquid (NAFL) (Pines, 1994) of Monthoux and Pines, to name a few. Since spin correlations (e.g., magnetism) are actually the result of Coulomb interaction (often with its covariant relativistic effects too) and the Pauli exclusion principle, the ECM approach of starting directly from the Coulomb potential and Pauli principle does not necessarily contradict to the spin-based models.

1.3 How to "Glue" Two Holes with an Exciton

Before delving into advanced (yet often tedious) calculations like the diagram techniques often associated with prior approaches in exciton superconductivity, can we understand from a simple yet intuitive picture, and from the most basic physics, how an exciton mediates an effective attraction to two like charges? It turns out to be quite lucid as shown below in Fig. 1.1.

The effective attraction effected by an exciton to two other holes is due to (1) geometry and (2) the inverse square Coulomb law.

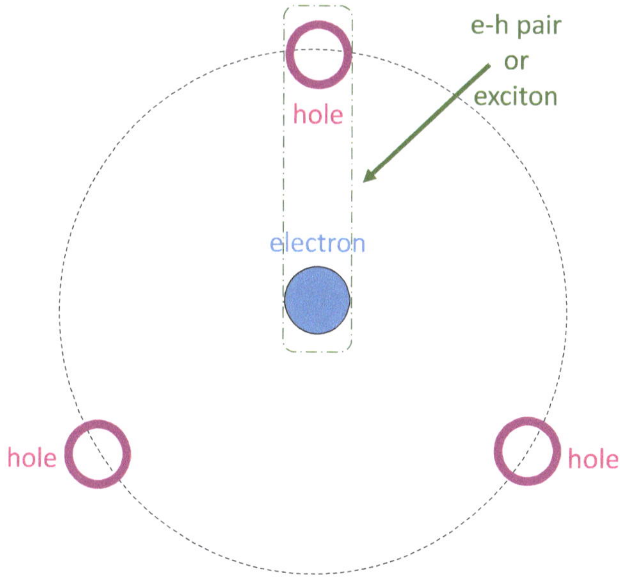

Figure 1.1 The geometry showing the interaction of two holes mediated by an exciton, as to why the inverse-square Coulomb law should give rise to an effective attraction. The green labeling indicates the exciton. The dashed circle is a guide to the eye. The attractive interaction is non-retarded. The exciton can similarly mediate attraction to two electrons as in Fig. 1.3.

It is well known in the superconductivity community that a highly polarizable media should strongly reduce the Coulomb repulsion of like charges. And over-screening is often mentioned in the literature too. However, Fig. 1.1 really explains why a neutral physical entity can *actually* result into *net attraction* to two like charges, purely due to the electrostatic Coulomb interaction. This is probably the simplest yet clear enough illustration of over-screening. Compared with ordinary phonons, the attraction is not retarded, and it can be termed purely electronic in that lattice dynamics does not play a (direct) role, at least in the first order. A valuable lesson gained from this example is that geometry should play important role in understanding the pairing mechanism, and in general, strong correlation effects.

In contrast, with the perturbative expansion of many-body diagram technique, much of the geometric information is lost. This is in part due to the use of a planewave basis set. Therefore, in the latter part of this Chapter we will formulate our theory based on real-space variational methods, which utilize correlated basis functions. This will be discussed in Section 1.10.2. Before we get fully involved in the details of variational calculation, a bit more of exposure to the cuprate phenomenology related to the key fortes and issues of the ECM and prior models, is highly helpful.

1.4 Exciton Mechanism in Its Prior Approach

Although the word "exciton" or alternatively "excitonic" is often used in the literature, there are slight differences among different theorists on the exact definition of such a mechanism. Little (Little, 1964), for example interprets it as an electronic excitation of an organic instead of relying on the Mott–Wannier or Frankel excitons, while the intermediate boson in the model (Allender et al., 1973) by Allender, Bray and Bardeen (ABB) are actually excitons in semiconductors, though being virtual. Here we discuss this mechanism roughly in line with ABB's interpretation although our excitons are not limited to those in semiconductors. More importantly, an exciton in a typical semiconductor is short-lived (Kittel, 1996), susceptible to recombination around the order of

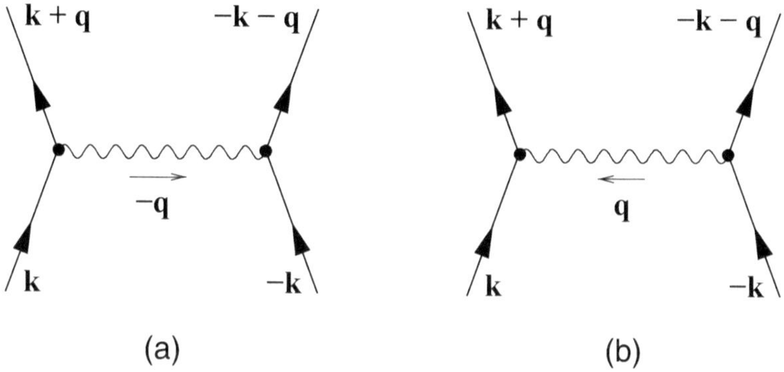

Figure 1.2 Feynman diagrams showing the interaction of two electrons mediated by an exciton, to the second order in the electron–exciton coupling constant.

submicrosecond (maybe up to roughly a millisecond depending on the band structure). The exciton in our excitonic Coulomb model, is an electron–hole bound state in a semimetal. It is stable against recombination due to reasons that we will explain later.

The formalism developed by the ABB paper basically follows that of BCS, with the phonon replaced by an exciton. To find out the attractive action mediated through the intermediate boson, BCS starts from the initial state with two electrons, respectively, in the eigenstates $\mathbf{k}$ and $-\mathbf{k}$, which are later scattered by an intermediate boson of momentum $\mathbf{q}$, into the final state made of $\mathbf{k}'$ and $-\mathbf{k}'$. One often resorts to diagrammatic techniques similar to Feynman diagrams shown in Fig. 1.2.

A second-order perturbation theory gives the effective interaction mediated by excitons

$$\begin{aligned} V_{\text{ex}} = \sum_q \frac{1}{2} \langle \mathbf{k}, -\mathbf{k} | \mathbf{H}_{\text{ex}} | \mathbf{k}', -\mathbf{k}, -\mathbf{q} \rangle \\ \frac{1}{E_{\mathbf{k},-\mathbf{k}} - E_{\mathbf{k}',-\mathbf{k},-\mathbf{q}}} \langle \mathbf{k}', -\mathbf{k}, -\mathbf{q} | \mathbf{H}_{\text{ex}} | \mathbf{k}', -\mathbf{k}' \rangle \\ + \frac{1}{2} \langle \mathbf{k}, -\mathbf{k} | \mathbf{H}_{\text{ex}} | \mathbf{k}, -\mathbf{k}', \mathbf{q} \rangle \\ \frac{1}{E_{\mathbf{k}',-\mathbf{k}'} - E_{\mathbf{k},-\mathbf{k}',\mathbf{q}}} \langle \mathbf{k}, -\mathbf{k}', \mathbf{q} | \mathbf{H}_{\text{ex}} | \mathbf{k}', -\mathbf{k}' \rangle, \end{aligned} \quad (1.1)$$

where $\langle \mathbf{k}, -\mathbf{k}|$ is the initial state in its conjugate form, with the $|\mathbf{k}', -\mathbf{k}'\rangle$ the final state. $|\mathbf{k}', -\mathbf{k}, -\mathbf{q}\rangle$ and $|\mathbf{k}, -\mathbf{k}', \mathbf{q}\rangle$ are two intermediate states. A reasonable and often used assumption is that the electron–exciton coupling is isotropic in $\mathbf{k}$, $\mathbf{k}'$, and $|\mathbf{k}| = |\mathbf{k}|$, matrix elements are now

$$\begin{aligned} &\langle \mathbf{k}, -\mathbf{k}|\mathbf{H}_{\text{ex}}|\mathbf{k}', -\mathbf{k}, -\mathbf{q}\rangle\langle \mathbf{k}', -\mathbf{k}, -\mathbf{q}|\mathbf{H}_{\text{ex}}|\mathbf{k}', -\mathbf{k}'\rangle \\ &= \langle \mathbf{k}, -\mathbf{k}|\mathbf{H}_{\text{ex}}|\mathbf{k}, -\mathbf{k}', \mathbf{q}\rangle\langle \mathbf{k}, -\mathbf{k}', \mathbf{q}|\mathbf{H}_{\text{ex}}|\mathbf{k}', -\mathbf{k}'\rangle \\ &= |M_q|^2. \end{aligned} \tag{1.2}$$

Also under these assumptions, $E_{\mathbf{k},-\mathbf{k}} - E_{\mathbf{k}',-\mathbf{k},-\mathbf{q}} = E_{\mathbf{k}',-\mathbf{k}'} - E_{\mathbf{k},-\mathbf{k}',\mathbf{q}} = -\hbar\omega_q$, where $\hbar\omega_q$ is the energy of exciton with wavevector $\mathbf{q}$. Therefore, we obtain

$$V_{\text{ex}} = \sum_q \frac{-2}{\omega_q} \frac{|M_q|^2}{\hbar}, \tag{1.3}$$

which is attractive because of the minus sign. After getting this attractive interaction in a perturbative way, a typical excitonic mechanism basically follows the mean field approach of BCS to get the gap function.

To get this attractive interaction by the procedure above, there are at least two assumptions. First, a perturbation method requires that the coupling is not very strong. Second, it also needs a standard Fermi liquid where k states near Fermi sphere are well defined. These two assumptions, however, do not fit cuprate superconductors. It is generally agreed that high-T_c materials are a strongly correlated electron system, and whether perturbative methods hold or not is an open question. The importance of non-perturbative approaches in high-T_c superconductivity was mentioned by Anderson (Anderson, 2000). Besides there is mounting evidence showing non-Fermi-liquid properties for the normal state of cuprates. It is not a surprise to understand a number of conflicts between the theory and the experiments. For example, the BCS formulation predicts that pair formation and phase coherence setting in together while experiments demonstrated otherwise (Uemura et al., 1989). There are numerous other issues with the BCS paradigm for high-T_c such as pairing symmetry, coherence length and so on, to just name a few.

1.5 The Case of High-T_c Superconductors

Yet there are other issues in general for the perturbative techniques in high-T_c. For example, is the momentum of the charge-carrying quasiparticle a good quantum number? It is not quite likely. In a typical low-T_c superconductor, the coherence length, which is roughly the size of the pair is much larger than the inverse of Fermi wavevector k_F of the underlying Fermi liquid. The momentum uncertainty δk due to pairing for each of the electrons are much smaller than k_F (Heisenberg uncertainty principle) thus momentum of an electron may still be a good quantum number. However, in cuprates, the coherence length ξ, of typically 20 Å, is on the same order of the inverse Fermi wavevector $1/k_F$. High-T_c is characterized by a fairly low charge-carrier density, which gives a small k_F. Thus the momentum uncertainty k_F due to pairing is of the same order of magnitude as the unperturbed momentum $k \sim k_F$. This place a severe restriction on the applicability of perturbation methods. (In the preceding discussion, we neglected the non-essential difference of 2π between wavenumber and wavevector; momentum and wavevector were used interchangeably too.)

The second problem with the BCS approach is that the electron in the exciton is not treated on the equal footing as the two electrons forming the Cooper pair. This was not a problem for the phonon mechanism but cannot be neglected here for an exciton mechanism because they are identical particles.

To remedy these problems, instead of considering two electrons plus an exciton, we cast the problem in a different perspective: three electrons plus one hole. By this doing, we are now putting the electrons on equal footing. We then drop the BCS notion that momentum is still a good quantum number k and instead focus on real space pairing as suggested by numerous experiments on high-T_c materials. By scattering three electrons and one hole in a two-dimensional plane we want to see if there is any way that one can get an effective attraction among the electrons. Here we have to resort to a favorable geometry.

A classical picture in which such a state of negative energy might be possible is given in Fig. 1.3, three electrons are equally spaced

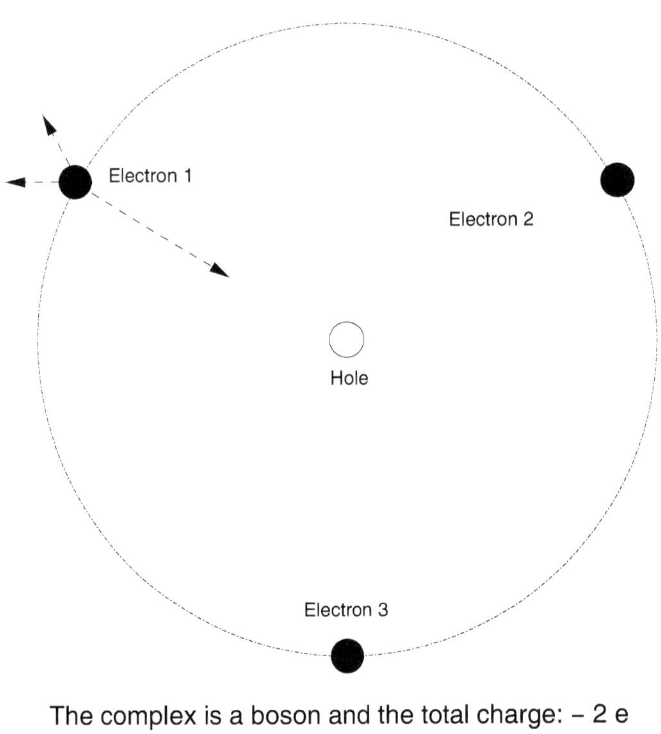

Figure 1.3 Three electrons and a hole arrange themselves to form a "bound" state in a classical picture. Arrows indicate the electrostatic forces whose magnitudes are given by arrows' lengths. The three electrons are at equal distance one to another. The dash-dotted circle traces the orbit of electrons.

on a circle, with one hole in the center. This is the direct opposite of Fig. 1.1, but both should lead to similar results in an SC. According to the simple knowledge of electrostatics, the force on the central hole is zero while those on the electrons are

$$\frac{(\sqrt{3}-1)e^2}{4\sqrt{3}\pi\epsilon r^2}, \tag{1.4}$$

radially inwards to the center, where r is the distance between the hole and any one of the electrons. The corresponding potential energy being negative is given by

$$-\frac{(3-\sqrt{3})e^2}{4\pi\epsilon r}. \tag{1.5}$$

In order not to fall into the center, the electrons may rotate around the hole in such a way that the electrostatic attraction is balanced by the centrifugal force. Intuitively a four-body system with a net charge of $-2e$ may not be stable because of the strong Coulomb repulsion. However, under the special geometry as shown, the peculiarity of Coulomb's law (an inverse square law) actually favors a bound state of three electrons and one hole ($3e + 1h$), at least within the framework of classical electrodynamics. Thus these electrons tend to move to the hole, giving rise to an effective binding among them.

The stability of such complexes against electron–hole recombination or annihilation actually is not a big concern: While externally excited electrons and holes recombine in a short time in typical semiconductors, equilibrium electrons and holes coexist peacefully inside semimetals without destroying one another. The stability is governed by the semimetallic energy band structures.

1.6 Stability against Electron–Hole Recombination

Whether electrons (e) and holes (h) should recombine or not depends on the band structure. Here we just give a very simple example: semimetals (Lovett, 1977). Electrons and holes coexist without annihilating one another, in semimetals (Lovett, 1977) such as Bi, As, Sb. The reason is as follows and shown in Fig. 1.4.

In typical semiconductors, the conduction and valence bands are separated by a positive energy gap as shown in Fig. 1.4a. By jumping from the conduction to the valence band which is at a lower energy (i.e., recombining with a hole), the thermally excited electron lowers its energy. This process thus occurs spontaneously to the e-h pair and makes it unstable. (The same process can be interpreted as a hole jumping up and note that in contrast to electron, the higher is the hole in the band diagram, the lower is its energy.) For semimetals with different band configurations, as typified in Fig. 1.4b,c, the situation is, however, totally different. If the electron were to jump to the hole state and then recombine,

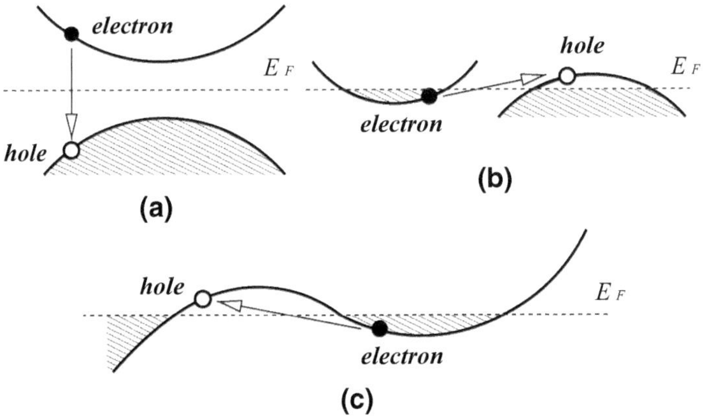

Figure 1.4 Mechanism for the stability of electron–hole pairs. (a) Typical band structure for semiconductors. Electron states have higher energy than hole states. As a result, electrons recombine with holes in a short time. This reasoning also applies to metal provided the relevant conduction band is partly filled. (b) and (c) Band structures of semimetals. Electron states have lower energy than hole states and the recombination of electrons and holes is unfavored in energy. In all figures, E_F is the Fermi energy.

as indicated by the arrows, it would have to take extra energy to reach there. This process is clearly unfavorable in energy and such a mechanism secures the stability of an electron–hole liquid against recombination in semimetals. Thus we expect the proposed $3e{+}1h$ complexes to find their best existing scenario in solids with semimetallic band structures.

In our discussion thereafter in this book, we imply that the electrons and holes are from a host conductor with semimetallic band-edges. We will also use interlayer Coulomb interaction and ECM to derive such an electronic structure even with single-layer cuprate like LSCO. Whether or not this semimetallic band-edge materializes in cuprate superconductors was not clearly answered nearly 20 years ago around the turn of the century. Still the author proposed (Luo, 2000b, 2001b) this even when the experimental data were lacking. The situation is much clearer now. Angel-resolved photoemission spectroscopy (ARPES) experiments have clearly established the electronlike portion of the Fermi surface in the p-type cuprates (Kaminski et al., 2006; Nakayama et al., 2006).

In other words, electrons and holes coexist in cuprates even if it is nominally called n-type or p-type as suggested by the type and valence of dopants.

1.7 Coexistence of Electron and Hole

We want to emphasize one more time: When coexistence of electron/hole is mentioned in this book, they are referred to in the sense of band electron and band hole, before turning-on of the (strong) superconducting pairing interaction. These are not to be confused with electronlike or holelike (Landau- or Bogoliubov-quasiparticle) excitations from a SC condensate or an ordinary electron Fermi gas (liquid).

First, take the archetypical single CuO_2 layer cuprate $La_{2-x}Sr_xCuO_{4-\delta}$ (LSCO) as the example. Once one understands the case in the single CuO_2 plane cuprates, others with two (or more) CuO_2 planes per unit cell will be even easier to comprehend: Interlayer (i.e., between the CuO_2 layers or planes) coupling splits the otherwise single Fermi-level-crossing band into two (or more) bands, one will be more holelike, with the other, however, more electronlike.

The previously often-held notion on LSCO is that it has only a hole pocket centered around (π, π). This view makes three oversimplifications. First, it neglects the three-dimensional nature of the cuprate Fermi surface. Second, it neglects the FS topological change upon increasing strontium doping. Third, it neglects that some portion of the Fermi surface could be electronlike even if the majority FS part is still holelike.

Such a notion, in the very minimum, neglects the k_z dispersion of the energy band, and the associated FS surface topological change. Although the band dispersion along k_z is weak, on the order of 0.1 eV at most, its effect on the FS topology is not trivial, and should not be neglected (Sahrakorpi et al., 2005). Aggravating the problem, is that the ARPES technique currently developed for cuprates, is not k_z sensitive. This contributes to the once much-debated issue as to whether some salient features represent electronlike (e-like) FS or not (Chuang et al., 1999; Fretwell et al., 2000; Luo, 2000b,

2001a). The presence of electronlike band quasiparticles were proposed before on a phenomenological base (Eagles, 1989; Eagles and Savvides, 1989; Hirsch and Marsiglio, 1991; Xing and Ting, 1988), and was emphasized in various occasions by the author due to its importance for excitonic pairing (Luo, 2000b; Luo and Miley, 2009).

Now it is well known that in LSCO, the FS becomes more electronlike with increasing doping of strontium. Still it is often forgotten that electrons and holes actually *coexist* around the optimal doping. This is a fact unambiguously demonstrated by both the latest ARPES and recent band-structure calculations. However, it would really help if more audience in the high-T_c community become aware of such a fact.

To better understand the FS topology, one needs to resort to electronic structure calculation. There have been numerous band structure studies for cuprates. Here we focus on the 3D Fermi surface topology. To the best of the author's knowledge, previous studies largely were on two-dimensional aspect of FS. For the sake of simplicity, only LSCO will be worked out as the example. The author has explored similar systems such as $Nd_{2-x}Ce_xCuO_4$ (NCCO as *e*-doped), Bi-2201 and 2212, with characteristically similar results. These studies will appear elsewhere in separate publications.

The band structure was studied by a first-principles electron-structure package (Blaha et al., 2001) utilizing a linearized augmented plane wave (LAPW) basis set, WIEN2k. The many-body correlation problem is treated within the framework of local density approximation. Correction due to electron density gradient abides by the general gradient approximation rules and the functional used was that of Perdew–Burke–Ernzerhof. Calculation was carried over 2000 k-points in the first Brillouin zone to ensure a high precision in the resulting band structure and Fermi energy. The $\mathbf{k}$ vectors were sorted out by their Fermi crossing and plotted in 3D (Kokalj, 1999).

Take Fig. 1.5 for example. Clearly the Fermi surface topology varies with k_z, the wavevector normal to the CuO_2 plane. One can imagine slicing the 3D FS with planes that are parallel to the k_x-k_y plane, but with different k_z. This gives a series of FS segments.

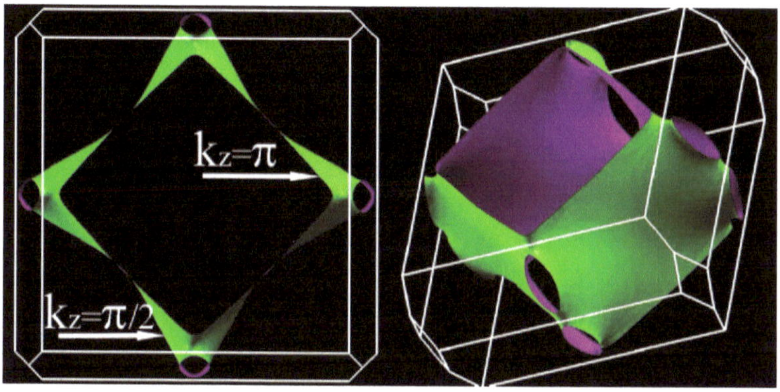

Figure 1.5 Three dimensional Fermi surface of LSCO near optimal doping $x \approx 0.16$. The arrow labeled as $k_z = \pi$ designates the largely electronlike FS section at $k_z = \pi$. The arrow labeling $k_z = \pi/2$ designates the still holelike FS section at $k_z = \pi/2$.

Near optimal doping, the Fermi segment at $k_z = \pi/2$ is still largely holelike while that at $k_z = \pi$ is already mostly electronlike. This demonstrates the coexistence of electron and hole in cuprates, at least for proper doping levels, and a band-edge similar to a semimetal, even with LDA/CGA calculations.

The ARPES techniques that have been developed for high-T_c research are not very sensitive in k_z. As the result, what we have seen for the supposed Fermi surface is merely its 2D projection along the k_z direction. The k_z dependence of the FS near $(\pi/2, \pi/2)$ is negligible. The ARPES FS is therefore strong and unambiguously holelike near $(\pi/2, \pi/2)$. The situation is complicated at $(\pi, 0)$ due to the strong k_z dependence. As a result, the ARPES response there is smeared out, forming the often seen "shaded" region around $(\pi, 0)$, which was interpreted as some sort of unknown electronic states. Of course, strong correlation further complicates the ARPES spectra, but the long neglected k_z dependence of FS are now becoming recognized.

The figure at left in Fig. 1.5 is a head-on view from the k_z direction. It is very similar to ARPES intensity plot (with only $k_\parallel$ resolved) which basically "collapses" the 3D FS along the k_z direction.

1.8 Stability Problem in Quantum Mechanics

The simple picture given in Fig. 1.3 becomes more involved in quantum mechanics. E. H. Lieb (Lieb, 1984a,b) actually proved rigorously that the system of three negative and one positive charges is not stable against the dissociation of one of the electrons into infinity *in vacuum*. (Here vacuum means there are no other particles and fields.) This is also the reason why we have seen singly charged negative hydrogen ion H^- (Hylleraas, 1947) but have never confirmed the doubly charged H^{--}. Nevertheless, a quasi-bound state of $3e+1h$ actually is not ruled out if it is not in vacuum and we have seen the very importance of geometric configuration in determining the ground state energy of strongly correlated electron system.

For the purpose of convenience, let us introduce a new term, *tetron*, to refer to the $3e+1h$ complex, in a sense similar to the term trion used for a bound state of two electrons and one hole.

Now we use a variational technique to put an upper limit on the ground-state energy of such system. We also require that the four particles are relatively localized or near one to another.

Our strategy for this part is as follows. We will first mention relevant 2D quantum mechanical results for the Coulomb potential in Section 1.9. Then as a warm-up exercise we try to find out the upper bound on the ground-state energy of a 2D trion in the heavy hole approximation, which is given in Section 1.10. Based on techniques developed for trions, we try a few fully antisymmetrized wavefunctions in Section 1.11 to show that states of negative energy indeed happen to the tetron.

1.9 Quantum Mechanics of an Electron in 2D Coulomb Potential

This problem has been worked out numerous times but we just repeat some relevant results here for quick reference.

The Hamiltonian for an electron in a 2D Coulomb potential reads

$$H = -\frac{\hbar^2}{2m}\nabla^2 - \frac{e^2}{4\pi\epsilon r}, \tag{1.6}$$

Table 1.1 The units chosen to make the Schrödinger equation dimensionless

Dimension	(energy) (length)^2	(energy) (length)	(length)	(energy)	(wavenumber)
Ordinary unit	$\frac{\hbar^2}{2m}$	$\frac{e^2}{4\pi\epsilon}$	$\frac{2\pi\epsilon\hbar^2}{me^2}$	$\frac{me^4}{8\pi^2\epsilon^2\hbar^2}$	$\frac{me^2}{2\pi\epsilon\hbar^2}$
Dimensionless unit	1	1	1	1	1

where ϵ is the permittivity, e the elementary charge and m the mass of the electron.

We try to solve the problem in a polar coordinate, where the gradient operator is given by

$$\nabla = \hat{\mathbf{r}}\frac{\partial}{\partial r} + \hat{\boldsymbol{\theta}}\frac{1}{r}\frac{\partial}{\partial \theta}, \tag{1.7}$$

with the corresponding Laplacian

$$\nabla^2 = \frac{1}{r}\frac{\partial}{\partial r}\left(r\frac{\partial}{\partial r}\right) + \frac{1}{r^2}\frac{\partial^2}{\partial \theta^2}. \tag{1.8}$$

The Schrödinger equation reads

$$\left\{-\frac{\hbar^2}{2m}\nabla^2 - \frac{e^2}{4\pi\epsilon r}\right\}\Psi = E\Psi. \tag{1.9}$$

We want to simplify Eq. (1.9) by making it dimensionless. Note that $\hbar^2/2m$ has a dimension of (energy)(length)^2 while $e^2/4\pi\epsilon$ is in (energy)(length). We select the units (or the energy and length scales) so that both $\hbar^2/2m$ and $e^2/4\pi\epsilon$ are 1. Under such a convention length will be measured in the unit $2\pi\epsilon\hbar^2/me^2$ while energy is measured in $me^4/8\pi^2\epsilon^2\hbar^2$. This unit system is summarized in Table 1.1.

Now the Schrödinger equation takes a simple form:

$$\left\{-\nabla^2 - \frac{1}{r}\right\}\Psi = E\Psi. \tag{1.10}$$

Equation (1.10) has the following solutions as well known:

$$\Psi_{nm}(r,\theta) = \frac{w_{nm}(r)}{\sqrt{r}}\frac{\exp(\pm im\theta)}{\sqrt{2\pi}}, \; m = 0, 1, 2, 3, \ldots, \tag{1.11}$$

where $w_{nm}(r)$ is given by

$$w_{nm}(r) = \frac{(2Kr)^{m+1/2}\exp(-Kr)}{(2m)!}\left[\frac{K(n+m-1/2)!}{n(n-m-1/2)!}\right]^{1/2}$$
$$\times {}_1F_1\left(-n+m+\frac{1}{2}; 2m+1; 2Kr\right). \tag{1.12}$$

Here ${}_1F_1$ is the confluent hypergeometric function, and

$$K = \frac{1}{2n},\ n = \frac{1}{2}, \frac{3}{2}, \ldots,\ m = 0, 1, \ldots, n - \frac{1}{2}. \tag{1.13}$$

The energy E_{nm} of eigenstate Ψ_{nm} is given by

$$E_{nm} = -\frac{1}{4n^2}. \tag{1.14}$$

For the purpose of convenience, we list the first four eigenstates as follows:

$$\Psi_{\frac{1}{2},0}(r,\theta) = \sqrt{\frac{2}{\pi}}\exp(-r),$$
$$\Psi_{\frac{3}{2},0}(r,\theta) = \sqrt{\frac{2}{27\pi}}\left[1-\frac{2}{3}r\right]\exp\left(-\frac{r}{3}\right),$$
$$\Psi_{\frac{3}{2},\pm 1}(r,\theta) = \sqrt{\frac{4}{243\pi}}\,r\exp\left(-\frac{r}{3}\right)\exp(\pm i\theta). \tag{1.15}$$

1.10 Upper Bound on Ground-State Energy of a 2D trion

We take this problem as a warm-up exercise. This is an illustrative example showing how to balance between the Coulomb and kinetic energy. It also gives us some intuition in choosing good trial wavefunctions and on how to deal with two-body Coulomb integrals. The basic line of reasoning follows that of H.A. Bethe (Bethe and Salpeter, 1957), E.A. Hylleraas (Hylleraas, 1947) and others (Lampert, 1958). Although there has been recent research (Stébé et al., 1997; Thilagam, 1997) in the same direction, our methods in dealing with the integrals are somehow different and hopefully easier to understand.

1.10.1 Introducing the Trion

The research on trions in 2D systems has shown many interesting results in recent years (Huard et al., 2000; Kheng et al., 1993; Stébé et al., 1997; Thilagam, 1997). In a broader sense, the trion problem, not limited to 2D, is linked to the existence and the stability of negative hydrogen ion H^- which has not only important implication for atomic physics and astrophysics (Bethe and Salpeter, 1957), but also vital application to other fields such as physical chemistry (Massey, 1938), accelerator engineering (Belchenko, 1993), and controlled fusion (Holmes, 1996).

In a trion, or charged exciton, two electrons are bound to one hole despite the strong Coulomb repulsion between electrons. (Likewise two holes can be bound to one electron as well but the discussion that follows is the same if "hole" is switched with "electron.") It is quite a marvel that the two electrons would not fall apart from the strong Coulomb repulsion between them. More surprising is that the ground-state energy of such a complex is *lower* than that of an exciton, at least in the limit of heavy hole (i.e., the mass of the hole is much larger than that of electrons, which effectively results into the H^- problem).

First let us try to understand in terms of classical mechanics why trion forms despite the strong Coulomb repulsion.

Suppose that two electrons are on two sides of one hole as shown in Fig. 1.6. The electrostatic forces felt by electron 1 (e1) are from the hole and electron 2 (e2). Because the forces are governed by the inverse square Coulomb's law, the repulsion exerted on e1 by e2 is only 1/4 that of the attraction given by the hole. Similar

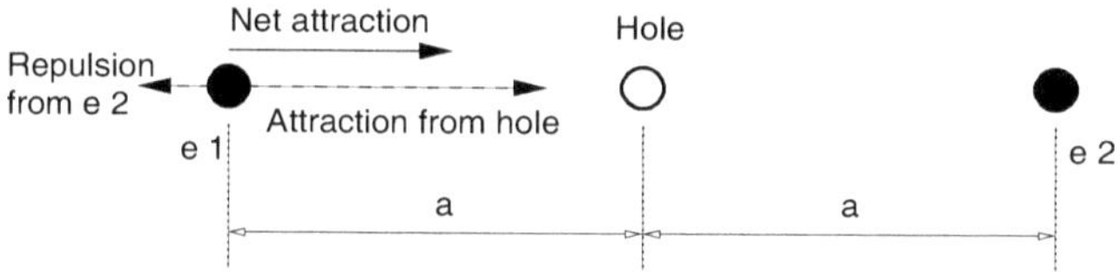

Figure 1.6 Two electrons and a hole arrange themselves to form the bound state of a trion in a classical picture of electrodynamics. Arrows indicate forces. Electrons e1 and e2 are shown as the shaded circles.

argument applies to e2 and therefore these two electrons tend to come together to form a bound state in spite of the strong repulsion among like charges. In order for electrons not to fall into the hole we can assume in a classical fashion that the electrons rotate around the hole in such a way that the inward electrostatic force is balanced by the centrifugal force. In quantum mechanics such tendency of electron's coming together will be stopped by the penalty in kinetic energy which increases as a^{-2}, where a is roughly the average distance between the hole and one of the electrons. Note that the gain in Coulomb energy varies at a slower pace approximately as a^{-1} so that it would be hard for the electrons to come near the hole without limit.

Therefore, the bound state of trion is not quite a surprise after all. It basically results from the inverse square Coulomb's law and a favorable geometric configuration of two electrons and one hole.

1.10.2 The Variational Treatment

For the purpose of simplicity, we take the mass of the hole as infinity. The Schrödinger equation for this problem is now clearly

$$\left\{-\nabla_1^2 - \nabla_2^2 - \frac{1}{r_1} - \frac{1}{r_2} + \frac{1}{r_{12}}\right\}\Psi = E\Psi. \tag{1.16}$$

where ∇_1^2 and ∇_2^2 are, respectively, short-hands for

$$\frac{1}{r_1}\frac{\partial}{\partial r_1}\left(r_1\frac{\partial}{\partial r_1}\right) + \frac{1}{r_1^2}\frac{\partial^2}{\partial\theta_1^2}, \tag{1.17}$$

and

$$\frac{1}{r_2}\frac{\partial}{\partial r_2}\left(r_2\frac{\partial}{\partial r_2}\right) + \frac{1}{r_2^2}\frac{\partial^2}{\partial\theta_2^2}. \tag{1.18}$$

Here subscripts refer to electrons such that r_1, θ_1 and r_2, θ_2 are the coordinates of e1 and e2, respectively.

We choose the following trial wavefunction:

$$\Psi = \exp[-\lambda(r_1 + r_2)][1 + k(r_1 - r_2)^2], \tag{1.19}$$

where λ and k are parameters to be determined later by variational principle. The first term in this wavefunction suggests that e1 and e2 are both in 1s states. Here 1s state are not exactly the $\Psi_{\frac{1}{2},0}$ given in

Eqs. (1.15) but rather that with a screening constant λ. The physical origin of λ is clear: The Coulomb field of the hole felt by one electron is reduced (screened) by the presence of the other. For this reason we also require that λ be positive and smaller than 1. The existence of the second term is to reduce the Coulomb interaction between e1 and e2. Note that this term is small when r_1 and r_2 are close which is another way of saying that electrons 1 and 2 tend to avoid each other. So why not just keep the second term only? The reason is that while this term is good for reducing the Coulomb energy it causes some penalty in kinetic energy when compared with the first one. Thus, the viable solution is to combine the merits of the two terms, which results in the trial wavefunction (1.19).

To facilitate our calculation we want better management on the trial wavefunction. Note that functions like $\exp[-\lambda(r_1 + r_2)]$ are not normalized and would be a headache for the algebra and calculus to come. The solution is to generalize the eigenstates of Eqs. (1.15) to accommodate the screening effect. The first four solutions now read

$$|1s\rangle = \sqrt{\frac{2}{\pi}}\lambda \exp(-\lambda r), \tag{1.20}$$

$$|2s\rangle = \sqrt{\frac{2}{27\pi}}\left[\lambda - \frac{2}{3}\lambda^2 r\right]\exp\left(-\frac{\lambda r}{3}\right), \tag{1.21}$$

$$|2p^{\pm}\rangle = \sqrt{\frac{4}{243\pi}}\lambda^2 r \exp\left(-\frac{\lambda r}{3}\right)\exp(\pm i\theta), \tag{1.22}$$

where we have used notations $1s$, $2s$, and $2p^{\pm}$ in place of the cumbersome Ψ. The meaning of these symbols should be self-evident given a background knowledge of hydrogen eigenstates in three-dimension (3D). It is not hard to verify that these wavefunctions are now normalized.

The trial wavefunction is then instead written as

$$\Psi = [1 + k(r_1 - r_2)^2]|1s_1 1s_2\rangle, \tag{1.23}$$

Table 1.2 Some useful integrals involving $|1s\rangle$ state

Integral	$\langle 1s\|r^{-1}\|1s\rangle$	$\langle 1s\|r\|1s\rangle$	$\langle 1s\|r^2\|1s\rangle$	$\langle 1s\|r^3\|1s\rangle$	$\langle 1s\|r^4\|1s\rangle$
Value	2λ	λ^{-1}	$\frac{3}{2}\lambda^{-2}$	$3\lambda^{-3}$	$\frac{15}{2}\lambda^{-4}$

where $|1s_1 1s_2\rangle = |1s_1\rangle|1s_2\rangle$ stands for

$$\sqrt{\frac{2}{\pi}}\lambda \exp(-\lambda r_1)\sqrt{\frac{2}{\pi}}\lambda \exp(-\lambda r_2), \tag{1.24}$$

and $|1s_1\rangle$ is the short-hand for

$$\sqrt{\frac{2}{\pi}}\lambda \exp(-\lambda r_1). \tag{1.25}$$

Similarly we define $|1s_2\rangle$.

This trial wavefunction is symmetric in the spatial coordinates and in order to make the total wavefunction antisymmetric we only need to multiply Ψ by a spin singlet state $\mathcal{S}^a_{12} = \frac{1}{\sqrt{2}}[|\uparrow_1\downarrow_2\rangle - |\downarrow_1\uparrow_2\rangle]$ where the subscripts are electron indices with the superscript a indicating antisymmetry. Note that the spin wavefunction in this case will not affect the calculation of either $\langle\Psi\mathcal{S}^a_{12}|\Psi\mathcal{S}^a_{12}\rangle = \langle\Psi|\Psi\rangle$ or $\langle\Psi\mathcal{S}^a_{12}|H|\Psi\mathcal{S}^a_{12}\rangle = \langle\Psi|H|\Psi\rangle$ thus we can drop it for now. We also tabulate some often used integrals in Table 1.2 to facilitate our calculation.

The Ritz variational principle states

$$E(\Psi) = \frac{\langle\Psi|H|\Psi\rangle}{\langle\Psi|\Psi\rangle} = \text{min.}, \tag{1.26}$$

where

$$\langle\Psi|H|\Psi\rangle = \langle\Psi| - \nabla_1^2 - \nabla_2^2 - \frac{1}{r_1} - \frac{1}{r_2} + \frac{1}{r_{12}}|\Psi\rangle. \tag{1.27}$$

First we need to calculate $\langle\Psi|\Psi\rangle$,

$$\begin{aligned}\langle\Psi|\Psi\rangle &= \langle 1s_1 1s_2|\{1 + 2k(r_1^2 + r_2^2 - 2r_1 r_2) \\ &\quad + k^2[r_1^4 + r_2^4 + 6r_1^2 r_2^2 - 4r_1^3 r_2 - 4r_1 r_2^3]\}|1s_1 1s_2\rangle \\ &= 1 + 2k\lambda^{-2} + 4.5k^2\lambda^{-4}.\end{aligned} \tag{1.28}$$

The first term on the right-hand-side (RHS) of Eq. (1.27) is

$$\langle\Psi| - \nabla_1^2|\Psi\rangle = \langle\nabla_1\Psi|\nabla_1\Psi\rangle, \tag{1.29}$$

where $|\nabla_1\Psi\rangle$ is to be understood as $\nabla_1|\Psi\rangle$, while $\langle\nabla_1\Psi|$, as a notation, is the Hermitian conjugate of $|\nabla_1\Psi\rangle$. This identity can be proved in rectangular Cartesian coordinate using integration by parts when Ψ is zero at infinity which our trial wavefunction satisfies.

Because

$$\nabla_1|\Psi\rangle = \left\{\hat{\mathbf{r}}_1\frac{\partial}{\partial r_1} + \hat{\boldsymbol{\theta}}_1\frac{1}{r_1}\frac{\partial}{\partial \theta_1}\right\}|1s_1 1s_2\rangle$$
$$= \left\{-\lambda[1 + k(r_1^2 + r_2^2 - 2r_1r_2)] + 2k(r_1 - r_2)\right\}\hat{\mathbf{r}}_1|1s_1 1s_2\rangle$$
$$= -\lambda\Psi\hat{\mathbf{r}}_1 + 2k(r_1 - r_2)\hat{\mathbf{r}}_1|1s_1 1s_2\rangle, \quad (1.30)$$

we have

$$\langle\nabla_1\Psi|\nabla_1\Psi\rangle = \lambda^2\langle\Psi|\Psi\rangle + 4k^2(3\lambda^{-2} - 2\lambda^{-2})$$
$$= \lambda^2 + 2k + 8.5k^2\lambda^{-2}. \quad (1.31)$$

Since Ψ is symmetric in r_1 and r_2 it should not be a surprise that the second term on RHS of Eq. 1.27, $\langle\Psi| - \nabla_2^2|\Psi\rangle$ is equal to $\langle\Psi| - \nabla_1^2|\Psi\rangle$.

The third term in $\langle\Psi|H|\Psi\rangle$ [i.e., Eq. (1.27)] is

$$-\langle\Psi|\frac{1}{r_1}|\Psi\rangle = -\langle 1s_1 1s_2|\frac{1}{r_1}\left\{1 + 2k\left(r_1^2 + r_2^2 - 2r_1r_2\right)\right.$$
$$\left. + k^2[r_1^4 + r_2^4 + 6r_1^2r_2^2 - 4r_1^3r_2 - 4r_1r_2^3]\right\}|1s_1 1s_2\rangle$$
$$= -(2\lambda + 4k\lambda^{-1} + 9k^2\lambda^{-3}), \quad (1.32)$$

to which the fourth term $-\langle\Psi|\frac{1}{r_2}|\Psi\rangle$ is equal because of the symmetry in r_1 and r_2.

The last term of $\langle\Psi|H|\Psi\rangle$ is a two-body Coulomb integral involving elliptic integrals, and

$$\langle\Psi|\frac{1}{r_{12}}|\Psi\rangle = \langle 1s_1 1s_2|\frac{1}{r_{12}}\{1 + 2k(r_1^2 + r_2^2 - 2r_1r_2)$$
$$+ k^2[r_1^4 + r_2^4 + 6r_1^2r_2^2 - 4r_1^3r_2 - 4r_1r_2^3]\}|1s_1 1s_2\rangle$$
$$= \langle 1s_1 1s_2|\frac{1}{r_{12}}\{1 + 2k(2r_1^2 - 2r_1r_2)$$
$$+ k^2[2r_1^4 + 6r_1^2r_2^2 - 8r_1^3r_2]\}|1s_1 1s_2\rangle, \quad (1.33)$$

where we have used identities

$$\langle 1s_1 1s_2|\frac{r_2^2}{r_{12}}|1s_1 1s_2\rangle = \langle 1s_1 1s_2|\frac{r_1^2}{r_{12}}|1s_1 1s_2\rangle,$$
$$\langle 1s_1 1s_2|\frac{r_2^4}{r_{12}}|1s_1 1s_2\rangle = \langle 1s_1 1s_2|\frac{r_1^4}{r_{12}}|1s_1 1s_2\rangle,$$
$$\langle 1s_1 1s_2|\frac{r_1r_2^3}{r_{12}}|1s_1 1s_2\rangle = \langle 1s_1 1s_2|\frac{r_1^3r_2}{r_{12}}|1s_1 1s_2\rangle, \quad (1.34)$$

which are easy to understand because of the symmetry of these integrals in r_1 and r_2.

As an example, we explicitly show in the following how to work out the first term on RHS of Eq. (1.33).

$$\begin{aligned}\langle 1s_1 1s_2 | \frac{1}{r_{12}} | 1s_1 1s_2 \rangle &= \left(\frac{2}{\pi}\right)^2 \lambda^4 \int_0^\infty r_1 dr_1 \int_0^\infty r_2 dr_2 \exp[-2\lambda(r_1 + r_2)] \\ &\times \int_0^{2\pi} d\theta_1 \int_0^{2\pi} d\theta_2 \frac{1}{\sqrt{r_1^2 + r_2^2 - 2r_1 r_2 \cos\theta_{12}}}, \end{aligned} \tag{1.35}$$

where $\theta_{12} = \theta_1 - \theta_2$.

We want to change the integration variables in the integral

$$\int_0^{2\pi} d\theta_1 \int_0^{2\pi} d\theta_2 \frac{1}{\sqrt{r_1^2 + r_2^2 - 2r_1 r_2 \cos\theta_{12}}} \tag{1.36}$$

from θ_1 and θ_2 to θ_{12} and θ_2. To understand how this is done, we refer to Fig. 1.7. Shown in Fig. 1.7 is a plane formed by θ_1 and θ_2. The integration limits of θ_1 and θ_2 in Eq. 1.36 or the integration domain corresponds to the square ABCD.

Because the integrand in integral (1.36)

$$\frac{1}{\sqrt{r_1^2 + r_2^2 - 2r_1 r_2 \cos\theta_{12}}} \tag{1.37}$$

is periodic in θ_1 with period 2π, the integration over triangle ABC is the same as that over triangle DCE. Note that DCE is the translate of ABC by 2π in the θ_1 direction. Thus we can now change the domain of integration from the square ABCD to the parallelogram ACED. Since the segments AC and DE, respectively, satisfy the relation $\theta_1 - \theta_2 = \theta_{12} = 0$ and $\theta_1 - \theta_2 = \theta_{12} = 2\pi$, if we transform integral variables to θ_{12} and θ_2 along with the change in integration domain, we will have

$$\begin{aligned}&\int_0^{2\pi} d\theta_1 \int_0^{2\pi} d\theta_2 \frac{1}{\sqrt{r_1^2 + r_2^2 - 2r_1 r_2 \cos\theta_{12}}} \\ &= \int_0^{2\pi} d\theta_2 \int_0^{2\pi} d\theta_{12} \frac{1}{\sqrt{r_1^2 + r_2^2 - 2r_1 r_2 \cos\theta_{12}}} \left| \frac{\partial(\theta_1, \theta_2)}{\partial(\theta_2, \theta_{12})} \right|, \end{aligned} \tag{1.38}$$

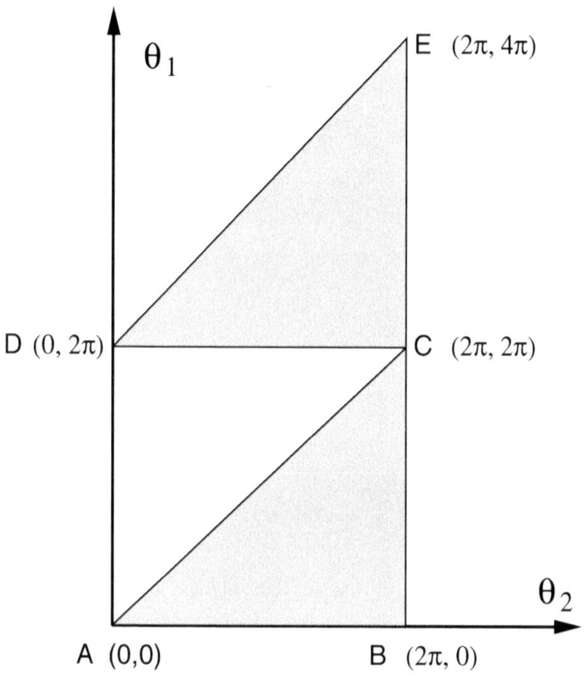

Figure 1.7 Change of variable from θ_1 to θ_{12} in integral (1.31).

where $\left|\frac{\partial(\theta_1,\theta_2)}{\partial(\theta_2,\theta_{12})}\right|$ is the absolute of Jacobian determinant

$$\frac{\partial(\theta_1,\theta_2)}{\partial(\theta_2,\theta_{12})} = \begin{vmatrix} \frac{\partial\theta_1}{\partial\theta_2} & \frac{\partial\theta_1}{\partial\theta_{12}} \\ \frac{\partial\theta_2}{\partial\theta_2} & \frac{\partial\theta_2}{\partial\theta_{12}} \end{vmatrix} = \begin{vmatrix} 1 & 1 \\ 1 & 0 \end{vmatrix} = -1, \tag{1.39}$$

which is identically 1 in its absolute value.

Thus integral (1.36) now satisfies

$$\int_0^{2\pi} d\theta_2 \int_0^{2\pi} d\theta_{12} \frac{1}{\sqrt{r_1^2 + r_2^2 - 2r_1r_2\cos\theta_{12}}}$$
$$= 2\pi \int_0^{2\pi} d\theta_{12} \frac{1}{\sqrt{r_1^2 + r_2^2 - 2r_1r_2\cos\theta_{12}}}. \tag{1.40}$$

Inserting Eq. (1.40) into Eq. (1.35), we get

$$\langle 1s_1 1s_2|\frac{1}{r_{12}}|1s_1 1s_2\rangle = \frac{8}{\pi}\lambda^4 \int_0^\infty dr_1 \int_0^\infty dr_2 \exp[2\lambda(r_1+r_2)]r_1 r_2$$
$$\int_0^{2\pi} d\theta_{12} \frac{1}{\sqrt{r_1^2+r_2^2-2r_1 r_2\cos\theta_{12}}}. \quad (1.41)$$

Equation (1.41) can be further simplified by the following change of variables:

$$r_1 = r,$$
$$r_2 = ur, \quad (1.42)$$

with corresponding Jacobian

$$\frac{\partial(r_1,r_2)}{\partial(r,u)} = \begin{vmatrix} \frac{\partial r_1}{\partial r} & \frac{\partial r_1}{\partial u} \\ \frac{\partial r_2}{\partial r} & \frac{\partial r_2}{\partial u} \end{vmatrix} = \begin{vmatrix} 1 & 0 \\ u & r \end{vmatrix} = r. \quad (1.43)$$

This transformation changes RHS of Eq. (1.41) to

$$\frac{8}{\pi}\lambda^4 \int_0^\infty udu \int_0^{2\pi} d\theta_{12} \frac{1}{\sqrt{1+u^2-2u\cos\theta_{12}}}$$
$$\int_0^\infty drr^2 \exp[2\lambda(1+u)r], \quad (1.44)$$

which can be first integrated with respect to r to give

$$\frac{8}{\pi}\lambda^4 2! \int_0^\infty \left[\frac{2}{\lambda(1+u)}\right]^3 udu \int_0^{2\pi} d\theta_{12} \frac{1}{\sqrt{1+u^2-2u\cos\theta_{12}}}. \quad (1.45)$$

Further evaluation of second integrand of Expression (1.45) gives

$$\int_0^{2\pi} d\theta_{12} \frac{u}{\sqrt{1+u^2-2u\cos\theta_{12}}} = \frac{4K\left(\frac{-4u}{1-2u+u^2}\right)}{\sqrt{1-2u+u^2}}, \quad (1.46)$$

after integration over θ_{12}, with **K** the complete elliptic integral of the first kind.

Integral (1.45) is therefore reduced to

$$\frac{16}{\pi}\lambda \int_0^\infty \left[\frac{2}{1+u}\right]^3 \frac{4u\mathbf{K}\left(\frac{-4u}{1-2u+u^2}\right)}{\sqrt{1-2u+u^2}} du, \quad (1.47)$$

which can be integrated numerically to give the final answer

$$\langle 1s_1 1s_2 \left| \frac{1}{r_{12}} \right| 1s_1 1s_2 \rangle = 1.1780972\lambda. \tag{1.48}$$

Similarly we work out other terms in Eq. (1.33):

$$2k\langle 1s_1 1s_2 \left| \frac{2r_1^2 - 2r_1 r_2}{r_{12}} \right| 1s_1 1s_2 \rangle = 1.0308351k\lambda^{-1}, \tag{1.49}$$

and

$$k^2\langle 1s_1 1s_2 \left| \frac{2r_1^4 + 6r_1^2 r_2^2 - 8r_1^3 r_2}{r_{12}} \right| 1s_1 1s_2 \rangle = 1.5186410k^2\lambda^{-3}. \tag{1.50}$$

Substituting Eqs. (1.48), (1.49) and (1.50) into (1.33) and combining terms from (1.31), (1.32), we find the expectation value of the Hamiltonian

$$\begin{aligned} \langle \Psi | H | \Psi \rangle = {} & 2\lambda^2 + 4k + 17k^2\lambda^{-2} - 2.8219028\lambda \\ & - 6.9691649k\lambda^{-1} - 16.4813590k^2\lambda^{-3}, \end{aligned} \tag{1.51}$$

where the first three terms are from kinetic energy and the last three are Coulomb ones.

We want to minimize the energy

$$\begin{aligned} E &= \frac{\langle \Psi | H | \Psi \rangle}{\langle \Psi | \Psi \rangle} \\ &= \frac{2\lambda^2 + 4k + 17k^2\lambda^{-2} - 2.8219028\lambda - 6.9691649k\lambda^{-1} - 16.4813590k^2\lambda^{-3}}{1 + 2k\lambda^{-2} + 4.5k^2\lambda^{-4}}, \end{aligned} \tag{1.52}$$

with respect to λ and k, which has numerical results as follows:

$$E_{\min} = -1.08147; k = 0.111270; \lambda = 0.693001. \tag{1.53}$$

This value, as an upper bound of the ground-state energy for our specific trion, is lower than that of an exciton with heavy hole, which is -1. The trion is therefore more stable than exciton and it forms spontaneously given one exciton and an extra electron.

Our calculation shows that the trion binding energy in two-dimension >0.08147 2D Rydberg, larger than that of 3D one which is ≈ 0.055 3D Rydberg. This is not a surprise because it is well known that a lower dimensionality favors the formation of bound states. As a result, pure excitons in 2D have a binding energy 4 times as large as that in 3D.

We are so far confined to the heavy hole limit. It is interesting to see quite some reports (Hylleraas, 1947; Lampert, 1958; Stébé et al., 1997; Thilagam, 1997) showing trion states more stable than those of exciton regardless of the electron–hole mass ratio although the author of this Chapter has not obtained such results using a slightly different yet more rigorous Hamiltonian. Not reaching a low energy after a few trials in variational methods does not necessarily rule out its actual existence and this problem will not be further probed here because it is not our main topic of this book. Such issues aside, we have been familiar with the techniques needed for further development and let us now investigate the possible negative-energy state of tetron or $3e + 1h$.

1.11 Variational Quantum Mechanics of 2D Tetron

1.11.1 The Hamiltonian

We assume that all particles have the same mass m. The full Hamiltonian is

$$\begin{aligned} H = &-\frac{\hbar^2}{2m}\left(\underline{\nabla}_0^2 + \underline{\nabla}_1^2 + \underline{\nabla}_2^2 + \underline{\nabla}_3^2\right) \\ &+\frac{e^2}{4\pi\epsilon}\left(\frac{1}{\underline{r}_{12}} + \frac{1}{\underline{r}_{23}} + \frac{1}{\underline{r}_{31}} - \frac{1}{\underline{r}_{10}} - \frac{1}{\underline{r}_{20}} - \frac{1}{\underline{r}_{30}}\right), \end{aligned} \tag{1.54}$$

where subscripts 1, 2, and 3 are indices to the three electrons while 0 designates the hole. $\underline{r}_{12}$ is the distance between particles 1 and 2. The meanings of $\underline{r}_{23}$, $\underline{r}_{31}$, $\underline{r}_{10}$, $\underline{r}_{20}$, and $\underline{r}_{30}$ are similarly defined. Using the underlined r instead of pure r is simply for notational purpose; we will in a moment transform the coordinates into a more convenient form. Under the current notation $\underline{\nabla}_0^2$ is given by

$$\underline{\nabla}_0^2 = \frac{1}{\underline{r}_0}\frac{\partial}{\partial \underline{r}_0}\left(\underline{r}_0\frac{\partial}{\partial \underline{r}_0}\right) + \frac{1}{\underline{r}_0^2}\frac{\partial^2}{\partial \underline{\theta}_0^2} \tag{1.55}$$

and $\underline{\nabla}_1^2$, $\underline{\nabla}_2^2$, and $\underline{\nabla}_3^2$ are understood similarly.

Performing the dimensionless unit transform same as made before in Section 1.9 , the Hamiltonian is now

$$H = -\left(\underline{\nabla}_0^2 + \underline{\nabla}_1^2 + \underline{\nabla}_2^2 + \underline{\nabla}_3^2\right) + \left(\frac{1}{\underline{r}_{12}} + \frac{1}{\underline{r}_{23}} + \frac{1}{\underline{r}_{31}} - \frac{1}{\underline{r}_{10}} - \frac{1}{\underline{r}_{20}} - \frac{1}{\underline{r}_{30}}\right). \quad (1.56)$$

We introduce a new coordinate system

$$\mathbf{r}_0 = \frac{\underline{\mathbf{r}}_0 + \underline{\mathbf{r}}_1 + \underline{\mathbf{r}}_2 + \underline{\mathbf{r}}_3}{4},$$
$$\mathbf{r}_1 = \frac{\underline{\mathbf{r}}_0 + \underline{\mathbf{r}}_1 - \underline{\mathbf{r}}_2 - \underline{\mathbf{r}}_3}{2},$$
$$\mathbf{r}_2 = \frac{\underline{\mathbf{r}}_0 - \underline{\mathbf{r}}_1 + \underline{\mathbf{r}}_2 - \underline{\mathbf{r}}_3}{2},$$
$$\mathbf{r}_3 = \frac{\underline{\mathbf{r}}_0 - \underline{\mathbf{r}}_1 - \underline{\mathbf{r}}_2 + \underline{\mathbf{r}}_3}{2}, \quad (1.57)$$

which are treated as vector transforms.

These transforms result in the following relations:

$$\mathbf{r}_0 - \mathbf{r}_1 = \underline{\mathbf{r}}_2 + \underline{\mathbf{r}}_3,$$
$$\mathbf{r}_0 - \mathbf{r}_2 = \underline{\mathbf{r}}_3 + \underline{\mathbf{r}}_1,$$
$$\mathbf{r}_0 - \mathbf{r}_3 = \underline{\mathbf{r}}_1 + \underline{\mathbf{r}}_2, \quad (1.58)$$

and

$$\mathbf{r}_1 - \mathbf{r}_2 = \underline{\mathbf{r}}_1 - \underline{\mathbf{r}}_2,$$
$$\mathbf{r}_2 - \mathbf{r}_3 = \underline{\mathbf{r}}_2 - \underline{\mathbf{r}}_3,$$
$$\mathbf{r}_3 - \mathbf{r}_1 = \underline{\mathbf{r}}_3 - \underline{\mathbf{r}}_1. \quad (1.59)$$

With the transformed coordinate system in place, the Hamiltonian now takes a new form,

$$H = -\left(\frac{1}{4}\nabla_0^2 + \nabla_1^2 + \nabla_2^2 + \nabla_3^2\right) + \left(\frac{1}{r_{12}} + \frac{1}{r_{23}} + \frac{1}{r_{31}} - \frac{1}{r_{1+2}} - \frac{1}{r_{2+3}} - \frac{1}{r_{3+1}}\right). \quad (1.60)$$

where for example

$$\nabla_0^2 = \frac{1}{r_0}\frac{\partial}{\partial r_0}\left(r_0\frac{\partial}{\partial r_0}\right) + \frac{1}{r_0^2}\frac{\partial^2}{\partial \theta_0^2}, \quad (1.61)$$

and

$$r_{12} = \sqrt{r_1^2 + r_2^2 - 2r_1r_2\cos\theta_{12}}, \quad (1.62)$$

while r_{1+2} is the short-hand for $|\mathbf{r}_1 + \mathbf{r}_2|$ or $\sqrt{r_1^2 + r_2^2 + 2r_1r_2\cos\theta_{12}}$.

$\mathbf{r}_0$ now refers to the center-of-mass (c.m.). After discarding the irrelevant first term on RHS of Eq. (1.60), we reduce the internal Hamiltonian in the c.m. frame to

$$\begin{aligned} H =& -\left(\nabla_1^2 + \nabla_2^2 + \nabla_3^2\right) \\ &+ \left(\frac{1}{r_{12}} + \frac{1}{r_{23}} + \frac{1}{r_{31}} - \frac{1}{r_{1+2}} - \frac{1}{r_{2+3}} - \frac{1}{r_{3+1}}\right), \end{aligned} \tag{1.63}$$

which is effectively a three-body problem.

1.11.2 Spin-Singlet Trial Wavefunction of $1s$, $2p^+$, and $2p^-$ States

It is easy to see that if the trial wavefunction for the three-body problem is antisymmetrized, the corresponding wavefunction for the our original four-body problem will be antisymmetric as well under the permutation of the original electron coordinates. Now we want to see how to make a three-body antisymmetrized wavefunctions.

The trial wavefunctions will be initially made only from the first four eigenstates for the purpose of simplicity.

$$\begin{aligned} |1s\rangle &= \sqrt{\frac{2}{\pi}}\lambda \exp(-\lambda r), \\ |2s\rangle &= \sqrt{\frac{2}{27\pi}}\left[\lambda - \frac{2}{3}\lambda^2 r\right] \exp\left(-\frac{\lambda r}{3}\right), \\ |2p^\pm\rangle &= \sqrt{\frac{4}{243\pi}}\lambda^2 r \exp\left(-\frac{\lambda r}{3}\right) \exp\left(\pm i\theta\right). \end{aligned} \tag{1.64}$$

which are copied from the previous section for quick reference here. As a notation convenience which will be extensively used later, we also define

$$|2p\rangle = \sqrt{\frac{4}{243\pi}}\lambda^2 r \exp\left(-\frac{\lambda r}{3}\right). \tag{1.65}$$

Note that $\langle 2p|2p\rangle = 1$ although $|2p\rangle$ is not necessarily orthogonal to other eigenstates.

We now try to make an antisymmetrized wavefunction Ψ out of Eqs. (1.64). Ψ should satisfy our main objective, $\langle\Psi|H|\Psi\rangle/\langle\Psi|\Psi\rangle < 0$.

The favorable symmetry property of the Hamiltonian (1.63) helps us rule out Ψ only made of s states. To see this, we need to understand that the kinetic terms in Hamiltonian are positive definite, or $\langle\Psi| - (\nabla_1^2 + \nabla_2^2 + \nabla_3^2)|\Psi\rangle/\langle\Psi|\Psi\rangle > 0$. Thus the only possibility left for a negative energy state is by a negative Coulomb term $\langle\Psi|\left(\frac{1}{r_{12}} + \frac{1}{r_{23}} + \frac{1}{r_{31}} - \frac{1}{r_{1+2}} - \frac{1}{r_{2+3}} - \frac{1}{r_{3+1}}\right)|\Psi\rangle/\langle\Psi|\Psi\rangle$. We only need to consider terms like $\langle\Psi|(\frac{1}{r_{12}} - \frac{1}{r_{1+2}})|\Psi\rangle/\langle\Psi|\Psi\rangle$ because the total Coulomb energy is just three times as much because of the symmetry of Ψ in $\mathbf{r}_1, \mathbf{r}_2, \mathbf{r}_3$. Now we show and exploit a further symmetry hidden in r_{12} and r_{1+2}.

Note that

$$r_{12} = \sqrt{r_1^2 + r_2^2 - 2r_1r_2\cos\theta_{12}}, \tag{1.66}$$

$$r_{1+2} = \sqrt{r_1^2 + r_2^2 + 2r_1r_2\cos\theta_{12}} \tag{1.67}$$

$$= \sqrt{r_1^2 + r_2^2 - 2r_1r_2\cos(\pi + \theta_{12})}. \tag{1.68}$$

Both r_{12} and r_{1+2} are periodic in θ_1 with a period 2π. By the same argument we made in Section 1.10 for two-body Coulomb integrals, we can integrate with respect to θ_{12} and θ_2, both from 0 to 2π. However, r_{1+2} is identical to r_{12} except for a phase shift of π in θ_{12}. Since $\cos\theta_{12}$ has a period of 2π in θ_{12}, $\langle\Psi|\frac{1}{r_{12}}|\Psi\rangle/\langle\Psi|\Psi\rangle$ would be equal to $\langle\Psi|\frac{1}{r_{1+2}}|\Psi\rangle/\langle\Psi|\Psi\rangle$ if Ψ does not have any angular dependency such as those given by pure s states. The total Coulomb energy would be 0 and we have no way whatsoever to obtain a negative energy. We will give a more general theorem later and exploit similar properties to simplify our calculation.

So one of the states chosen in Ψ has to be $|2p^+\rangle$ or $|2p^-\rangle$. From our trial-and-error experience, we find that it is generally preferable to have two $2p$ states and one s state.

As the first trial we want to test wavefunctions made out of $|1s\rangle$, $|2p^+\rangle$ and $|2p^-\rangle$.

By using the variational approach, we also want to investigate the spin structure of such 3E+1H complex. Since there is no spin-orbit coupling term in the Hamiltonian, we need only to specify the total spin of three electrons for any trial wavefunction. Basically there are two possible options for three electrons: Either they are all lined up in the same direction or one of them is against the other two. The hole is also a fermion, whose spin tends to be opposite to at least one of the electrons. Thus the total spin of this complex is either 0 or $\hbar$, corresponding to spin singlet or triplet states. We would consider this explicitly in our calculation.

First let us start with three orthogonal states $|1s\ \updownarrow\rangle$, $|2p^+\ \uparrow\rangle$ and $|2p^-\ \downarrow\rangle$. The arrows designate spin states. The complex is now in a spin singlet state because the mixed spin wavefunction $|\ \updownarrow\rangle = \frac{1}{\sqrt{2}}|\uparrow\rangle + \frac{1}{\sqrt{2}}|\downarrow\rangle$ should combine with that of the hole to give a singlet for one electron–hole pair. Writing $|\ \updownarrow\rangle$ also means that we do not care about the exact spin state; it can be either up or down.

The corresponding Slater determinant is

$$\sqrt{\frac{1}{6}}\begin{vmatrix} |1s_1\ \updownarrow_1\rangle & |2p_1^+\ \uparrow_1\rangle & |2p_1^-\ \downarrow_1\rangle \\ |1s_2\ \updownarrow_2\rangle & |2p_2^+\ \uparrow_2\rangle & |2p_2^-\ \downarrow_2\rangle \\ |1s_3\ \updownarrow_3\rangle & |2p_3^+\ \uparrow_3\rangle & |2p_3^-\ \downarrow_3\rangle \end{vmatrix}$$

$$= \sqrt{\frac{1}{6}}\begin{vmatrix} |2p_2^+\ \uparrow_2\rangle & |2p_2^-\ \downarrow_2\rangle \\ |2p_3^+\ \uparrow_3\rangle & |2p_3^-\ \downarrow_3\rangle \end{vmatrix}|1s_1\ \updownarrow_1\rangle$$

$$+\sqrt{\frac{1}{6}}\begin{vmatrix} |2p_1^+\ \uparrow_1\rangle & |2p_1^-\ \downarrow_1\rangle \\ |2p_2^+\ \uparrow_2\rangle & |2p_2^-\ \downarrow_2\rangle \end{vmatrix}|1s_3\ \updownarrow_3\rangle$$

$$-\sqrt{\frac{1}{6}}\begin{vmatrix} |2p_1^+\ \uparrow_1\rangle & |2p_1^-\ \downarrow_1\rangle \\ |2p_3^+\ \uparrow_3\rangle & |2p_3^-\ \downarrow_3\rangle \end{vmatrix}|1s_2\ \updownarrow_2\rangle$$

$$\begin{aligned} = \sqrt{\frac{1}{6}}\{&|2p_2 2p_3\rangle\{\exp[i(\theta_2-\theta_3)]|\uparrow_2\downarrow_3\rangle \\ &-\exp[i(\theta_3-\theta_2)]|\uparrow_3\downarrow_2\rangle\}|1s_1\ \updownarrow_1\rangle \\ &+|2p_1 2p_2\rangle\{\exp[i(\theta_1-\theta_2)]|\uparrow_1\downarrow_2\rangle \\ &-\exp[i(\theta_2-\theta_1)]|\uparrow_2\downarrow_1\rangle\}|1s_3\ \updownarrow_3\rangle \\ &-|2p_1 2p_3\rangle\{\exp[i(\theta_1-\theta_3)]|\uparrow_1\downarrow_3\rangle \\ &-\exp[i(\theta_3-\theta_1)]|\uparrow_3\downarrow_1\rangle\}|1s_2\ \updownarrow_2\rangle\}, \end{aligned} \tag{1.69}$$

where for example $\updownarrow_1$ indicates a mixed spin state for electron 1 while $2p_2^+$ says that electron 2 is in a $2p^+$ state and so on so forth. Subscripts 1, 2, and 3 as usual are electron indices.

There will be 6 terms in this antisymmetrized trial wavefunction, which would render follow-on manipulation difficult. Here we do a small rearrangement by adding to the above the following Slater determinant, which is similar to Eq. (1.69) except for the switch between $2p^+$ and $2p^-$ states:

$$\sqrt{\frac{1}{6}}\begin{vmatrix} |1s_1 \updownarrow_1\rangle & |2p_1^- \uparrow_1\rangle & |2p_1^+ \downarrow_1\rangle \\ |1s_2 \updownarrow_2\rangle & |2p_2^- \uparrow_2\rangle & |2p_2^+ \downarrow_2\rangle \\ |1s_3 \updownarrow_3\rangle & |2p_3^- \uparrow_3\rangle & |2p_3^+ \downarrow_3\rangle \end{vmatrix}$$

$$= \sqrt{\frac{1}{6}}\begin{vmatrix} |2p_2^- \uparrow_2\rangle & |2p_2^+ \downarrow_2\rangle \\ |2p_3^- \uparrow_3\rangle & |2p_3^+ \downarrow_3\rangle \end{vmatrix} |1s_1 \updownarrow_1\rangle$$

$$+\sqrt{\frac{1}{6}}\begin{vmatrix} |2p_1^- \uparrow_1\rangle & |2p_1^+ \downarrow_1\rangle \\ |2p_2^- \uparrow_2\rangle & |2p_2^+ \downarrow_2\rangle \end{vmatrix} |1s_3 \updownarrow_3\rangle$$

$$-\sqrt{\frac{1}{6}}\begin{vmatrix} |2p_1^- \uparrow_1\rangle & |2p_1^+ \downarrow_1\rangle \\ |2p_3^- \uparrow_3\rangle & |2p_3^+ \downarrow_3\rangle \end{vmatrix} |1s_2 \updownarrow_2\rangle$$

$$\begin{aligned} &= \sqrt{\frac{1}{6}}\{|2p_2 2p_3\rangle\{\exp[i(\theta_3-\theta_2)]| \uparrow_2\downarrow_3\rangle \\ &- \exp[i(\theta_2-\theta_3)]| \uparrow_3\downarrow_2\rangle\}|1s_1 \updownarrow_1\rangle \\ &+|2p_1 2p_2\rangle\{\exp[i(\theta_2-\theta_1)]| \uparrow_1\downarrow_2\rangle \\ &- \exp[i(\theta_1-\theta_2)]| \uparrow_2\downarrow_1\rangle\}|1s_3 \updownarrow_3\rangle \\ &-|2p_1 2p_3\rangle\{\exp[i(\theta_3-\theta_1)]| \uparrow_1\downarrow_3\rangle \\ &- \exp[i(\theta_1-\theta_3)]| \uparrow_3\downarrow_1\rangle\}|1s_2 \updownarrow_2\rangle\}. \end{aligned} \tag{1.70}$$

The result is the following after normalization:

$$\begin{aligned} \Psi = \frac{2}{\sqrt{6}}\{&|2p_1 2p_2 1s_3\rangle \cos\theta_{12}|\mathcal{S}^a_{12,3}\rangle \\ &+|2p_2 2p_3 1s_1\rangle \cos\theta_{23}|\mathcal{S}^a_{23,1}\rangle \\ &+|2p_3 2p_1 1s_2\rangle \cos\theta_{31}|\mathcal{S}^a_{31,2}\rangle\}, \end{aligned} \tag{1.71}$$

where the spin wavefunction $|\mathcal{S}^a_{12,3}\rangle = \frac{\sqrt{2}}{2}(| \uparrow_1\downarrow_2\rangle - | \uparrow_2\downarrow_1\rangle)| \updownarrow_3\rangle$ which is antisymmetric in the first two indices 1 and 2, and note that

$$\langle\mathcal{S}^a_{12,3}|\mathcal{S}^a_{23,1}\rangle = \langle\mathcal{S}^a_{12,3}|\mathcal{S}^a_{31,2}\rangle = \langle\mathcal{S}^a_{23,1}|\mathcal{S}^a_{31,2}\rangle = -\frac{1}{2}. \tag{1.72}$$

Ψ is already normalized and thus $\langle\Psi|\Psi\rangle = 1$. The kinetic energy of this trial wavefunction $\langle\Psi| -\nabla_1^2 - \nabla_2^2 - \nabla_3^2|\Psi\rangle$ is easy to work out

$$\langle\Psi| -\nabla_1^2 - \nabla_2^2 - \nabla_3^2|\Psi\rangle = \frac{11}{9}\lambda^2. \tag{1.73}$$

Actually this can be figured out without explicit calculation. We have three particles in $1s$, $2p^+$ and $2p^-$ states whose kinetic energy are, respectively, λ^2, $\lambda^2/9$ and $\lambda^2/9$, thus the total kinetic energy should be the sum, which gives $11\lambda^2/9$.

As for the Coulomb energy, we need only to calculate this term

$$\langle\Psi|\frac{1}{r_{12}}-\frac{1}{r_{1+2}}|\Psi\rangle, \tag{1.74}$$

because

$$\langle\Psi|\frac{1}{r_{12}}-\frac{1}{r_{1+2}}|\Psi\rangle=\langle\Psi|\frac{1}{r_{23}}-\frac{1}{r_{2+3}}|\Psi\rangle=\langle\Psi|\frac{1}{r_{31}}-\frac{1}{r_{3+1}}|\Psi\rangle, \tag{1.75}$$

considering symmetry in permutation of the electrons.

We then write explicitly

$$\begin{aligned}
&\langle\Psi|\frac{1}{r_{12}}-\frac{1}{r_{1+2}}|\Psi\rangle\\
&=\frac{4}{6}\Big\{\langle 2p_1 2p_2 1s_3|\cos^2\theta_{12}\left(\frac{1}{r_{12}}-\frac{1}{r_{1+2}}\right)|2p_1 2p_2 1s_3\rangle\\
&\quad+\langle 2p_2 2p_3 1s_1|\cos^2\theta_{23}\left(\frac{1}{r_{12}}-\frac{1}{r_{1+2}}\right)|2p_2 2p_3 1s_1\rangle\\
&\quad+\langle 2p_3 2p_1 1s_2|\cos^2\theta_{31}\left(\frac{1}{r_{12}}-\frac{1}{r_{1+2}}\right)|2p_3 2p_1 1s_2\rangle\\
&\quad-\langle 2p_2 2p_3 1s_1|\cos\theta_{12}\cos\theta_{23}\left(\frac{1}{r_{12}}-\frac{1}{r_{1+2}}\right)|2p_1 2p_2 1s_3\rangle\\
&\quad-\langle 2p_3 2p_1 1s_2|\cos\theta_{12}\cos\theta_{31}\left(\frac{1}{r_{12}}-\frac{1}{r_{1+2}}\right)|2p_1 2p_2 1s_3\rangle\\
&\quad-\langle 2p_3 2p_1 1s_2|\cos\theta_{23}\cos\theta_{31}\left(\frac{1}{r_{12}}-\frac{1}{r_{1+2}}\right)|2p_3 2p_2 1s_1\rangle\Big\}.
\end{aligned} \tag{1.76}$$

For obvious identities from symmetry

$$\begin{aligned}
&\langle 2p_2 2p_3 1s_1|\cos^2\theta_{23}\left(\frac{1}{r_{12}}-\frac{1}{r_{1+2}}\right)|2p_2 2p_3 1s_1\rangle\\
&\quad=\langle 2p_3 2p_1 1s_2|\cos^2\theta_{31}\left(\frac{1}{r_{12}}-\frac{1}{r_{1+2}}\right)|2p_3 2p_1 1s_2\rangle,\\
&\qquad\langle 2p_2 2p_3 1s_1|\cos\theta_{12}\cos\theta_{23}\left(\frac{1}{r_{12}}-\frac{1}{r_{1+2}}\right)|2p_1 2p_2 1s_3\rangle\\
&\quad=\langle 2p_3 2p_1 1s_2|\cos\theta_{12}\cos\theta_{31}\left(\frac{1}{r_{12}}-\frac{1}{r_{1+2}}\right)|2p_1 2p_2 1s_3\rangle,
\end{aligned}$$

Equation (1.76) is simplified to

$$\begin{aligned}
&\langle\Psi|\frac{1}{r_{12}}-\frac{1}{r_{1+2}}|\Psi\rangle \\
&=\frac{4}{6}\Big\{\langle 2p_1 2p_2 1s_3|\cos^2\theta_{12}\left(\frac{1}{r_{12}}-\frac{1}{r_{1+2}}\right)|2p_1 2p_2 1s_3\rangle \\
&\quad+2\langle 2p_2 2p_3 1s_1|\cos^2\theta_{23}\left(\frac{1}{r_{12}}-\frac{1}{r_{1+2}}\right)|2p_2 2p_3 1s_1\rangle \\
&\quad-2\langle 2p_2 2p_3 1s_1|\cos\theta_{12}\cos\theta_{23}\left(\frac{1}{r_{12}}-\frac{1}{r_{1+2}}\right)|2p_1 2p_2 1s_3\rangle \\
&\quad-\langle 2p_3 2p_1 1s_2|\cos\theta_{23}\cos\theta_{31}\left(\frac{1}{r_{12}}-\frac{1}{r_{1+2}}\right)|2p_2 2p_3 1s_1\rangle\Big\}.
\end{aligned} \tag{1.77}$$

The first term on RHS of Eq. (1.77) will be zero. To understand this, we prove the following theorem:

Theorem: Suppose that $f_1(r_1, r_2, r_3)$ and $f_2(r_1, r_2, r_3)$ are functions only of r_1, r_2, and r_3 (i.e., do not depend on θ_1, θ_2, and θ_3). Also assume that $\langle f_1|\frac{1}{r_{12}}|f_2\rangle$ exists where f_1 and f_2 are short-hands for $f_1(r_1, r_2, r_3)$ and $f_2(r_1, r_2, r_3)$. Then we have

$$\begin{aligned}
&\langle f_1|\cos^n\theta_{12}\left(\frac{1}{r_{12}}-\frac{1}{r_{1+2}}\right)|f_2\rangle \\
&=\begin{cases} 0, & \text{if } n \text{ is even;} \\ \langle f_1|2\cos^n\theta_{12}\left(\frac{1}{r_{12}}\right)|f_2\rangle, & \text{if } n \text{ is odd.}\end{cases}
\end{aligned} \tag{1.78}$$

Proof:

$$\begin{aligned}
&\langle f_1|\cos^n\theta_{12}\left(\frac{1}{r_{12}}-\frac{1}{r_{1+2}}\right)|f_2\rangle \\
&=\int_0^\infty r_1dr_1\int_0^\infty r_2dr_2\int_0^\infty r_3dr_3\int_0^{2\pi}d\theta_1\int_0^{2\pi}d\theta_2 \\
&\qquad\int_0^{2\pi}d\theta_3\, f_1 f_2\cos^n\theta_{12}\left(\frac{1}{r_{12}}-\frac{1}{r_{1+2}}\right) \\
&=\int_0^\infty r_1dr_1\int_0^\infty r_2dr_2\int_0^\infty r_3dr_3\int_0^{2\pi}d\theta_3 \\
&\qquad\int_0^{2\pi}d\theta_2\int_0^{2\pi}d\theta_{12}\, f_1 f_2\cos^n\theta_{12}\left(\frac{1}{r_{12}}-\frac{1}{r_{1+2}}\right)
\end{aligned} \tag{1.79}$$

where in the second step we have changed variables from θ_1 to θ_{12} for the same reason that we discussed before in Section 1.10. We consider first the integration with respect to θ_{12} in Eq. (1.79), or the following expression:

$$\begin{aligned}\int_0^{2\pi} & d\theta_{12}\cos^n\theta_{12}\left(\frac{1}{r_{12}}-\frac{1}{r_{1+2}}\right)\\ &= \int_0^{2\pi} d\theta_{12}\cos^n\theta_{12}\left(\frac{1}{\sqrt{r_1^2+r_2^2-2r_1r_2\cos\theta_{12}}}\right.\\ &\quad\left.-\frac{1}{\sqrt{r_1^2+r_2^2-2r_1r_2\cos(\pi+\theta_{12})}}\right).\end{aligned} \tag{1.80}$$

The last term on RHS of Eq. (1.80) is

$$-\int_0^{2\pi} d\theta_{12}\cos^n\theta_{12}\frac{1}{\sqrt{r_1^2+r_2^2-2r_1r_2\cos(\pi+\theta_{12})}}. \tag{1.81}$$

Changing variable from θ_{12} to an arbitrary $x=\pi+\theta_{12}$ in this integral, we have

$$\begin{aligned}-\int_0^{2\pi} & d\theta_{12}\cos^n\theta_{12}\frac{1}{\sqrt{r_1^2+r_2^2-2r_1r_2\cos(\pi+\theta_{12})}}\\ &= -\int_\pi^{3\pi} dx\cos^n(x-\pi)\frac{1}{\sqrt{r_1^2+r_2^2-2r_1r_2\cos x}}\\ &= -\int_\pi^{3\pi} dx(-1)^n\cos^n x\frac{1}{\sqrt{r_1^2+r_2^2-2r_1r_2\cos x}}.\end{aligned} \tag{1.82}$$

Because the integrand on RHS of Eq. (1.82) is a function of x with a period 2π, integrating from 0 to π is equal to integrating from 2π to 3π, or

$$\begin{aligned}-\int_{2\pi}^{3\pi} & dx(-1)^n\cos^n x\frac{1}{\sqrt{r_1^2+r_2^2-2r_1r_2\cos x}}\\ &= -\int_0^{\pi} dx(-1)^n\cos^n x\frac{1}{\sqrt{r_1^2+r_2^2-2r_1r_2\cos x}},\end{aligned} \tag{1.83}$$

or alternatively

$$\begin{aligned}-\int_{\pi}^{3\pi} & dx(-1)^n\cos^n x\frac{1}{\sqrt{r_1^2+r_2^2-2r_1r_2\cos x}}\\ &= -\int_0^{2\pi} dx(-1)^n\cos^n x\frac{1}{\sqrt{r_1^2+r_2^2-2r_1r_2\cos x}}.\end{aligned} \tag{1.84}$$

After Eq. (1.84) is substituted in Eq. (1.82),

$$
\begin{aligned}
&-\int_0^{2\pi} d\theta_{12}\cos^n\theta_{12}\frac{1}{\sqrt{r_1^2+r_2^2-2r_1r_2\cos(\pi+\theta_{12})}}\\
&=-\int_0^{2\pi} dx(-1)^n\cos^n x\frac{1}{\sqrt{r_1^2+r_2^2-2r_1r_2\cos x}}\\
&=-\int_0^{2\pi} d\theta_{12}(-1)^n\cos^n\theta_{12}\frac{1}{\sqrt{r_1^2+r_2^2-2r_1r_2\cos\theta_{12}}},
\end{aligned}
\tag{1.85}
$$

where in the second step variable x is relabeled as θ_{12}.

Inserting Eq. (1.85) into (1.80), we have

$$
\begin{aligned}
&\int_0^{2\pi} d\theta_{12}\cos^n\theta_{12}\left(\frac{1}{r_{12}}-\frac{1}{r_{1+2}}\right)\\
&=\int_0^{2\pi} d\theta_{12}\cos^n\theta_{12}\left[\frac{1}{\sqrt{r_1^2+r_2^2-2r_1r_2\cos\theta_{12}}}\right.\\
&\qquad\left.-\frac{(-1)^n}{\sqrt{r_1^2+r_2^2-2r_1r_2\cos\theta_{12}}}\right]\\
&=\begin{cases}0, & \text{if } n \text{ is even;}\\ \int_0^{2\pi} d\theta_{12}\frac{2\cos^n\theta_{12}}{r_{12}}, & \text{if } n \text{ is odd.}\end{cases}
\end{aligned}
\tag{1.86}
$$

Then we substitute Eq. (1.86) back to Eq. (1.79) to give the theorem. This concludes our proof.

The second term on RHS of Eq. (1.77) or $2\langle 2p_2 2p_3 1s_1|\cos\theta_{23}^2(\frac{1}{r_{12}}-\frac{1}{r_{1+2}})|2p_2 2p_3 1s_1\rangle$ has the factor

$$
\begin{aligned}
\cos^2\theta_{23} &= [\cos\theta_2\cos\theta_3+\sin\theta_2\sin\theta_3]^2\\
&=\cos^2\theta_2\cos^2\theta_3+\sin^2\theta_2\sin^2\theta_3\\
&\quad+2\cos\theta_2\cos\theta_3\sin\theta_2\sin\theta_3,
\end{aligned}
\tag{1.87}
$$

where the last term on the right gives 0 after integration with respect to θ_3, while the first two terms, combined with $\frac{1}{r_{12}}-\frac{1}{r_{1+2}}$ are 0 after integration with respect to θ_{12} from our theorem, because there is no odd power of $\cos\theta_{12}$ in the integrand.

The third term on RHS of Eq. (1.77) is 0 after integration with respect to θ_3 and we have only the last term left.

Because

$$\begin{aligned}\cos\ \theta_{23}\cos\theta_{31}\\ &= [\cos\theta_2\cos\theta_3+\sin\theta_2\sin\theta_3][\cos\theta_3\cos\theta_1+\sin\theta_3\sin\theta_1]\\ &= \cos\theta_2\cos\theta_1\cos^2\theta_3+\sin\theta_2\sin\theta_1\sin^2\theta_3\\ &\quad+\cos\theta_3\cos\theta_1\sin\theta_2\sin\theta_3+\cos\theta_2\cos\theta_3\sin\theta_3\sin\theta_1, \end{aligned} \tag{1.88}$$

which after integration with respect to θ_3 gives

$$\pi(\cos\theta_2\cos\theta_1+\sin\theta_2\sin\theta_1)=\pi\cos\theta_{12}. \tag{1.89}$$

Thus

$$\begin{aligned}&-\langle 2p_3 2p_1 1s_2|\cos\theta_{23}\cos\theta_{31}\left(\frac{1}{r_{12}}-\frac{1}{r_{1+2}}\right)|2p_3 2p_2 1s_1\rangle\\ &= -\langle 2p_1 1s_2|\frac{\cos\theta_{12}}{r_{12}}|2p_2 1s_1\rangle, \end{aligned} \tag{1.90}$$

which is worked out in the same fashion as integral (1.35) in section 1.10 to give

$$-\langle 2p_1 1s_2|\frac{\cos\theta_{12}}{r_{12}}|2p_2 1s_1\rangle = -0.0172573\lambda. \tag{1.91}$$

The total Coulomb energy is thus -0.0345146λ and we have

$$\begin{aligned}E &= \frac{\langle\Psi|H|\Psi\rangle}{\langle\Psi|\Psi\rangle}\\ &= \frac{11\lambda^2}{9}-0.0345146\lambda\\ &= \frac{11}{9}(\lambda-0.0141196)^2-0.00019936, \end{aligned} \tag{1.92}$$

from which the upper bound for ground-state energy is therefore -0.00019936.

This confirms our conjecture that $3e$+$1e$ indeed forms a quasi-bound state. Here quasi-bound merely means that the 4 particles are relatively localized or near one to another and the total energy of this system is *lower* than when they are infinitely apart. Note that the 4 particle as a whole could still move freely with an effective mass of $4m$.

We proceed on to try other wavefunctions and want to improve our result.

1.11.3 Spin-Triplet Trial Wavefunction of $1s$, $2p^+$, and $2p^-$ States

The next logic choice is still $1s$, $2p^+$, and $2p^-$ states but with a spin-triplet configuration:

$$\Psi = \frac{2}{\sqrt{6}}\{|2p_1 2p_2 1s_3\rangle \sin\theta_{12}|\mathcal{S}^s_{1,23}\rangle + |2p_2 2p_3 1s_1\rangle \sin\theta_{23}|\mathcal{S}^s_{23,1}\rangle + |2p_3 2p_1 1s_2\rangle \sin\theta_{31}|\mathcal{S}^s_{31,2}\rangle\}, \qquad (1.93)$$

where the normalized spin wavefunction $|\mathcal{S}^s_{12,3}\rangle$ is symmetrized in indices 1 and 2, and describes a spin triplet state for electrons 1 and 2. $|\mathcal{S}^s_{123}\rangle$ may be written $|\uparrow_1\uparrow_2\updownarrow_3\rangle = \frac{\sqrt{2}}{2}|\uparrow_1\uparrow_2\rangle(|\uparrow_3\rangle + |\downarrow_3\rangle)$ but can also take other forms, for example $|\downarrow_1\downarrow_2\updownarrow_3\rangle$ or $|\frac{1}{\sqrt{2}}(\uparrow_1\downarrow_2 + \uparrow_2\downarrow_1)\updownarrow_3\rangle$. Three forms of $|\mathcal{S}^s_{12,3}\rangle$ all satisfy $\langle\mathcal{S}^s_{12,3}|\mathcal{S}^s_{12,3}\rangle = 1$ and $\langle\mathcal{S}^s_{12,3}|\mathcal{S}^s_{23,1}\rangle = 1/2$ and so on.

This trial wavefunction gives exactly the same result as the spin singlet trial wavefunction,

$$E = \frac{\langle\Psi|H|\Psi\rangle}{\langle\Psi|\Psi\rangle} = \frac{11\lambda^2}{9} - 0.0345146\lambda = \frac{11}{9}(\lambda - 0.0141196)^2 - 0.00019936, \qquad (1.94)$$

which gives $E_{\min} = -0.00019936$. We are not yet able to tell whether the spin singlet or the triplet state is more stable.

1.11.4 Spin-singlet Trial Wavefunction of Modified $1s$, $2p^+$, and $2p^-$ States

In such a $3e$+$1h$ complex, the strong repulsive Coulomb force is a penalty that like charges need to avoid, thus it would be beneficial if the trial wavefunction is small when two electron coordinates are close one to the other. Take the spin-singlet wavefunction for example:

$$\Psi = \frac{2}{\sqrt{6}}\{|2p_1 2p_2 1s_3\rangle \cos\theta_{12}|\mathcal{S}^a_{12,3}\rangle + |2p_2 2p_3 1s_1\rangle \cos\theta_{23}|\mathcal{S}^a_{23,1}\rangle + |2p_3 2p_1 1s_2\rangle \cos\theta_{31}|\mathcal{S}^a_{31,2}\rangle\} \qquad (1.95)$$

If we can modify Ψ so that it is small when two electrons are close, the total energy then might be lower. So we choose the following:

$$\begin{aligned}\Psi = \frac{2}{\sqrt{6}}\{&|2p_1 2p_2 1s_3\rangle(1-\cos\theta_{12})|S^a_{12,3}\rangle\\ &+|2p_2 2p_3 1s_1\rangle(1-\cos\theta_{23})|S^a_{23,1}\rangle\\ &+|2p_3 2p_1 1s_2\rangle(1-\cos\theta_{31})|S^a_{31,2}\rangle\},\end{aligned} \tag{1.96}$$

where the first term on RHS will be zero when $\theta_{12} = 0$, or $\theta_1 = \theta_2$, or in other words, when electrons 1 and 2 are near.

Note that after this modification, the three particles are no longer in pure $1s$, $2p^+$, and $2p^-$ states, but we may still use these words for notational purpose, and call them modified $1s$, $2p^+$, and $2p^-$ states. Then

$$\begin{aligned}\langle\Psi|\Psi\rangle &= 2\langle 2p_1 2p_2 1s_3|1-2\cos\theta_{12}+\cos^2\theta_{12}|2p_1 2p_2 1s_3\rangle\\ &\quad -2\langle 2p_1 2p_2 1s_3|1-\cos\theta_{12}-\cos\theta_{23}\\ &\quad +\cos\theta_{12}\cos\theta_{23}|2p_2 2p_3 1s_1\rangle\\ &= 3-2\{\langle 2p_1 1s_3|2p_3 1s_1\rangle\\ &\quad +\langle 2p_1 2p_2 1s_3|\cos\theta_{12}\cos\theta_{23}|2p_2 2p_3 1s_1\rangle\},\end{aligned} \tag{1.97}$$

where we have used identities

$$\begin{aligned}&\langle 2p_1 2p_2 1s_3|1-2\cos\theta_{12}+\cos^2\theta_{12}|2p_1 2p_2 1s_3\rangle\\ &\quad = \langle 2p_2 2p_3 1s_1|1-2\cos\theta_{23}+\cos^2\theta_{23}|2p_2 2p_3 1s_1\rangle\\ &\quad = \langle 2p_3 2p_1 1s_2|1-2\cos\theta_{31}+\cos^2\theta_{31}|2p_3 2p_1 1s_2\rangle,\end{aligned} \tag{1.98}$$

and

$$\begin{aligned}&\langle 2p_1 2p_2 1s_3|1-\cos\theta_{12}-\cos\theta_{23}+\cos\theta_{12}\cos\theta_{23}|2p_2 2p_3 1s_1\rangle\\ &\quad = \langle 2p_2 2p_3 1s_1|1-\cos\theta_{23}-\cos\theta_{31}+\cos\theta_{23}\cos\theta_{31}|2p_3 2p_1 1s_2\rangle,\end{aligned} \tag{1.99}$$

and the like.

The exact value of $\langle\Psi|\Psi\rangle$ is found to be 90/32.

Next we want to find out the kinetic energy

$$\begin{aligned}\langle\Psi|-\nabla_1^2-\nabla_2^2-\nabla_3^2|\Psi\rangle &= 3\langle\Psi|-\nabla_1^2|\Psi\rangle\\ &= 3\langle\nabla_1\Psi|\nabla_1\Psi\rangle,\end{aligned} \tag{1.100}$$

where

$$\nabla_1 = \hat{\mathbf{r}}_1\frac{\partial}{\partial r_1}+\hat{\boldsymbol{\theta}}_1\frac{1}{r_1}\frac{\partial}{\partial\theta_1}. \tag{1.101}$$

Because

$$\nabla_1|\Psi\rangle = \frac{2}{\sqrt{6}}\left\{\left(\frac{1}{r_1}-\frac{\lambda}{3}\right)|2p_1 2p_2 1s_3\rangle(1-\cos\theta_{12})\mathcal{S}^a_{12,3}\right.$$
$$-\lambda|2p_2 2p_3 1s_1\rangle(1-\cos\theta_{23})\mathcal{S}^a_{23,1}$$
$$\left.+\left(\frac{1}{r_1}-\frac{\lambda}{3}\right)|2p_3 2p_1 1s_2\rangle(1-\cos\theta_{31})\mathcal{S}^a_{31,2}\right\}\hat{\mathbf{r}}_1$$
$$+\frac{2}{\sqrt{6}}\frac{\hat{\boldsymbol{\theta}}_1}{r_1}\left\{|2p_1 2p_2 1s_3\rangle\sin\theta_{12}\mathcal{S}^a_{12,3}\right.$$
$$\left.-|2p_3 2p_1 1s_2\rangle\sin\theta_{31}\mathcal{S}^a_{31,2}\right\} \quad (1.102)$$

we have

$$\langle\nabla_1\Psi|\nabla_1\Psi\rangle$$
$$=\frac{4}{6}\left\{2\langle 2p_1 2p_2|\left(\frac{1}{r_1^2}-\frac{2\lambda}{3r_1}+\frac{\lambda^2}{9}\right)(1-2\cos\theta_{12}+\cos^2\theta_{12})|2p_1 2p_2\rangle\right.$$
$$+\langle 2p_2 2p_3|\lambda^2\left(1+\cos^2\theta_{23}-2\cos\theta_{23}\right)|2p_2 2p_3\rangle$$
$$-2\langle 2p_1 1s_3|\left(\frac{\lambda^2}{3}-\frac{\lambda}{r_1}\right)|1s_1 2p_3\rangle$$
$$-\langle 2p_1 2p_2 1s_3|\left(\frac{1}{r_1^2}-\frac{2\lambda}{3r_1}+\frac{\lambda^2}{9}\right)|2p_1 2p_3 1s_2\rangle$$
$$\left.+2\langle 2p_1 2p_2|\frac{\sin^2\theta_{12}}{r_1^2}|2p_1 2p_2\rangle\right\}, \quad (1.103)$$

where we have used symmetry identities similar to Eqs. (1.98) and (1.99) to merge some terms. Those terms which go to 0 after integration of one variable, say θ_2, are dropped as well.

After somewhat tedious but straightforward calculation, the total kinetic energy is $\frac{1405}{432}\lambda^2$.

The Coulomb energy of the trial wavefunction is calculated as usual though with added complexity. We just give the result here, -0.566523λ.

Thus

$$E = \frac{\langle\Psi|H|\Psi\rangle}{\langle\Psi|\Psi\rangle}$$
$$= 1.15638\lambda^2 - 0.201430\lambda$$
$$= 1.15638(\lambda-0.0870951)^2 - 0.00877178, \quad (1.104)$$

which gives $E_{\min} = -0.00877178$, a much improved result over previous cases.

Also recall that we have spin-singlet state in this case. In contrast, the spin-triplet trial wavefunction of Eq. (1.95) cannot be so modified as Eq. (1.96) because the first term on RHS of Eq. (1.95) is already small when electrons 1 and 2 are close. This suggests that antiparallel spins be energetically more favorable than parallel ones. We will hereafter focus our attention on the spin-singlet state only.

1.11.5 Spin-Singlet Trial Wavefunction of Modified $2s$, $2p^+$, and $2p^-$ States

The next trial wavefunction and the final one in this Chapter is made of modified 2s, $2p^+$, and $2p^-$ states:

$$\begin{aligned}\Psi = \frac{2}{\sqrt{6}}\{ &|2p_1 2p_2 2s_3\rangle(1-\cos\theta_{12})|\mathcal{S}^a_{12,3}\rangle \\ &+|2p_2 2p_3 2s_1\rangle(1-\cos\theta_{23})|\mathcal{S}^a_{23,1}\rangle \\ &+|2p_3 2p_1 2s_2\rangle(1-\cos\theta_{31})|\mathcal{S}^a_{31,2}\rangle\}. \qquad (1.105)\end{aligned}$$

The treating of such Ψ is similar to the previous case, and we just list the final results as follows:

$$\begin{aligned}E &= \frac{\langle\Psi|H|\Psi\rangle}{\langle\Psi|\Psi\rangle} \\ &= 0.632997\lambda^2 - 0.168735\lambda \\ &= 0.632997(\lambda - 0.133283)^2 - 0.0112447. \qquad (1.106)\end{aligned}$$

Then the upper bound for ground-state energy is now improved to -0.0112447.

This largely analytical treatment of the tetron problem gives us a proper start to understand the wavefunction symmetry, in both the orbital and the spin part. The relatively transparent expression in both the wavefunction and energy has the added benefit of removing any doubts, which might arise with a full numerical yet often opaque computational method. However, it should not be a surprise to see that a single parameter variation model is not able to answer the question of meta-stable state in vacuum, because the 2D tetron has essentially 6 degrees of freedom. A numerical method for high precision study is therefore needed and will be discussed later. But before any numerical details, we also want a clear picture of what is at stake for the stability of tetron against dissociation.

1.12 Stability against Dissociation

As mentioned before, a tetron is not stable against the dissociation of one of its electrons into infinity in *vacuum*. The theorem by Lieb states that the ground-state energy of a $3e+1h$ complex must be higher than that of a trion plus an electron, which is assumed to be infinitely far away, having arbitrarily small kinetic energy. Therefore, if a quasi-bound state of $3e+1h$ is placed in vacuum, the loosely "bound" complex tends to lower its energy by moving one of the electrons away to minimize the energy. This corresponds to the dissociation (detachment) of one electron as shown in Fig. 1.8. However, if the tetron is not in vacuum, for example, if many other electrons and holes are present nearby in a form of Fermi sea, then there is hardly any gain in energy for either the tetron or the whole system even if an electron dissociates from one tetron; this electron would simply increase it kinetic energy with other electrons of Fermi sea due to the Pauli principle. Further, we can

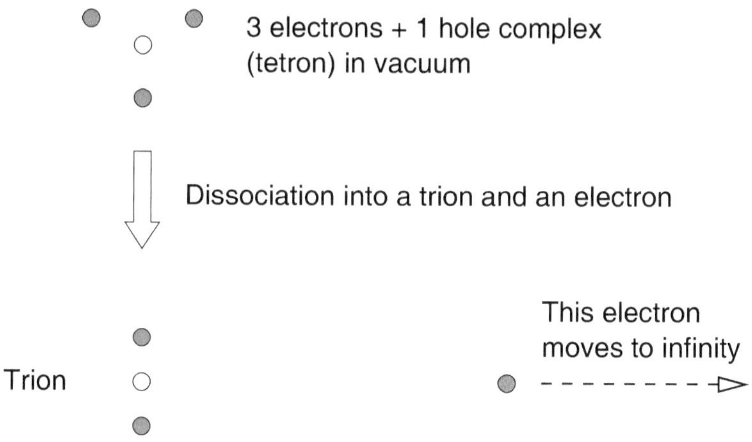

Figure 1.8 The reason why the tetron ($3e+1h$) is not stable in vacuum.

model the effect of the electrons of the Fermi sea that assumes a Fermi energy E_F, which is dominated by kinetic energy under the Pauli exclusion principle. Take the independent electron gas model then the Coulomb energy is zero. The effect of dissociating one of the three electrons in the tetron complex then corresponds to adding the electron to the Fermi sea neighboring the tetron. Without the loss of generality, one can assume such a two-dimensional electron gas occupies anywhere except the circle roughly defined by the radius (r_{te}) of the tetron. Then the whole problem is reduced to the square well case that we are to encounter in Section 1.12.1, with the potential depth $-V$ replaced by Fermi level (chemical potential) E_F at absolute zero $T = 0$. Then with a suitable choice of the electron density n that governs the E_F we can have a range of n in which the tetron is stable against losing one of its electrons to the neighboring Fermi sea. The detailed analyses are to be given in Section 1.13.

When the densities of electron and hole are low, the situation is more like a vacuum and therefore tetrons are not stable. With an increasing charge density, however, the situation could be increasingly different as we see in the preceding paragraph. Actually Nature may have already suggested a similar mechanism for us: Doubly charged negative ions until recently have not been seen in gas phase, or in isolation. In contrast, they have long been known to exist in condensed matter systems such as liquids and solutions (Scheller et al., 1995).

If the densities are to further increase by too much, the energy of the tetron state will no longer be negative after a second threshold. The reason is simple, the gain in Coulomb energy increases as a^{-1} while the penalty paid for the kinetic energy scales as a^{-2}, where a is the average distance between particles. Then we would see a system close to the ordinary metal because a bound state is no longer possible.

To be more precise some numerical variational methods are needed as the next step. We will show that, with a square potential well of certain depth and radius, the tetron will *not* dissociate into a trion plus an electron. Therefore, the tetron state is the *true* ground state, rather than the trion, under such a circumstance.

1.12.1 Numerical Results on the Tetron

The stochastic variational methods (Suzuki and Varga, 1998) (SVM) utilizes the correlated Gaussian orbital as the basis function:

$$\exp\left[-\frac{1}{2}\sum_{k<l}^{N}\alpha_{kl}(\mathrm{r}_k-\mathrm{r}_l)^2-\frac{1}{2}\sum_{k=1}^{N}\beta_k\mathrm{r}_k^2\right]=\exp\left(-\frac{1}{2}\mathrm{r}A\mathrm{r}\right), \tag{1.107}$$

where r_k, r_l are notational as the 2D coordinates for the k-th and l-th particle. N is the total number of electrons and holes. The explicit presence of the term $(\mathrm{r}_k - \mathrm{r}_l)^2$ makes the basis strongly correlated, and is regulated by the nonlinear parameters α_{ij} to variationally change the pair correlation between the particles i and j. This explains the meaning of "correlated Gaussian": For example, a large α_{ij} implies that the wavefunction is only significant when particles i and j are nearby. Part of this feature was already seen in the trion and tetron wavefunctions of preceding sections. α_{kl} and β_k are then matrix elements of A, as the notation is compact. The second term

$$\exp\left(-\frac{1}{2}\sum_{k=1}^{N}\beta_k\mathrm{r}_k^2\right) \tag{1.108}$$

is a product of independent single-particle Gaussian orbitals that describe the individual motion of particle k. β_k then relates to its kinetic energy.

RHS of the equation above is written in a matrix form with A the matrix, given explicitly by

$$\mathrm{r}A\mathrm{r}=\sum_{i,j=1}^{N}A_{ij}\mathrm{r}_i\mathrm{r}_j. \tag{1.109}$$

For two dimensions, the wavefunction is much simplified with r written as a complex $x + iy$, and with explicit account of each particle's angular momentum m_k. The correlated Gaussian basis function is written compact as

$$\Phi^A_{MS}(\mathrm{r})=\mathcal{A}\left\{\prod_{k=1}^{N}(x_k+iy_k)^{m_k}\exp\left\{-\frac{1}{2}\mathrm{r}A\mathrm{r}\right\}\chi_{SM_S}\right\}, \tag{1.110}$$

where $\mathcal{A}$ antisymmetrizes terms within {} . $M = (m_1, \ldots, m_N)$ gives the angular momentum of individual particles, and $\mathrm{r}A\mathrm{r}$ is already given by Eq. (1.109). The spin function χ_{SM_S} is a linear combination of single-particle spin functions that sum up to a total spin quantum number S and the corresponding projection M_S. For the orbital momentum that is indexed by integer m, the following identities prove helpful:

$$(x+iy)^m = \begin{cases} (x+iy)^m & m \geq 0 \\ (x-iy)^{-m} & m < 0 \end{cases} \tag{1.111}$$

when the complex's modulus is one.

Note that a similar basis is often utilized for correlated electron states in the Fractional Quantum Hall Effect (FQHE) (Laughlin, 1983), initially adopted by Laughlin. The main advantage of this basis is that its matrix elements can be calculated analytically. The method of Ref. Varga (2008) is adapted to this numerical study, with modifications to taking into account the loosely bound state of tetron. The original code approximates some potential matrix element integrals $\int V(r) r^{2n} \exp\{-\frac{c}{2}r^2\}$ as constants when the square well potential is defined at large r. This is a good approximation for tightly bound states. For the loosely bound tetron here, such an assumption is no longer valid and hence the code was modified so that a double precision numerical integration could be carried out.

Suppose that a tetron is thrown to a 2D square potential well of depth V and radius a. Note that because of the presence of the external potential, the tetron is no longer in a strict vacuum, and the Lieb theorem is no longer applicable. In reality such a case exists in a broad range of subfields in condensed matter physics, besides the cuprate superconductivity. For example, it may manifest as quantum dots (QD) in a quasi-2D quantum well, irregularities in a QW due to random variations in the composition or doping, in some defects at interfaces, or due to the presence of other electrons nearby. Therefore, the techniques well developed for the semiconductor quantum well physics can be borrowed to help to better understand the tetron physics. To simply the numerical procedure, we neglect any non-essential difference that the well potential might present separately to the electron and the hole. Instead the same potential is assumed for both. This is a good approximation for many quantum wells with local variations in the chemical composition. In practice, the difference in the offsets of the conduction and the valence band can be compensated by the so-called modulation doping, to give a same effective confining potential to both the electron and the hole.

The potential $V(r)$ is defined as

$$V(r) = \begin{cases} -V & \text{if } r < a; \\ 0 & \text{if } r \geq a. \end{cases} \tag{1.112}$$

First we ran the numerical code to find out the ground state energy of trion E_{tr} in the square well. The code was carefully benchmarked in the well-known limiting case of 2D trion to get $E_{tr} \sim -1.12$, in vacuum. We also changed the size of the basis $\mathcal{K}$ (see Ref. (Varga, 2008)) from 200 to 600, with the binding energy improved from 0.12036247 to 0.12130221. Afterwards, the code found the ground state energy of tetron E_{te} in the square well. Similarly the computation was also well benchmarked in vacuum to give an E_{te} of tetron approaching to that of trion. E_{te}, however, could never be lower than E_{tr} in vacuum. Therefore, the energy drop in each numerical iteration of E_{te} merely depicts the physical process of a tetron dissociating into a trion plus an electron, with the electron going to infinity. Now with the square well potential $V(r)$ turned on, we want see if tetron is more stable. Then we need to compare E_{te} and $E_{tr} + E_e$ where E_e is the electron energy when it is far away. If one of the electrons escapes to infinity as mandated by the dissociation, E_e is positive definite because $\langle \psi_e | V(r) | \psi_e \rangle \rightarrow 0$ when $r \rightarrow \infty$ and the kinetic energy is positive definite. Therefore, if $E_{te} < E_{tr}$, then one must have $E_{te} < E_{tr} + E_e$, which means the tetron is the true ground state and it is stable against dissociation into a trion.

The trial wavefunction that we chose for the trion is in a spin doublet and all three particles are in the s-orbit. The trial wavefunction for the tetron is in a singlet state and the orbital part starts with the hole and one electron in the s-orbit and the other two electrons in the $p\pm$ orbits.

The SVM randomly chooses new wavefunctions in the basis set, and once it finds a lower energy, the basis is updated. This process is iterated and the iteration keeps going on until a predetermined number of iteration is reached. The trion is a three-body problem which converges very fast with the SVM. It in general only takes a

Table 1.3 Comparison of ground state energy for trion, E_{tr}, and for tetron, E_{te}, in square wells of different strength and radius

Size a	28	28	28	14	10	7	7	5
Depth V	−0.5	−0.1	−0.04	−0.1	−0.15	−0.25	−0.30	−0.35
E_{te}	−3.070	−1.427	−1.2103	−1.4197	−1.5593	−1.8368	−2.0286	−2.0278
E_{tr}	−2.620	−1.418	−1.2003	−1.4163	−1.5608	−1.8470	−1.9963	−2.1172
Stable?	y	y	y	y	i	i	y	n

Note: y: tetron is more stable; n: tetron is less stable; i: inconclusive.

Table 1.4 Continuation of Table 1.3

Size a	5	5	5	3	3	3	3	3
Depth V	−0.43	−0.44	−0.47	−1.0	−1.02	−1.03	−1.1	−1.18
E_{te}	−2.35492	−2.39490	−2.50369	−3.88114	−3.95555	−3.99243	−4.25357	−4.52510
E_{tr}	−2.35519	−2.38492	−2.47419	−3.90227	−3.96120	−3.98977	−4.19711	−4.43272
Stable?	n	y	y	n	i	y	y	y

few hours on a computer with an Intel Core (~3 GHz) series CPU to converge to the fifth digit after the decimal point, with a $\mathcal{K} = 600$. The convergency of the four-body tetron energy is generally slower in terms of time, often by one order of magnitude. The E_{te} also starts at a higher energy than the E_{tr}. However, its drop in energy is typically steeper on a per-iteration base, compared to the trion. E_{te} often approaches E_{tr}, and if the condition is right, may become lower. Therefore, if E_{te} is found to be significantly lower than E_{tr}, say starting at the second or third digit after the decimal point, then there leaves little doubt that the tetron is more stable than the trion. On the contrary, if E_{te} is found only slightly different from E_{tr}, say in the fifth digit after the decimal point, we still cannot draw a conclusion as to which is more stable. We then label it as inconclusive, to be on the conservative side, although the tetron energy has more variables to optimize and its slower convergency also means more room to drop.

Tables 1.3 and 1.4 summarize the result. In many cases, there is no doubt that the tetron is the true ground state for the type of potential outlined in Eq. (1.112). One important observation is that the confinement potential does not need to be very strong to stabilize the tetron state. In fact a mere potential of -0.043 suffices when radius of the potential well is significantly larger than the rough size of tetron (~7), for example 28. The purpose of many data in Tables 1.3 and 1.4 is to illustrate that the a-V plane actually forms a phase diagram where a line will separate the two relative stability phases of trion and tetron. The exact position of this boundary is subject to high precision numerical studies in the future. The numbers listed in Tables 1.3 and 1.4 though should give a description that is quantitative enough. The most profound discovery of this numerical result is that *tetron is the true ground state*, with a square well deep and wide enough. For example, the large energy difference between E_{te} and E_{tr} in the second column of Table 1.3 is beyond any possible numerical error, and testifies without any doubt that a well radius of 28 and depth of -0.5 make the tetron the true ground state.

While the current result does not answer whether or not a meta-stable state exists for tetron in vacuum, it definitely shows such a state exists for tetron under a proper potential. To see this, just tune the strength of the $V(r)$ of Eq. (1.112) so that $E_{tr} = E_{te}$.

Then by definition, the tetron is either in a meta-stable state or in a degenerate ground state. To destabilize the degenerate ground state, one decreases the magnitude of V in $V(r)$ by an arbitrarily small amount, then the tetron ends up in a meta-stable state. Notice that we still cannot give a definite answer as to the lifetime, but the proof of its existence in an external field is a step forward. The implication of tetron is not limited to the condensed matter physics. For atomic physics, for example, it indicates that the H^{--} likely stabilizes in external potentials, in both experiment and theory. This should not come as a surprise after all since many ions are now studied in ion traps anyway, especially for the magnetic type.

1.13 Phase Diagram for the Tetron Stability

Tables 1.3 and 1.4 should give a boundary that defines the stability of tetron versus trion if a is plotted against the potential V. By curve fitting the "critical" points (not in a strict sense), such as the one at $a = 28$, $|V| \approx 0.04$, the relative stability regions of the trion and tetron are defined.

The points are curve fitted to give a nonlinear power (allometric) form:

$$V = 5.16a^{-1.484}. \tag{1.113}$$

It is worth noting that the expression of Eq. (1.113) is only meaningful for a moderate a. If a is too small, say less than 1 Bohr radius, then it implies a high density, or a small Wigner-Seitz radius r_s, when one tries to confine the four particles of a tetron within the potential well. In a practical discussion of electron–hole interaction in cuprates, if the density is high the correlation effects are then reduced: An *e-h* liquid (of non-equal number) then results. Furthermore, it is not meaningful to study wavelike correlations with long-range Coulomb interaction when restricted to a short distance. Therefore, we are only concerned with a from a few 2D Bohr radius $r_{B,2D}$ up to a few hundred. We need a few more cautions on the 2D Bohr radius,

$$r_{B,2D} = \frac{2\pi\epsilon\hbar^2}{me^2}, \tag{1.114}$$

where m is understood as the reduced mass of the exciton. The normal (3D) Bohr radius (in atomic physics) is defined as $4\pi\epsilon_0\hbar^2/m_e e^2$ for vacuum permittivity and rest mass m_e of the electron. Here in the physics of excitonic complexes, the permittivity is now the dielectric response of the parent insulator and the mass replaced by the reduced mass of the exciton. One more difference between the 3D and the 2D Bohr radius is the factor of 2, reflecting the enhanced confinement in the 2D Coulomb problem, which favors the formation of bound-states. A further difference between the 2D and 3D Coulomb problem is the binding energy: The 2D value is 4 times that of its 3D counterpart; and the most probable radius for 2D ground state, is one half of the 2D Bohr radius. The importance of two-dimensionality and its connection to high T_c in the cuprates were emphasized by Leggett (Leggett, 2006).

Equation (1.113) defines a line separating two phases where trions and tetrons are, respectively, more stable. The line is given in Fig. 1.9 outlining the boundary of the trion and tetron phase. Figure 1.9 therefore becomes the actual phase diagram for the relative stability of the trion and tetron.

This a-V phase diagram is of central importance when we make a connection with the pairing mechanism of cuprate superconductors, in Section 1.14. But before we make such a bold move, we should ask ourselves: Does the tetron really exist in the physical world? To answer this, we keep in mind that trions are detected often with optical spectroscopy, like excitons. The semimetallic tetron (e.g., in cuprates) is stable against recombination yet difficult to characterize with the standard optical means. In semiconductors the situation is reversed and therefore existing experiments there should help shed a light on tetron's behaviors. A short digress into semiconductor quantum wells and quantum dots could reveal some delightful inspiration.

1.13.1 Evidence from QW and QD

The very fact that the binding energy of tetron can be higher than that of trion in a square-well may help explain a long existing over-binding problem in quantum well and quantum dot physics (Bracker et al., 2005; Peeters et al., 2001). The experimental

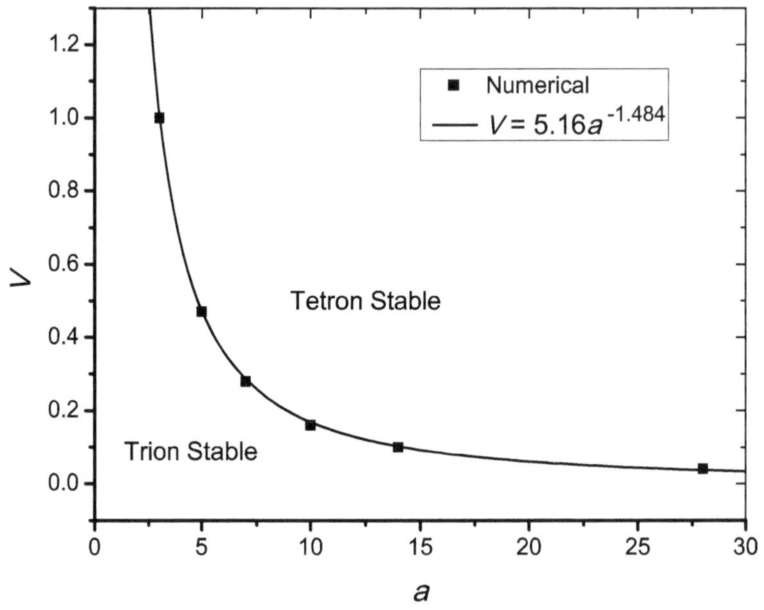

Figure 1.9 Phase diagram (a-V) of relatively stable region for the tetron and trion. Square data points are from Tables 1.3 and 1.4. The solid line is a nonlinear power fitting.

trion binding energy is known to be significantly higher than the theoretical one. One proposed solution to this problem is the possible interfacial, doping or compositional irregularities (disorders) in the QW (Dacal et al., 2002). The effect of such irregularity is manifested in the numerical results. We can assume that the potential defined in Eq. (1.112) as due to a local anomaly in the chemical composition. For example a local excess of the indium composition x in a GaAs/$Ga_{1-x}In_xAs$/GaAs QW in a circle defined by $r < a$ can give a potential for electrons and holes as described by Eq. (1.112). Another example which is more feasible and relevant to the experiment is a local QW width increase in an $Al_xGa_{1-x}As$/GaAs/$Al_xGa_{1-x}As$ QW. Without the loss of generality, let us assume $a = 28$ and $V = -0.1$ as from Table 1.3. Although the irregularity due to the potential is only 0.1, the increase in the binding energy of the trion is nearly twice as much. More significant increase lies in the tetron, which is nearly 3 times of V compared to a tetron in vacuum in its dissociation limit. Such a trend is profound in understanding why local disorders play such a prominent role in the experimental over-binding of charged excitonic complexes, which should also include the doubly charged types. Experimentally, it is also natural to ask if some of exceedingly large binding energies are actually due to the tetron. Moreover, if one considers local disorder traps of a QW as another form of QD, the tetron then likely plays a role in the QD.

In this section, we briefly comment on experimental circumstances for observing such excitonic complexes. This depends on electronic structure of the underlying material. If the material is of a semiconductor type, optical spectroscopic techniques like photoluminescence might have already revealed its existence although a clear separation from the trion peaks is not simple due to the small difference in the binding energy. Also the tetron's existence is somehow regulated by doping. If the QW is properly (modulation) doped so that there is a 3 to 1 ratio in the two carriers (e and h) at thermal equilibrium, its observation is much more likely.

Because the tetron is a boson, it might be revealed by quantum statistics, much like the highly sought Bose–Einstein condensation (BEC) of excitons (Eisenstein and MacDonald, 2004). In this case, however, semiconductors might not be a good venue for the experiment. Recall the much debated Bose–Einstein (BE) condensation of excitons (Eisenstein and MacDonald, 2004). The excitons are so light in a III-V semiconductor and hence in principle its BE condensation should be relatively easy to materialize. However, only until recently in the unexpected bilayer QW structure and under an external magnetic field, was the evidence starting to emerge. The likely culprit, often neglected at the first glance is the electron–hole recombination that quickly annihilates the exciton before a robust condensation can form. A similar problem exists here for ordinary semiconductor micro-structures, with the probable exception of quantum-wells with a misaligned Type-II hetero-interface. In ordinary III-V quantum wells, the exciton radiative recombination time is a very short, on the scale of 40 ps (picosecond) or less (Finkelstein et al., 1998; Vinattieri et al., 1994). To see the condensate, one can define a de Broglie time scale given by $\tau_d = \lambda_d/v = h/pv = h/mv^2 = h/3kT$ where λ_d is the de

Broglie wavelength. The de Broglie time scale can be roughly taken as the time to achieve phase coherence in the condensate, barring scatterings due to phonons and impurities. For a temperature of 1 K, $\tau_d = 16$ ps, and so it is on the same order of magnitude as the exciton lifetime for 1 K, but should be much larger with additional scatterings due to defects. Therefore, the recombination indeed hinders the BE condensation in semiconductors, with probably the exception of Type-II hetero-structures.

As mentioned before the stability of electron–hole liquid against recombination is given by thermodynamics in semimetals. Similar stability should exist in hetero-structures with semimetal band-edges. A good example is the Type II hetero-structure with a misaligned bandgap, i.e., the GaSb/InAs/GaSb quantum well (Chang and Ploog, 2012). Further example should include Type III hetero-structures, especially if the carrier concentration can be controlled.

The terminology needs to clarify one more time here. When the *e*-*h* bound state is discussed for semiconductor, it is referred to as excitonic, literally like an exciton as an excited state. For semimetallic materials, when "excitonic" is mentioned, it is in the strict sense of an attractive Coulomb bound state, and does not imply any excitation from the ground state. The typical semimetals, like As, Sb and Bi, have too high an electron/hole density and therefore screening is excessive, and excitonic structures do not seem to be significant there. However, there exist diluted semimetals which typically happen near a metal-insulator(semiconductor) transition, where the carrier density is often controllable by doping (Luo and Miley, 2009). The Type II and III hetero-structures could be a good venue for doubled charged excitonlike complexes if the carrier mass is not too small.

There exists one additional twist to be emphasized one more time: The stable exciton-like complex of diluted semimetal actually makes its observation difficult with the standard optical spectroscopy, from which most of the our experimental knowledge on the semiconductor exciton is obtained. With the stable exciton-like Coulomb bound state in semimetals, the traditional spectroscopic methods are largely ineffective. But one thing is clear: It forms an ideal playground for condensation physics. Therefore, many of the techniques utilized in the study of BE condensation of cold atoms, like the laser optical probe (Bradley et al., 1995), may be adapted to the tetron.

1.14 Connect the Tetron to the Cooper Pair

Now we are ready to make a formal connection from the tetron to the Cooper pair (Cooper, 1956): (1) Both are double charged; (2) both obey the Bose–Einstein (BE) statistics. This connection is particularly attractive due to facts about the nature of Cooper pair in cuprates: (1) Cuprate superconductivity is largely (at least from the underdoped to the optimal doping) due to the BE condensation (BEC) of preformed pairs; (2) The pairing is local in real space with the pair size ($\sim$ coherence length) ξ from 15 Å to some 30 Å (in-plane); (3) Pairs form below the pseudogap temperature T^*. We can check that the tetron's wavefunction gives a size of around 30 Å at $\epsilon_r \sim 15$ and single band hole (electron) mass of $m = 2m_e$, and a critical temperature up to $\sim$300 K, with formulae in later sections. Therefore, it is reasonable, natural and logical to relate the tetron to the Cooper pair, for the cuprate. Also keep in mind that the effective mass of the SC condensate is around 8 m_e (6 m_e to 12 m_e for example) from the London penetration depth λ_L (see Section 5.2), significantly higher than what is inferred from a two-body entity, i.e., 2-electron (or 2-hole) Cooper pair. In contrast, it reasonably fits into the mass of a four-body complex as the tetron.

In summary we are ready to make a bold yet educated statement: The Cooper pair in the (p-type) cuprates is actually the *tetron*, made of 3 holes and 1 electron.

In the following sections, we derive the well-known phase diagram of the cuprate superconductors from the tetron/trion/exciton physics. Since cuprate systems vary in details, to get the generic phase diagram we need a model cuprate, whose critical physical parameters represent the average material. We use the following: (1) lattice constant $a_0 = 3.80$ Å; (2) $c_0 = 6.80$ Å (unless otherwise specified); (3) close-coupled copper oxide spacing (in multiplane cuprates) 3.2 Å (Shaked, 1994); (4) band mass (of electron $m_e^b \approx$ that of hole m_h) $= 2m_e$; (5) undoped low temperature high frequency dielectric constant: 18 (Chen et al., 1989, 1991b; Samara et al.,

1990; Varyukhin and ZAkharov, 1991; Viskadourakis et al., 2015). The dielectric constance (or more precisely permittivity) ϵ and band mass m play a critical role in the phase diagram via the Coulomb interaction. We need a clarification on the in-plane ϵ of the familiar cuprate systems. From Ref. (Samara et al., 1990) $\epsilon_r = 14.7$ is the lowest among all data so far surveyed and this is from YBCO (Y-123). Data from other cuprates like LSCO are generally higher and we take a median value of 18, as the representative low-T high-frequency limit. Because ϵ is really temperature and frequency related, even for undoped cuprates, the low-T high-frequency limit represents an intrinsic property of the undoped cuprate. This is not difficult to understand: There are always defect-related charge carriers, though not very mobile, still can move via hopping and thermal agitation. This causes the T and f dependence in the effective ϵ. In the doped regime, the exciton complexes should contribute significantly to the low-T permittivity: These polarize readily in an applied electric field, adding additional ϵ beyond possible with the undoped parent matrix alone. The out-of-plane (c-axis) $\epsilon_\perp$ is typically a few times lower, reflecting anisotropic electrostatics and electrodynamics. The three known doping regimes of the cuprate, underdoped, optimal-doping and overdoped, will be in various occasions referred to, respectively, as UD, OP, OD.

1.15 Doping Dependent Phase Diagram of Tetron (Cooper Pair)

As mentioned before, a tetron should dissociate into a trion plus a hole (or an electron) in the crystal vacuum. Here "vacuum" really means undoped cuprate crystal, if in Eq. (1.114) m is the reduced band mass and ϵ the permittivity of undoped matrix. Therefore, the tetron is not the true ground state if doping density approaches zero. However, the situation changes when more holes are added to the system. We first borrow and reformulate our doping model from Eq. (2.12) of Section 2.6.1.

$$\begin{cases} n_p = 1.5x/a_0^2 \\ n_e = 0.5x/a_0^2 \end{cases} \quad ; \quad \text{when} \quad x \leq 0.2, \qquad (1.115)$$

where the density is understood as areal, and a_0 is the basal lattice constant (neglecting non-essential complication of orthorhombic distortions and so on). x is the doping fraction. This doping model is verified in part by high precision DFT electron-structure computation of Section 4.5.2.

At low doping the free carrier densities are $1.5x/a_0^2$ for holes and $0.5x/a_0^2$ for electrons were it not for excitonic Coulomb bound states. Instead, the ground state is made of positive trions and holes of same density $0.5x/a_0^2$ due to the excitonic action at the very dilute limit. If x and thus the density of holes further increase tetron should start forming.

Let us imagine putting a tetron in the hollow center of a Fermi sea made of holes and trions. Also assume that densities are not too high to screen out the intra-tetron Coulomb force. For the tetron to lose a hole to a Fermi sea of other holes, the supposedly leaving hole would have to pay a penalty in kinetic energy of E_F plus additional Coulomb energy by getting close to the hole sea of density $0.5x/a_0^2$.

The hole sea of $0.5x/a_0^2$ has by itself a Fermi energy of

$$E_F = n_p \frac{\pi \hbar^2}{m_h} = \frac{0.5x\pi \hbar^2}{m_h a_0^2}, \qquad (1.116)$$

where m_h is the hole band mass and is close to 2 times the bare mass of a free electron, or $2m_e$. The band mass of electron m_e^b is so labeled as to differentiate from the bare rest mass of a free electron m_e. n_p is the density of free holes and the subscript simply reflects the convention of referring to holes as p-type, to be consistent with labeling in other chapters. One might label it as n_h instead if this makes the symbology self-evident.

Besides the hole sea, there is the admixed fermion sea of trions with similar k space occupation. This leaves a lowest occupied state twice as high in energy versus Eq. (1.116). The leaving hole then would fight a multi-component Fermi energy of

$$E_{F,m} = n_p \frac{\pi \hbar^2}{m_h} = \frac{2 \times 0.5x\pi \hbar^2}{m_h a_0^2}. \qquad (1.117)$$

If the leaving hole needs to join the hole sea of $0.5x/a_0^2$ then it has to fight the Coulomb force from there by being closer to the sea. If the Coulomb energy U_C is not screened, it relates to free hole E_F

by

$$\frac{U_C}{E_F} = \frac{r_s}{r_{B,2D}} = \sqrt{\frac{1}{\pi n_p} \frac{1}{r_{B,2D}}}, \tag{1.118}$$

where r_s is the 2D Wigner-Seitz radius. However, screening does exist and the screening vector q_s is on par with the inverse of average inter-hole distance. The q_s should differ from the well-known Thomas–Fermi q_{tf} derived from the 3D case due to (1) modified electrostatics/electrodynamic in anisotropic quasi-2D space; and (2) the quasi-2D density of states (DOS) being step-like, rather than a constant as would from a pure 2D. The 2D Lindhard formula which gives an n-independent screening hence does not apply due to its neglecting such effects. Instead, a workable notion is still the screening q_s as the inverse of the average inter-hole distance $q_s = \sqrt{n_p}$, or

$$U_C(r) = \frac{1}{4\pi\epsilon r} e^{-q_s r} = \frac{1}{4\pi\epsilon r} e^{-\sqrt{n_p} r}, \tag{1.119}$$

which gives the above formula an additional $e^{-1/\sqrt{\pi}}$ factor at r_s, or

$$\frac{U_C}{E_F} = \frac{e^{-1/\sqrt{\pi}}}{\sqrt{\pi n_p} r_{B,2D}} = \sqrt{\frac{2}{\pi x}} \frac{e^{-1/\sqrt{\pi}} a_0}{r_{B,2D}} = \sqrt{\frac{2}{\pi x}} \frac{0.5688 a_0}{r_{B,2D}}. \tag{1.120}$$

In general, the relative magnitude of r_s and $r_{B,2D}$ determines the relative importance of the Coulomb and kinetic energy. This is also the case for the quasi-2D physics of cuprates. If r_s is significantly larger than one (in the Bohr unit) the Coulomb dominates. However, if r_s is too big, there is a propensity to forming the Wigner crystal, and the physics is not of much relevance to superconductivity. It is the region where r_s is a few times larger than $r_{B,2D}$ that is most interesting to the SC and superfluidity field of strongly correlated physics. In other words, the Coulomb interaction energy (in magnitude) on par and up to a few times larger than the kinetic energy (KE) is the focus of our study. The Coulomb energy (CE) is larger in absolute terms than the KE, still it wants to maximize the attractive part and minimize the repulsive ones. This may actually result in the reduction of the total CE in its magnitude. Then this hurts its ability to dominate the KE making the system still fluid-like. If one considers the screening of the Coulomb, the actual CE should come even closer to the KE. Take our model cuprate for example, one can estimate the relative strength of CE/KE in the underdoped regime with x from 0.01 to 0.15 from Eq. (1.120). The results are 3.631 and 0.937, respectively. This verifies our assertion that the Coulomb is general on par or larger than the corresponding KE.

Then the total energy that a hole has to pay to join a hole sea plus a trion sea of $0.5x/a_0^2$ each is

$$E_t = E_{F,m} + U_C = \frac{0.5x\pi\hbar^2}{m_h a_0^2}\left(2 + \sqrt{\frac{2}{\pi x}}\frac{0.5688a_0}{r_{B,2D}}\right)$$
$$= C(x)\frac{0.5x\pi\hbar^2}{m_h a_0^2}, \quad (1.121)$$

where $C(x)$ is the Coulomb + KE factor to give the total energy penalty for tetron to dissociate into a multi-component Fermi sea. It is given by

$$C(x) = 2 + \sqrt{\frac{2}{\pi x}}\frac{0.5688a_0}{r_{B,2D}}. \quad (1.122)$$

Because the supposedly leaving hole has to fight the positive Fermi and Coulomb energy of the sea of holes plus trions, it is equivalent to being set in the step potential well of Eq. (1.113) where V is now given by $E_t = E_{F,m} + U_C$. The radius of the potential well a is conveniently given by

$$a = \frac{a_0}{\sqrt{0.5x}}. \quad (1.123)$$

From the phase diagram in Fig. 1.9, in order for the tetron to form, the "applied" potential, essentially $E_{F,m} + U_C$, has to be higher than the line defined by Eq. (1.113), or

$$E_t = C(x)\frac{0.5x\pi\hbar^2}{m_h a_0^2} > V = 5.16a^{-1.484}. \quad (1.124)$$

Inserting Eq. (1.123) into Eq. (1.124) above, we obtain

$$C(x)\frac{0.5x\pi\hbar^2}{m_h a_0^2} > V = 5.16\left(\frac{a_0}{\sqrt{0.5x}}\right)^{-1.484}. \quad (1.125)$$

In other words, the following inequality has to be satisfied:

$$(0.5x)^{0.258} > \frac{5.16m_h}{\pi\hbar^2}C(x)^{-1}a_0^{0.516}. \quad (1.126)$$

Note that the term $m_h/\hbar^2$ is the inverse of $\hbar^2/2m_\mu$, since $m_h \sim 2m_\mu$ where m_μ is the reduced mass of an exciton. $\hbar^2/2m_\mu$ has a dimension of (energy)(length)^2, and can be made to equal 1 when a natural unit system is adopted. In this system, distance is measured in $r_{B,2D}$ and energy in exciton's binding energy. The equation above is then much simplified to

$$0.5x > \left[\frac{5.16}{\pi}C(x)^{-1}\right]^{3.876}\left(\frac{a_0}{r_{B,2D}}\right)^2. \tag{1.127}$$

The 2D Bohr radius $r_{B,2D} = 2\pi\epsilon\hbar^2/me^2$ varies with the permittivity and thus the relative dielectric constant. Given an $\epsilon_r = 14.7$ from YBCO a 2D Bohr radius of 3.89 Å is calculated, slightly larger than the ~3.80 Å lattice constant. For the model cuprate of $\epsilon_r = 18$, we got $r_{B,2D} = 4.76$ Å, and therefore $a_0/r_{B,2D} \approx 1/1.25$.

Note though that $r_{B,2D}$ thus defined only constitutes a lower bound. For one thing doping significantly increases the effective ϵ and $r_{B,2D}$. The doped exciton bound state can have a dielectric constant double the pristine value. All this should affect the numerical results of Eq. (1.127).

The LHS of Eq. (1.127) is a faster varying function of x than the RHS factor of $C(x)$. This factor does not hold for a too small x, because that corresponds to the classical limit and the notion of a Fermi gas/liquid is no longer valid. It is not very accurate for the overdoped regime either, where correlation is weak. Therefore, we focus on the mid-range x from 0.01 to ~ 0.15 and take an average $\bar{C}(x)$. The average is a bit arbitrary (e.g., arithmetic versus geometric, over x or over C). $C(x)$ typically gives numerical values centered between 3 and 4, as discussed before. We then resort to the simplest approach, and start by taking the geometric average of x first, which gives the average $1/\bar{x}$,

$$\frac{1}{\bar{x}} = \left[\sqrt{\frac{1.5}{10}\times\frac{1}{100}}\right]^{-1} = 25.82. \tag{1.128}$$

$1/\bar{x}$ is then substituted in Eq. (1.122) to give

$$C(\bar{x})^{-1} = \left[2 + \sqrt{\frac{2}{\pi\bar{x}}}\left(\frac{0.5688a_0}{r_{B,2D}}\right)\right]^{-1} = 3.845^{-1}. \tag{1.129}$$

The numerical result of Eq. (1.129) is inserted into Eq. (1.127) to give

$$0.5x > \left[\frac{5.16}{3.845\pi}\right]^{3.876}\left(\frac{3.8}{4.76}\right)^2 = 0.0238. \tag{1.130}$$

In other words, $x > 0.0476$ has to be satisfied in order to form tetrons. Or only after $x > 0.0476$ is the onset of superconductivity possible.

The aforementioned derivation, although simple, is concise enough to outline the essential physics. The message is clear, at $x \sim 0.05$, the tetron becomes more stable than the trion. Considering variation of intrinsic dielectric constant ϵ, we can reasonably put the

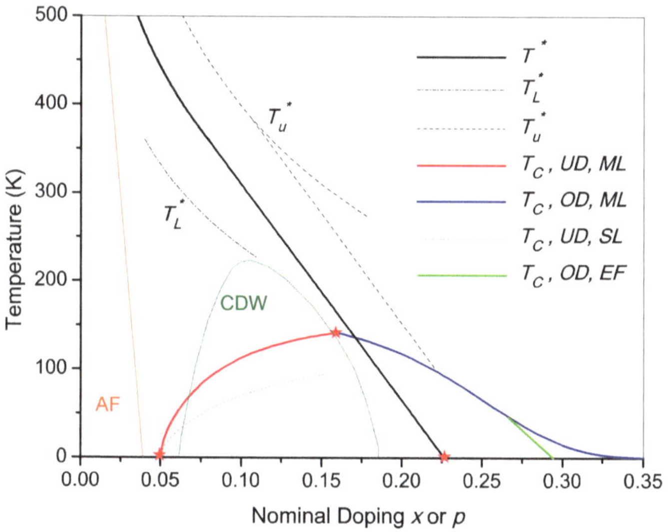

Figure 1.10 Calculated phase diagram of the model cuprate. The T_c from $x \sim 0.05$ up to 0.16 is calculated with the tetron model and the screened excitonic binding energy of Sections 1.15 and 1.17. The T_c from $x = 0.16$ upward is obtained with a BCS-like theory with excitonic attractive action of Section 1.19. T^*: pseudogap temperature; T_L^*: lower pseudogap temperature; T_U^*: upper pseudogap temperature; T_c, UD, ML: critical temperature for underdoped multi-layer cuprate; T_c, OD, ML: critical temperature for overdoped multi-layer cuprate; T_c, UD, SL: critical temperature for underdoped single-layer cuprate; T_c, OD, EF: critical temperature for overdoped cuprates based on Fermi surface change. The green dotted curve outlines the approximate regime where an incipient CDW order is energetically favored, as estimated in Section 1.20. The CDW may manifest as a dip in T_c, as in the lanthanum barium copper oxide (LBCO), but could also be of a fleeting nature, without appreciable effects on SC properties. Therefore, only the CDW's approximate domain is denoted, without an explicit consideration of its effects on T_c. Such effects are specific to a particular material, which are not considered in our model cuprate for the sake of simplicity.

boundary where pairs (tetrons) are to preform at $x = 0.05$. This registers the first point in the phase diagram of Fig. 1.10, as the red star located at $x = 0.05$.

1.16 Phase Boundary for the Pseudogap

In the preceding section we derived the correct doping level for the onset of superconductivity. This reinforces our confidence that the trion/tetron should indeed be relevant to cuprates. A natural extension is now to make the correspondence between the pseudogap and the exciton binding energy E_b. Alternatively the pseudogap can be thought of as the total binding energy of the trion E_{tr} as well. Since E_b and E_{tr} differ by roughly 10%, the two identifications are mostly very similar. In other words, we have

$$k_B T^* = E_b = \frac{c_{\text{dim}} e^4 m_\mu}{32\pi^2 \epsilon^2 \hbar^2} e^{-2q_s r_{B,2D}/2} = \frac{c_{\text{dim}} e^4 m_\mu}{32\pi^2 \epsilon^2 \hbar^2} e^{-q_s r_{B,2D}}, \quad (1.131)$$

where c_{dim} is a factor between 1 and 4 to account for the different binding energy of exciton in 3D and 2D, and the screening q_s, as defined before. The factor of 2 before q_s means that the screening factor is a reduction in the effective dielectric constant (permittivity) ϵ, and since $E_b \propto \epsilon^{-2}$, a factor 2 then appears before q_s in the first line of Eq. (1.131). The 2 in the denominator of $r_{B,2D}/2$ reflects the fact that the most probable orbit radius of a 2D exciton is $r_{B,2D}/2$ in its ground state. The underdoped cuprate approaching the MIT boundary from the conducting side near 0 K has strong 2D properties. However, it is increasingly 3D when approaching the T^* and with an increasing doping level. Therefore, c_{dim} is actually a function of n_p and T too. The reduced T^* with doping is actually understandable from the Coulomb screening, which increases with charge density n_p. The Thomas–Fermi screening strength q_s is on par with the reciprocal inter-particle distance (roughly given by $\sqrt{\pi}/r_s$), which then contributes to the exponential factor in Eq. (1.131).

The first ground state energy of an exciton having the Yukawa potential can only be numerically obtained and there does not exist an analytical solution yet. At small screening, however, an analytical approximation can be used. Actually the physical meaning is clear: At weak screening when electron/hole densities are low, the potential is not that much from the Coulomb, especially at the most probable radius of the ground state. Therefore, the exponential factor is to increase the effective dielectric constant, and the binding energy is

$$k_B T^* = E_b = \frac{c_{\text{dim}} e^4 m_\mu}{32\pi^2 \epsilon_0^2 \epsilon_r^2 \hbar^2} e^{-\sqrt{x} r_{B,2D}/a_0}. \qquad (1.132)$$

The n_h implicit through x in Eq. (1.132) is understood now as an effective mobile charge density, higher than the low-temperature free hole limit of $0.5x/a_0^2$, plus a trion density of similar value, if trion+hole is the stable phase. Alternatively, n_h is higher than double-charge density of $0.5x/a_0^2$, if tetron is stable. Either way, the effective charge density is higher than x/a_0^2, because at higher $T \sim T^*$ of a few hundred K, the excitons are for most part already excited to free holes/electrons. Similarly the weakly bound hole of a trion has largely detached too. Therefore, a realistic effective n is now given by $n_h + n_e = 2x/a_0^2$: Higher screening then emerges around T^*.

Taking the geometric average of c_{dim} at its maximal and minimal due to different dimensionality, giving $c_{\text{dim}} = 2$. This is substituted in Eq. (1.132) , with the new effective density $2x/a_0^2$ to give

$$k_B T^* = E_b(n) = 0.084 \times e^{-1.25\times 2\times 1.414\sqrt{x}}, \qquad (1.133)$$

in electron volt, where the factor $\sqrt{2}$ accounts for the higher effective mobile charge density of $2x/a_0^2$. The 2 in the exponent is the factor to describe the higher ϵ_r at T^*. The effective ϵ_r can be calculated with Clausius–Mossoti relation, with an example given in Section 1.19.

Equation (1.133) is then converted into the expression for temperature to give

$$T^* = 974 \times e^{-3.535\sqrt{x}}, \qquad (1.134)$$

in Kelvin.

These two equations (1.133) and (1.134) are valid when x is not too small, i.e., outside the anti-ferromagnetic phase, and in the underdoped region up to $x \sim 0.10$.

The curve given by Eq. (1.134) is plotted as the upper part of the thick black line in Fig. 1.10 to give the often quoted pseudogap T^* phase boundary in the underdoped regime. The lower part of T^*, which is approximated by the thick black straight line in Fig. 1.10, will be given later. Before that let us first make a few general remarks on the often ill-defined T^*.

Different experiments have various sensitivity to the dimensionality and other factors. Thus the curve of Eq. (1.134) is an average T^*, given by a geometric average of $c_{\bar{\text{dim}}} = 2$. In fact, recent experiments often start to define two pseudogap temperatures in a variety of notations (Barišić et al., 2013; Honma and Hor, 2015), and the concept of upper and lower pseudogaps starts to emerge. One specific probe may have a c_{dim} of, say 3/2, which is different from c_{dim} of another, for example, 5/2. The scattering may be due to the difference in binding energy for exciton and trion, too. The overall effects can be summarized in $c_{\text{dim}} \sim 1.5$ or 2.5 or other values in-between. We proceed with defining

$$T_L^* = \frac{3}{4}T^* = 731 \times e^{-3.535\sqrt{x}} \tag{1.135}$$

$$T_U^* = \frac{5}{4}T^* = 1218 \times e^{-3.535\sqrt{x}}. \tag{1.136}$$

The lower extreme is given by the black dash-dotted curve in Fig. 1.10. The upper part of the T_U^* is given by the black dashed curve, and more on the straight dashed part is deferred later. The broad region defined between T_L^* and T_U^* is where various T^* or T^{**} are reported. The feature of scattered data explains in part the vast variation in T^*. Another reason why T^* is not well defined is that the pseudogap has both k and T dependence, making its identification with a single value impractical. This is to be explained in Chapter 3 and specifically in Section 3.8.

Strictly speaking Eqs. (1.134) and (1.135) are valid only when the total charge density n is not too high. The screened Coulomb, or the Yukawa potential, is defined similar to Eq. (1.119) as

$$U_Y(r) \propto -\frac{e^{-q_s r}}{r} = -\frac{1}{4\pi\epsilon r}e^{-\sqrt{n}r}, \tag{1.137}$$

where the total charge density n replaces the more restricted n_p in Eq. (1.119).

The Yukawa (screened negative Coulomb) potential has a critical screening strength beyond which the bound state vanishes. If the n increases the q_s in the exponent $e^{-q_s r}$ also increases according to $q_s = 1.25\sqrt{n} = 1.25\sqrt{x}$ since at very low T, $n \sim x/a_0^2$. If

$1.25\sqrt{x} > 0.595$ (Edwards et al., 2017), then the bound exciton state no longer exists. The pseudogap therefore disappears at the often quoted quantum critical point (QCP). $1.25\sqrt{x} = 0.595$ then gives 0.2265 which is a very good result considering the simplicity of the model cuprate. $x = 0.2265$ is denoted by another red star in Fig. 1.10.

One consequence of the QCP is that T^* has to disappear through it and hence Eq. (1.133) is no longer valid at moderate doping beyond the optimal. A detailed numerical study of the ground state energy of a screened Coulomb (Yukawa) potential, especially in 2D, is highly desirable. We are, however, restricted with the limited scope in this book. It should constitute important future research since we essentially interpret T^* as the ground state binding energy of Eq. (1.137). For now, we resort to qualitative physics with simplified procedures as follows.

Note that $x = 0.2265$ represents the low-temperature limit of the pseudogap phase boundary. It is experimentally around 0.23 and the exact location is material related to some degree. Because T^* is measured with different physical properties which has different sensitivity to dimensionality (and the related binding energy difference), we expect variation in T^* anyway. It is therefore legitimately approximated by a straight line through $x = 0.2265$ and tangent to the curve given by Eq. (1.134). This gives the lower and straight part of the solid curve denoting T^* in Fig. 1.10. Note that the straight part is a rough approximation of T^* only. The actual numerical result may show appreciable curvature of a concave shape. A similar dashed line is drawn to match the curved T_U^* at doping less than $x \sim 0.11$. We kept the curved dash up to $x \sim 0.18$, to give an idea of discrepancy of the two. In reality both straight parts of T^* and T_U^* should have appreciable curvature approaching QCP, merging and disappearing from there.

Immediately below the pseudogap temperature, pairs (tetrons) form but superconductivity is not present yet due to the low pair density. This is called pairing without superconductivity at low carrier density, which was discussed long before the era of cuprate superconductors. See, for example, Eagles (1969), for a more general discussion.

1.17 The SC Boundary

Upon formation of the tetron at $x > 0.05$ we start accumulate supercarrier density n_s, also referred to as the SC condensate density. They are charged bosons with double charge. The transition critical temperature is given by the well-known formula for BEC,

$$k_B T_c = 3.31 \frac{n_{te}^{2/3} \hbar^2}{m^*}, \tag{1.138}$$

where the tetron density n_{te} has replaced the n_s, with m_s^* the effective mass of the supercarrier.

The formula applies to 3D. In pure 2D of infinite size, BEC is absent for a system with short-ranged interactions and a continuous Hamiltonian symmetry due to the Mermin-Wagner theorem. Still for quasi-2D cuprates, we expect a formula similar to Eq. (1.138), where the exponent could be different due to modified density of states.

The onset of SC in ECM has similarity to some boson-fermion high-T_c models where there is a thermodynamic balance regarding the supercarrier density and that of the fermion (Friedberg and Lee, 1989b). R. Friedberg and T. D. Lee (Friedberg and Lee, 1989a) demonstrated a critical density that to a certain degree is resembled by the logic here. Besides, we expect a continuous binding energy variation from the a-V phase diagram for the trion and tetron, barring some pathological conditions. Therefore, the supercarrier density near $x = 0.05$ at low T should be proportional to $(x - 0.05)$ up to an exponent. The exact doping x-dependent supercarrier density n_{te} (with te indicating the tetron) in the underdoped cuprate has to be calculated from a detailed mapping of relative energetics of trion versus tetron as a function of a-V. In other words, complete continuous surfaces in the 3D space spanned by E_{te} (or E_{tr}) vs. a-V, are highly desired. The simple diagram of Fig. 1.9 is not informative enough. Still we can elk out some important characteristics based on simple logic reasoning. Let us start with a crude assumption on $n_{te} = (x - 0.05)/2$. This is the simplest that one can get, essentially saying no tetrons below $x = 0.05$, and that across the tetron/trion boundary line immediately near $x = 0.05$, tetron and trion have the same chance of formation. Both are simple and fair assumption.

Substituting $n_{te} = (x - 0.05)/2$ into Eq.(1.138) gives

$$T_c \propto (x - 0.05)^{2/3}. \tag{1.139}$$

Equation (1.138) gives a characteristically correct SC boundary at the *very* underdoped regime near $x = 0.05$. The concave section of $T_c(x)$ is qualitatively right in Fig. (1.10), colored in either red or magenta.

In general there exists a power law $T_c \propto (x - 0.05)^\kappa$ where κ is the exponent that is generally positive but less than 1. The exact value should depend on dimensionality and other factors, subject to further exploration.

Equation (1.139), however, does not give good enough physics when x increases. Since the tetron density n_t really depends on the pseudogap, it has to be weighted down by the exponent in the T^*. In other words, if the pseudogap disappears, the real-space paired four-body entity is no longer in existence. Then the tetron density in the underdoped regime is given by

$$n_{te} = (x - 0.05)e^{-3.535\sqrt{x}}, \tag{1.140}$$

where the factor of 2 like that in $n_{te} = (x - 0.05)/2$ no longer exists because at higher x, the tetron is more stable: Most of the mobile fraction is made of tetron.

Inserting Eq. (1.140) into Eq. (1.138), and adopting parameters of the model cuprate, we obtain

$$T_c = 1083(x - 0.05)^{2/3}e^{-2.357\sqrt{x}}, \tag{1.141}$$

with the temperature given in Kelvin. This is plotted as dotted magenta line in Fig. 1.10. It reaches a high of sub 100 K (96.8K) near the optimal doping of 0.16.

The T_c formula of Eq. (1.141), though very simple, should give a maximal T_c after $x > 0.15$. Equation (1.140) is oversimplified though due to the fact we do not have yet an analytic expression for the T^* in the optimal-overdoped regime. Still this very simple expression gives good enough description of T_c vs. x in the underdoped-optimal regime, barring any additional depression due to the CDW order.

This, though might appear accidental, actually well represents the maximal T_c in single layer cuprates. We have not mentioned the CDW so far and the possible dent of T_c in the underdoped regime. It will be qualitatively discussed separately in Section 1.20.

1.18 Maximal Critical Temperature

In the optimal doped regime, we have two characteristic temperatures that are associated with the SC state. The first is the critical temperature given by the BEC

$$k_B T_c^{BEC} = 3.31 \frac{n_{te}^{2/3} \hbar^2}{m_s^*}$$
$$= 3.31 \frac{n_{te}^{2/3} \hbar^2}{4m_h}, \quad (1.142)$$

where m_s^* is the supercarrier mass and n_{te} its density. m_h is band mass of the hole.

In the tetron model, $m_s^* \approx 8m_e \approx 4m_h$, to the first order, without considering additional mass renormalization effects. Assuming $n_{te} = (x - 0.05)/a_0^2 c_0$ for convenience, we just need a first-cut estimate. The 3D supercarrier density can be calculated to be $0.585 \times 10^{27}\text{m}^{-3}$ and $0.635 \times 10^{27}\text{m}^{-3}$ for LSCO and Y-123 (only planar contribution), respectively. Note that this likely leads to an over-optimistic estimate because the tetron is in a thermodynamic equilibrium with fermion excitations like trions, holes, electrons, etc. This serves only as an upper limit guide for cuprates. In other words, an average supercarrier density of $6 \times 10^{26}\text{m}^{-3}$ for LSCO and Y-123 at OP (optimal doping). In other p-type cuprates similar values are expected, differing by a factor of 2~3 at most.

This n_{te} then leads to an upper limit 250 K to T_c^{BEC} given by Eq. (1.142). This critical temperature T_c^{BEC} indeed accommodates the maximum of 164 K experimentally achieved in cuprates. Notice that though, this serves as an upper limit for the current model, with the current understanding of the band-mass between 1 and $\sim 2\, m_e$. Also keep in mind that this 250 K is higher than the ~98 K T_c that is given in Eq. (1.141) in the preceding section. A $T_c^{BEC} = 250$ K neglected the reduction of the pseudogap due to screening from charge-carrier density, which becomes prominent when approaching maximal T_c. This introduces a second characteristic temperature T^*, the pseudogap temperature, which is associated with screened binding energy.

Equation (1.142) shows that T_c^{BEC} is inversely proportional to m_h. Reducing band mass of the hole hence increases T_c^{BEC}. This is in contrast with the pseudogap temperature T^* given by

$$T^* = \frac{c_{\text{dim}} e^4 m_\mu}{32\pi^2 \epsilon_0^2 \epsilon_r^2 \hbar^2} \frac{e^{-\sqrt{n} r_{B,2D}/a_0}}{k_B}$$
$$= \frac{c_{\text{dim}} e^4 m_h}{64\pi^2 \epsilon_0^2 \epsilon_r^2 \hbar^2} \frac{e^{-\sqrt{n} r_{B,2D}/a_0}}{k_B}, \tag{1.143}$$

where reduced mass $m_\mu \approx m_e$ of the exciton is derived from band mass of electron $m_e^b \approx 2m_e$ and hole $m_h \approx 2m_e$. The b indicates "band," and m_e is the rest mass of a bare electron in vacuum.

Hence T^* is proportional to m_h. Since in our pairing model T^* has to be satisfied first to get the preformed Cooper pair (tetron), T^* is a necessary condition for superconductivity. Combined with the demands on T_c^{BEC}, the actual (final) superconducting T_c is given by

$$T_c \leq \min(T_c^{BEC}, T^*). \tag{1.144}$$

T_c^{BEC} increases with x but T^* decreases. The conflicting requirement is the main physical basis that gives rise to the SC *dome* in the cuprate T-p (or T-x) phase diagram.

T_c^{BEC} decreases with m_h but T^* increases. This forms conflicting demands on the band mass. Its role in the search for new high-T_c materials will be discussed in Section 5.6. If one wants to increase T_c due to BEC, a small in-plane mass is desired. However, from a Coulomb bound state perspective, a heavy mass is favored. It appears that anisotropy plays a role in enhancing T_c, evident in the quasi-2D nature of the cuprate.

The anisotropy is manifested in various aspects, with two major parameters worth mentioning: the mass tensor and the dielectric constant tensor. First, as for the mass tensor, $m_\parallel \ll m_\perp$. The small $m_\parallel$ helps increase T_c^{BEC} while $m_\perp$ helps maintain (quasi) two-dimensionality. Although the interplane exciton's binding energy is not directly proportional to $m_\perp$, large $m_\perp$ helps reduce the c-axis kinetic energy due to confinement.

Second, $\epsilon_{r\perp} > \epsilon_{r\parallel}$, and this anisotropy in dielectric constant of the undoped parent matrix can be modeled by alternating slabs of different but isotropic ϵ_r, ϵ_1 and ϵ_2, for example.

1.18.1 Anisotropic Electrostatics

For simplicity equal thickness for Layer 1 (ϵ_1) and Layer 2 (ϵ_2) is assumed. Then the dielectric constant normal to, and parallel to the layer interface, $\epsilon_\perp$ and $\epsilon_\|$, are derived by using Gauss's Law,

$$\frac{2}{\epsilon_\perp} = \frac{1}{\epsilon_1} + \frac{1}{\epsilon_2},$$
$$\epsilon_\| = \frac{1}{2}(\epsilon_1 + \epsilon_2). \tag{1.145}$$

Solving the equations above for ϵ_1 and ϵ_2, one derives

$$\epsilon_2 = \epsilon_\| + \sqrt{\epsilon_\|^2 - \epsilon_\|\epsilon_\perp},$$
$$\epsilon_1 = \epsilon_\| - \sqrt{\epsilon_\|^2 - \epsilon_\|\epsilon_\perp}. \tag{1.146}$$

For cuprates the general relation holds as $\epsilon_\| \sim 3\epsilon_\perp$ when undoped, then we get simple relation as

$$\epsilon_2 = \left(3 + \sqrt{6}\right)\epsilon_\perp,$$
$$\epsilon_1 = \left(3 - \sqrt{6}\right)\epsilon_\perp. \tag{1.147}$$

Now we may identify the CuO_2 layer as the layer with ϵ_2 due to the higher dielectric constant. We consider the electrostatic potential effected from another nearby close-coupled CuO_2 layer via the insulating (Ca, Y) layer of ϵ_1. We put the source charge inside the ϵ_1 layer on top of a CuO_2 layer as in Fig. 1.11(b), the distance being ~3.2 Å. The source charge e should border the other CuO_2 on the further top. But for now let us neglect the complication because the effect can be simulated with an effective ϵ_1 that can be modified later on. We define that the origin is at the interface and the interface divides space into two halves, the upper and lower half space.

By using the image charge method in electrostatics, the potential in the lower half is given by

$$\epsilon_0 V_{es}(\mathbf{r}) = \frac{\mathbf{e}}{4\pi\epsilon_1\mathbf{r}} + \frac{\mathbf{e}'}{4\pi\epsilon_1\mathbf{r}'}, \tag{1.148}$$

where the subscript *es* indicates "electrostatic," to differentiate the V_{es} from similar notations of other chapters. e' is the image charge whose magnitude is given by

$$e' = -e\frac{\epsilon_2 - \epsilon_1}{\epsilon_2 + \epsilon_1}. \tag{1.149}$$

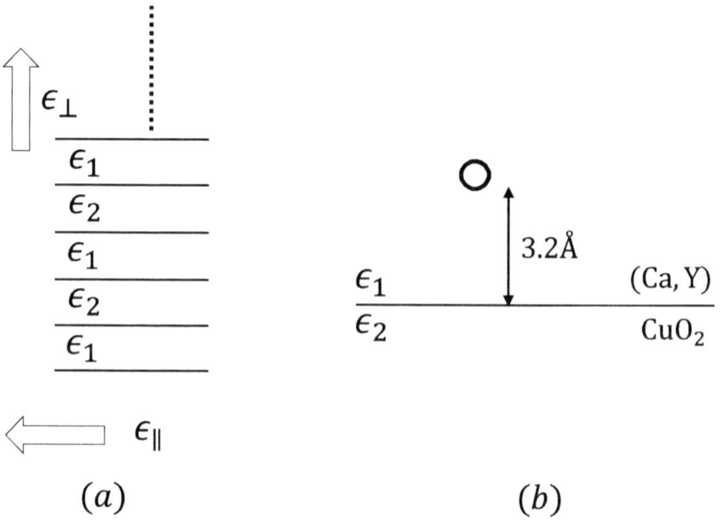

Figure 1.11 Geometry used in the study of anisotropic electrostatics relevant to the cuprate. (a) Alternating slabs of dielectrics of different ϵ. (b) A test charge is placed 3.2 Å away from the edge of a CuO_2 layer; the potential distribution is to be found.

Plugging Eq. (1.149) into Eq. (1.148) gives

$$\epsilon_0 V_{es}(\mathbf{r}) = \frac{\mathbf{e}}{4\pi\mathbf{r}} \frac{\mathbf{2}}{\epsilon_1 + \epsilon_2}. \tag{1.150}$$

1.18.2 Interlayer Coulomb and Why It Enhances T_c

The potential given by Eq. (1.150) turns out to be *higher* than the in-plane ones given by $e/(4\pi\epsilon_0\epsilon_\parallel\mathbf{r})$. This might appear counter-intuitive at the first glance but is indeed of perfect logic of the interlayer coupling theory. The Coulomb enhancement due to the reduced effective dielectric constant is a simple $(3+\sqrt{6})/3$ or 1.8615. However, the out-of-plane distance is indeed longer at $\sim\sqrt{3.2^2+4^2}$, rather than the ground-state exciton radius of $\sim$4 Å (due to higher effective $\epsilon_2 \sim 2\epsilon_\parallel$). This results in overall enhancement of 42% in the effective excitonic binding energy. Since the effective ϵ_2 could be less than what is given by Eq. (1.147) due to close proximity to Layer 1, the enhancement might be lower than quoted above for bilayer cuprates. However, in the trilayer cuprate, the effect is offset by further inter-layer coupling; therefore, T_c enhancement of some 42% is a good reference for multilayer cuprate, without considering other factors such as hydrostatic pressure. This results in improved binding energy, and enhancement in T_c. The SC boundary thus defined at underdoped regime (UD) is given as the solid red curve in Fig. 1.10.

Up to this point, important parameters of the phase diagram in Fig. 1.10 have been obtained without parametric fitting. All that we utilized are lattice constants, band-mass and dielectric constants (in-plane and out-of-plane).

We have not explicitly considered the additional polarizability enhancement along the c-axis. Still, we see the importance of interlayer Coulomb coupling, which was emphasized by Leggett (Leggett, 1999a), not long after the discovery of trilayer cuprates.

With the out-of-plane dielectric constant in mind, the Coulomb interaction energy to the undoped limit could approach the limit set by

$$\frac{e^2}{4\pi\epsilon_0\epsilon_\perp r_c}, \tag{1.151}$$

where r_c is the interplane separation, ranging from 3.2Å in some multilayer cuprates to ~6.8Å in many single-layer types. This should give a characteristic energy from 0.35 ~0.75 eV. If the in-plane mass is large like that of antiferromagnetic phase, the interplane energy may dominate the binding behavior of very underdoped cuprates. Of course, this is only an upper limit, but its large magnitude nevertheless indicates why interlayer Coulomb interaction is important in cuprate superconductivity. Also be reminded that the often discussed 0.4 eV MIR energy scale is right within this energy window.

It is well known that the homologous series 22(n-1)n reaches maximal T_c when $n = 3$ (or 4 but mostly 3). This has long been a puzzle and stimulated many papers. We are now ready to give a simple geometric interpretation for this, resorting to the interlayer excitonic Coulomb interaction.

A tetron comprises 4 particles, i.e., 3 holes and 1 electron. As illustrated in Fig. 1.12, the electron, shown as a sphere, resides in

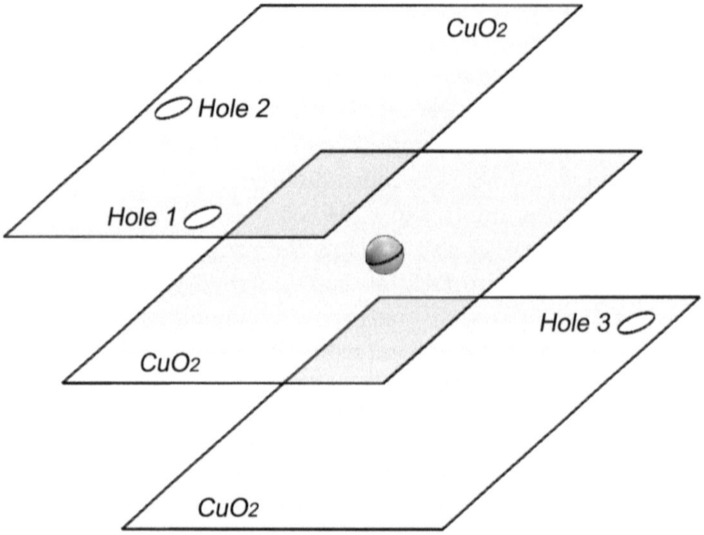

Figure 1.12 A tetron straddles across three closely coupled CuO_2 layers in a tri-layer cuprate, to maximize the binding energy. This explains why the highest T_c occurs to tri-layer or tetra-layer cuprates.

the center plane. In this way the three pair-wise attractive Coulomb forces are maximized, while the two repulsive types between h1-h3 and h2-h3 are minimized due to increased distance. The Coulomb repulsion between h1 and h2 is also reduced due to enhanced in-plane screening from the large in-plane dielectric constant. The interplane spacing is greatly exaggerated (by a factor ~3) for clarity. The distance between h1 and h2 can be calculated to be around 15 Å. This is very close to the inplane coherence length found in multilayer cuprates.

Additional consequences can be inferred from the aforementioned pairing mechanism. In many cases when the T_c is limited by the supercarrier density n_s, it is not due to the hole density n_p but rather due to the electron density n_e. For *p*-type cuprates, one wants a density ratio of 3 to 1 in n_p/n_e, to optimize the T_c given the same total carrier density. If the actual electron density is significantly less than 1/3 of n_h, then $n_s \approx n_e$ and T_c is suppressed. The electron density is sensitive to local band structure, particularly near the $(0, \pi)$ saddle point, and details in the interlayer Coulomb interaction.

In the oversimplified single carrier doping model, doping of a fraction x gives x holes per unit cell. if the doping were this simple, one would expect the SC condensate density directly proportional to x. One then would *NOT* see much difference in n_s and hence T_c. A notable counter example is Bi-2201 versus Bi-2212.

The pairing model also explains the pressure-dependent T_c in many cuprates. Hydrostatic pressure in solid-state physics often increases the hopping of quasiparticles among neighboring sites giving more dispersion or band width and thus lowers the effective band mass. This should increases T_c when underdoped. The Hg based cuprates are known to have the highest T_c, often under hydrostatic pressure, and the Hg-cuprates are known to be often underdoped.

From $0.05 < x < 0.20$, due to the pseudogap, the effective carrier density of p-cuprates increases with T. Near T^* we can safely put the total mobile charge on the order of the doping level. From the Hall study and many other spectroscopic pieces of evidence as shown in Section 2.5 and 2.6, we know that electrons are doped in as well. An ideal doping model is that x dopants generate $1.5x$ holes, and $0.5x$ electrons, as already mentioned early in this Chapter. In comparison, the traditional doping model states that x dopants generate $1x$ holes only. Such over-simplified models have been shown to have severe problems, particularly in understanding the temperature-dependent transport coefficients. See Section 2.6.1 for further information.

Now with this modified doping model, one can satisfactorily explain the temperature dependence of most transport coefficients below T^*, for the underdoped and optimal doping regimes. The T dependence is merely the manifestation of thermal activation process against the exciton binding energy, which defines T^*. In the significantly overdoped regime, the exciton binding energy disappears, and many of the transport properties, like R_H, S, which are closely associated with the carrier density, lose their significant temperature dependence above T_c up to the T^*.

In general we can summarize the p-type doping as in the underdoped and optimal-doped regimes with

$$\begin{cases} n_p = (1 + a_e)x/v_f \\ n_e = a_e x/v_f \end{cases} ; \quad \text{when} \quad x \leq 0.2, \tag{1.152}$$

where a_e denotes the fraction of electron that is generated for every dopant added. v_f is the volume of the formula unit. For divalent-trivalent replacement, each dopant is counted as one; with oxygen, each is counted as ~2. For optimal doping to maximize T_c, we need $a_e = 0.5$, which will give a 3:1 ratio in densities of hole and electron. However, many cuprates should deviate from this value, and we know for sure such relation does not hold in the significantly overdoped regime. If the a_e is too low, $T_c \approx 0$. a_e can be calculated with state-of-the-art computational tools. Section 4.5.2 has found $a_e \approx 0.31$ for LSCO near $x = 0.125$.

From the underdoped to overdoped regime via the optimal doping, the cuprate undergoes a BEC to BCS cross-over, for which various studies (Eagles and Leggett, for example) emerged (Eagles, 1969; Leggett, 1980) long before modern approaches (for example (Chen et al., 2005)) of the high-T_c era.

1.19 Overdoped Regime

Now we reach the final portion of the phase diagram, the overdoped regime beyond the QCP. There is no excitonic Coulomb bound state, at least deep into the OD. Still the highly polarizable *e*-*h* liquid can mediate an effective attraction, and the superconductivity proceeds more in line with the BCS scheme. We have

$$k_B T_c = 1.13\hbar\omega_c e^{-1/N(0)\mathcal{V}}, \tag{1.153}$$

where $\mathcal{V}$ is so written to differentiate it from the V in the a-V phase diagram, and is the supposedly constant matrix element $\langle k|V_i|k'\rangle$ of the attractive pairing potential V_i. The assumption of a constant $\mathcal{V}$ actually is not outrageous for OD cuprates evaluated at the Fermi surface. However, the matrix element is more involved than the standard screened Coulomb. This is due to the following: (1) The Coulomb between two holes are repulsive. The attractive interaction has to go through the *e-h* liquid, even if without excitonic bound states; (2) due to the relevant length scales of OD, the effective attraction is more of a dipole nature:

$$V_d = \frac{\mathbf{p}\cdot\hat{\mathbf{r}}}{4\pi\epsilon r^2}. \tag{1.154}$$

This dipole potential can be negative or attractive when **p** is anti-parallel to **r**. The polarization **p** is due to Coulomb induction by one of the holes to be paired. It turns out that the magnitude of **p** is proportional to the screened Coulomb. And the magnitude of this dipolar interaction is lower than a screened Coulomb. Since there is no longer any excitonic bound state, we would evaluate the Coulomb interaction at a characteristic distance, still reasonably given by the Bohr radius. Hence $\mathbf{p} \propto \exp(-r_{B,2D}q_s)/\epsilon$. In the overdoped regime, besides the TF-like screening, the effective reduction of the Coulomb via an increased dielectric constant cannot be overlooked either. Note that the Coulomb interaction energy scale is a monotonically decreasing function of increasing doping x (or p). At high density limit, the kinetic energy dominates the coulomb at r_s, and the pairing will be a cross-over from the local pair to a more BCS-like picture. Therefore, it is sound to evaluate the Coulomb at $r_{B,2D}$. If the effective ϵ doubles due to more mobile charge, $r_{B,2D}$ also doubles. The interaction energy then scales as 1/4. Also notice that in the OD regime, ξ is still rather short on the 15~40 Å scale. Therefore, the distance r is still not too high to be dominated by the inverse r^2 potential of the dipole field. Hence we may use an effective interaction potential of the form

$$V_i = \frac{-e^2}{4\pi\epsilon_{\text{eff}}^2/\epsilon_u r} e^{-q_s r_{B,2D}} e^{-q_s r}, \tag{1.155}$$

where ϵ_u is the undoped dielectric permittivity, taken for our model cuprate to be 18 times the vacuum value. $\epsilon_{\text{eff}}^2/\epsilon_u$ is then the overall effective pairing permittivity. The square in ϵ_{eff} means the interaction first go from a hole to the *e-h* liquid, to polarize the liquid and generate **p** and then from the induced **p** of the *e-h* liquid to a second hole. The dielectric medium is traversed twice in sequence and the effective pairing interaction is proportional to $\epsilon_{\text{eff}}^{-2}$.

Equation (1.155) can be further re-arranged and simplified as

$$V_i = \frac{e^{-q_s r_{B,2D}}}{\epsilon_{\text{eff}}^2/\epsilon_u^2} \frac{-e^2}{4\pi\epsilon_u r} e^{-q_s r}$$
$$= -\frac{e^{-q_s r_{B,2D}}}{\epsilon_{\text{eff}}^2/\epsilon_u^2} 2E_b \frac{r_{B,2D}}{r} e^{-q_s r}$$
$$= -2E_b \frac{e^{-q_s r_{B,2D}}}{\epsilon_{\text{eff}}^2/\epsilon_u^2} \frac{e^{-q_s r}}{r}, \tag{1.156}$$

where in the last step we switch to the Bohr unit in r. $2E_b$ is the Coulomb energy evaluated at $r_{B,2D}$, and gives 0.168 eV for $\epsilon = 18$.

The factor

$$\frac{e^{-q_s r_{B,2D}}}{\epsilon_{\text{eff}}^2/\epsilon_u^2} \tag{1.157}$$

is a function of x only and therefore can be factored out in evaluation of the matrix element $\langle k|V_i|k'\rangle$.

Since ϵ_{eff} can be calculated by the Clausius–Mossoti relation via the polarizability of weakly localized carrier, it is typically a few times larger than the matrix ϵ of 18 at the very overdoped case. Therefore, we can parameterize the $\epsilon_{\text{eff}}/\epsilon$ using a very simple relation $\epsilon_{\text{eff}}/\epsilon = 1 + 8(x - 0.20)$, which will give a conservative ϵ_{eff} twice as much as 18 at the overdoping boundary where SC approaches disappearance. As the result, the x-dependent V_i is approximately given as

$$V_i = -\frac{2E_b}{[1 + 8(x - 0.2)]^2} e^{-1.25\times 1.414\times[1+8(x-0.2)]\times\sqrt{x}} \frac{e^{-q_s r}}{r}. \tag{1.158}$$

The only r-dependent portion in V_i is $\exp(-q_s r)/r$ and its matrix element in 2D is $2\pi/(q^2+q_s^2)^{1/2}$, where $q = |\mathbf{q}| = |\mathbf{k}-\mathbf{k}'|$. The factor $1/(q^2 + q_s^2)^{1/2}$ actually gives a length on the scale of interparticle distance. When combined with $r_{B,2D}$ it gives an area of the unit cell $\sim a_0^2$, for overdoped cuprates. Therefore, $\mathcal{V} \sim -4\pi E_b a_0^2$. For 3D, a factor of 4π appears in the integral but the longer lattice constant in c-axis weighs it down and the relation still holds as $\mathcal{V} \sim 2\pi \times 2E_b \times$ [volume of unit cell].

The density of states $N(0)$ can be calculated by LDA/GGA (see Chapter 4) in the overdoped regime with good confidence, as shown in Fig. 1.13. The DOS read from the peak near the Fermi energy when

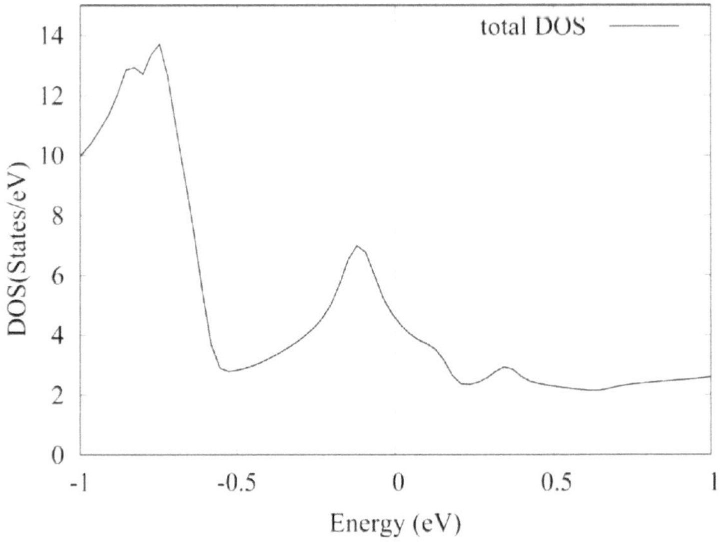

Figure 1.13 Numerical density of states in LCO, calculated with the WIEN2k package. The Fermi level is at the 0 energy. The DOS is that for every CuO_2 site, or per formula unit.

hole-doped is ~7.42, in the unit of state per eV per formula-unit. Then we have

$$N(0)\mathcal{V} = \frac{7.8}{[1+8(x-0.2)]^2} e^{-1.768[1+8(x-0.20)]\times\sqrt{x}}. \tag{1.159}$$

The corresponding expression for T_c is given by

$$\begin{aligned} T_c &= 155 \times \exp(-1/N(0)\mathcal{V}) \\ &= 155 \times \exp\left[\frac{-[1+8(x-0.2)]^2 e^{1.768[1+8(x-0.2)]\times\sqrt{x}}}{7.8}\right]. \end{aligned} \tag{1.160}$$

Equation (1.160) is plotted as the solid blue line in Fig. 1.10. The coefficient of 155 K is given by requirement to match the underdoped T_c at $x = 0.16$. It is also loosely given by the T^* at that doping. In other words, $1.13\hbar\omega_c$ of Eq. (1.153) can be replaced by pseudogap $k_B T^*$. The pseudogap energy near optimal (OP) then plays the role of Debye energy. T^* in Fig. 1.10 gives a value of 165 K, very closely matching the value given in Eq. (1.160). This gives us added confidence as to the ECM pairing's self-consistency.

The tail extends beyond $x = 0.30$ as given in some materials for example Tl-2201. This part is much materials-related as reported in (Rourke et al., 2010) of the Hussey group. In LSCO for example, at $n \sim 0.235$ from LDA/GGA calculation (trustworthy at high doping) topological change is seen in the FS, as in Fig. 2.8. The change in Fermiology signifies the loss of hole as a major carrier of electric current. Then at $n = 0.25$ and above, most of the quasiparticles are electronlike, the overwhelming repulsion then overcomes the marginal attraction that can be mediated by the small fraction of holes, and hence superconductivity is lost near $n = 0.3$. This behavior is given by the green solid line in Fig. 1.10.

Up to this point, the full SC dome of the cuprate T-x diagram has been derived. Different orders exist varying with the change of doping. The calculated dome closely follows the experimental and reinforces our confidence that attributing tetron as the Cooper pair to the UD and OP phases is the right designation. Generally speaking, there are many different orders in condensed matter physics. The tetron order is a geometric type that is different from the ordinary crystal planewave type long range order. Near a center of mass of a particular tetron, there is locally a geometric correlation in a shape of triangle. Such local patterns are often neglected in conventional thermodynamics textbooks, but definitely affect the energy and entropy of the system. A notable example is in the local vorticity of the Berezinskii–Kosterlitz–Thouless transition (Kosterlitz and Thouless, 1973). These examples demonstrated the importance of geometric configuration for strongly correlated systems.

1.20 Charge Density Wave

The competition and balance of repulsive and attractive interaction in the ECM model so far have shown many interesting physics. The attractive interaction is also well known to cause charge density wave in the condensed matter. To see this we calculate the dielectric susceptibility using the linear response theory. The generalized susceptibility $\chi(q, \omega)$ is given by

$$\mathbf{P} = \chi(\mathbf{q}, \omega)\epsilon_0 \mathbf{E}. \tag{1.161}$$

Assuming that the **E** field is small the susceptibility can be evaluated with the second-order perturbation theory to give

$$\chi(\mathbf{q},\omega) = \frac{4\pi}{\mathbb{V}} \int d^3\mathbf{k} \frac{\langle \mathbf{k}+\mathbf{q}|e^{-i\mathbf{q}\mathbf{r}}|\mathbf{k}\rangle^2 f(E(\mathbf{k})+\mathbf{q}) - f(E(\mathbf{k}))}{E(\mathbf{k}+\mathbf{q}) - E(\mathbf{k}) - \hbar\omega + i\alpha}, \tag{1.162}$$

where $\mathbb{V}$ is the volume of the system, and $f(E)$ is the Fermi function. The χ thus defined is evaluated at the $\alpha \to 0$ limit. Equation (1.162) holds only for independent-fermion systems.

If an interaction (U) between the charged fermions is turned on, the susceptibility is renormalized with the following approximation:

$$\chi^*(\mathbf{q},\omega) = \frac{\chi(\mathbf{q},\omega)}{1 - U(\mathbf{q})\chi(\mathbf{q},\omega)}, \tag{1.163}$$

where the denominator is actually the dielectric function given by

$$\epsilon(\mathbf{q},\omega)/\epsilon_0 = 1 - U(\mathbf{q})\frac{4\pi}{\mathbb{V}} \int d^3\mathbf{k} \frac{f(E(\mathbf{k}+\mathbf{q})) - f(E(\mathbf{k}))}{E(\mathbf{k}+\mathbf{q}) - E(\mathbf{k}) - \hbar\omega + i\alpha}. \tag{1.164}$$

Since $\chi(\mathbf{q},\omega)$ is given by

$$\chi(\mathbf{q},\omega) = \frac{4\pi}{\mathbb{V}} \int d^3\mathbf{k} \frac{f(E(\mathbf{k}+\mathbf{q})) - f(E(\mathbf{k}))}{E(\mathbf{k}+\mathbf{q}) - E(\mathbf{k}) - \hbar\omega + i\alpha}, \tag{1.165}$$

if we consider the static limit $\omega = 0$,

$$\chi(\mathbf{q},0) = \frac{4\pi}{\mathbb{V}} \int d^3\mathbf{k} \frac{f(E(\mathbf{k}+\mathbf{q})) - f(E(\mathbf{k}))}{E(\mathbf{k}+\mathbf{q}) - E(\mathbf{k})}. \tag{1.166}$$

Since the Fermi function $f(E)$ is a monotonically decreasing function of E, $\chi(\mathbf{q},0)$ is in general negative.

$\chi(\mathbf{q},\omega)$ can be further evaluated for a specific fermion system where the Hamiltonian may give special nesting effects near the Fermi surface which generate large or diverging χ, for special $\mathbf{q}$ vectors. For cuprates since the DOS near $(0,\pi)$ is high and with the saddle point extended, $\chi(\mathbf{q},\omega)$ can be quite large for a $\mathbf{q}$ vector that is smaller than the BZ. Then with an attractive interaction like the screening Coulomb between the electron and hole, or between two like charges with effective attraction mediated by a screened dipole field, $U(\mathbf{q})$ is negative which leads to a zero dielectric function or a divergent susceptibility according to Eqs. (1.163) and

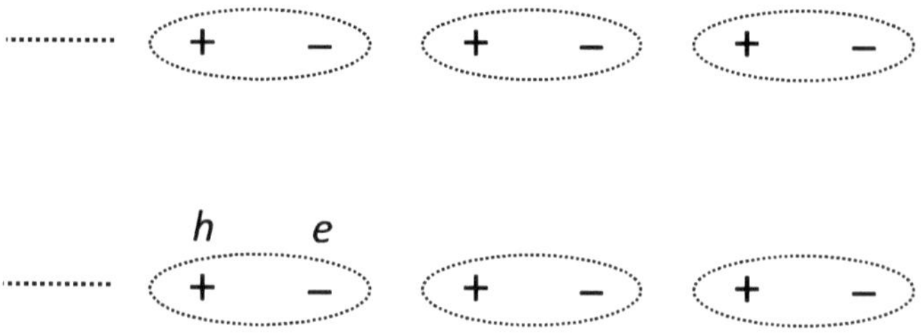

Figure 1.14 How excitonic Coulomb bound states may form a CDW.

(1.164). The resulting spontaneous polarization then gives rise to a CDW.

An exact solution in this approach is really model and parameter dependent, and especially considering the interlayer coupling forming the excitonic Coulomb state, a quantitative description is beyond the scope of this book. Still we can qualitatively understand the *e*-*h* excitonic interaction as a natural candidate scenario for the CDW in cuprates.

At proper density, exciton molecules could form in a direct analogy to hydrogen molecules. In the two-dimension, the increased confinement favors bound states and increases the likelihood of exciton molecules. The simplest exciton molecule is formed from two excitons and such a complex is also called biexciton in the literature. The biexciton is made of 2 electrons and 2 holes altogether and the total ground state energy is ~ -2.2 (Singh et al., 1996) in the $m_e^b = m_h$ limit. We utilized the SVM to verify too and the result is -2.192. Since each exciton has a ground state energy of -1, the binding energy for the two exciton is hence $\sim -2 - (-2.19) = 0.19$. In other words, there is actually an attractive interaction between the two excitons although each one of them is electrically neutral. We want to find out why, and it against turns out to be the result of geometric factors and the inverse squared Coulomb law. Consider the instantaneous polarization of each of the exciton. They can align the dipole moments up to make an effective attraction as given in Fig. 1.14. In fact, this is not much different from the Van der Waals force between neutral molecules.

If the two excitons gain the potential energy by polarizing and aligning-up, the end result is a periodic charge density modulation in the aligned direction. More excitons can join this so that the total energy is further reduced: A density wave is clearly formed. The periodicity of the density wave, can be estimated from the size of a trion. We take the trial wavefunction $\psi(r_1, r_2)$ of Eq. (1.23) with the variational parameters given by Eq. (1.53).

We plot the $(r_1 - r_2)^2\psi(r_1, r_2)^2$ to find out the mean square value of the separation between r_1 and r_2. When r_1 is small, the maximum appears when r_2^2 is 2.79587 and vice versa when r_2 is swapped with r_1. The most probable r_1 may then be defined as $\sqrt{2.796} = 1.672$ 2D Bohr radius. Alternatively, we may define the most probable inter-particle distance squared as $\langle\psi|(r_1 - r_2)^2|\psi\rangle$, with proper normalization. This then gives 2.893 and the average r_{12} is then 1.7. This size happens when the center hole/electron is fixed, or $m \approx \infty$. For trion of similar mass in electron and hole, the size is nearly doubled, and we would have 3, all in 2D Bohr radius for underdoped cuprates ~4 Å. The periodicity of the CDW is then ≈12 Å, fairly close to experimental results (Blanco-Canosa et al., 2014; Chang et al., 2016; Croft et al., 2014; Miao et al., 2017).

The CDW has a favorable periodicity closely matching the linear size of exciton molecule. This then poses a restriction on the density of doping. When doping is too low, the CDW periodicity then would be too large to take the full advantage of Coulomb correlation. Instead, the excitons might just oligomerize into small exciton molecules or droplets, forming some carrier rich puddles without the long range order implicated by CDW. Meanwhile, the CDW is in this paradigm interpreted as an excitonic Coulomb phenomenon and so it is to be bounded by the T^*. This then leads to the dome-shaped phase boundary outlined in olive short dot-dashed curve, again given in Fig. 1.10.

In general, one needs to keep in mind that CDWs in cuprates are of a fleeting and incipient nature, due to the competition with other orders.

CDW's stability in cuprates is controlled by doping behavior. If the hole/electron ratio is 3:1, CDW is then not favored. This contributes to CDW's ephemeral presence in UD cuprates. If instead, the ratio is close to 2:1 or lower, CDW is then enhanced at price of the reduced SC order.

Of course, the CDW is but one phase that an excitonic complex may condense into. The charge order comes in at a price in the kinetic energy. This phase is in competition (or, delicate balance) with other orders, like the tetron liquid, trion/hole liquid. The double charged tetron complex may further combine with another exciton to form a $4h+2e$ boson, which may co-BE-condense with the tetron in the superconducting state. Clearly the physics of excitonic Coulomb states is indeed very rich and covers a wide range of interdisciplinary areas in condensed matter physics. Since the relative stability of different phases is very sensitive to local electron structure, temperature and pressure, the CDW order does not extend to large correlation length/volume, rather it often tops out at a few tens of Å.

The CDW periodicity (inverse of the Q wavevector) varies due to factors like effective mass and dielectric constants. If it happens to be commensurate with the periodicity of the lattice, locking or strong coupling to various phonon modes may result. This locking may pin down the CDW, degrade transport coefficients and suppress SC. It is tempting to see if this helps to resolve the well-known 1/8 anomaly, best manifested in LBCO.

1.21 Conclusion

In this chapter, we started with few-body quantum mechanics and have shown analytically that the exciton can give attractive pairing to two holes, in the form of a tetron. The high precision and well-benchmarked numerical study leads to an $a-V$ phase diagram on the stability of tetron. Tetron was then identified as the Cooper pair in cuprates, and the ubiquitous T-doping phase diagram of cuprates was derived. Satisfactory matches to experiments were found, essentially free of parameter fitting. This enables us to further identify the pseudogap as the excitonic binding energy of the (screened) Coulomb bound state, and the QCP as the manifestation of critical screening where bound state disappears in a screened Coulomb potential. The tetron picture in the ECM explains in geometrical terms why multilayer cuprates has a higher T_c and why the highest T_c is achieved at $n = 3$.

The major derivation of this Chapter is also based on the premise of electrons in p-type cuprates and a doping behavior not given in naively simple doping models. In Chapter 2, we will demonstrate the experimental support, and the preliminary theoretical understanding of such a behavior, namely the existence of electrons in p-type cuprates.

Chapter 2

Transport and the Evidence for Electrons in *P*-type Cuprates

2.1 Introduction

Anomalies in the normal-state transport are some of the earliest evidence indicating the pseudogap phenomenon. Peculiarities in electronic transport, magneto-transport, and thermal properties clearly reveal the significant presence of electronlike quasiparticles in *p*-type cuprates, which possess the highest critical temperature obtainable at ambient conditions.

This chapter, written by Dr. Nie Luo, explains why the linear T dependence of resistivity and T^2 dependence in the cotangent of Hall angle can be neatly understood from a single relaxation rate, due to pure electronic (*e*-*e* and *e*-*h*) scatterings: The effective carrier density scales linearly with T within a proper doping and T domain. The variable carrier density is just a manifestation of the T and k dependent pseudogap. The pseudogap T^* itself is accommodated by the framework of excitonic Coulomb binding energy. The sign reversals of many transport coefficients encountered in the mixed state (i.e., vortex state), is mainly due to the formation of excitonic

High-T_c Superconductivity: The Excitonic Coulomb Perspective
Nie Luo and George H. Miley

ISBN 978-981-4800-57-0 (Hardcover), 978-0-429-27870-9 (eBook)
www.jennystanford.com

complexes, either trions or tetrons. The minority electrons "freeze" out the majority charges, with a ratio of 1:2 or 1:3, making the sign reversal possible.

The presence of electrons in *p*-type cuprates naturally explain the preformed pairs inferred from many experiments, notably the muon spin rotation (μSR) and microwave surface reflectance. When T cools down to T^*, on the energy scale of the exciton or trion binding energy, trions and/or tetrons start to form, with proportions governed by doping x and T. The tetron portion is then counted as Cooper pairs, as proposed and identified in Chapter 1.

Below T_c, ODLRO starts to set in. As the supercarrier, the tetron starts to lose contribution to the Hall voltage and the thermopower S. The tetron is also significantly more massive, resulting in much less mobility in the normal state. What are left to still significantly contribute to normal transport coefficients are residue charges according to n_p-$3n_e$. Because of strong fluctuation effects in the mixed state of cuprates, an abrupt change is not likely seen in either the Hall or S. Instead, there exists a balance between the trion, tetron, exciton, free e and h. If $n_p < 3n_e$ initially at high T, then changes in the sign of the Hall and S are likely at temperature cold enough, down into the mixed state. Because the ambipolar Hall coefficient also depends on the relative mobility of electron and hole, the sign reversal in Hall is relatively rare to see. Suffice it to say that the sign change of S in the mixed state is a very clear sign of electron–hole co-existence.

Even if without ODLRO to the tetron, sign reversal of transport coefficient is still possible in that the tetron is heavy resulting in a much reduced mobility. Its "positive" contribution to the transport is then much reduced, leaving room for the left-over electrons to dominate the normal or mixed state transport coefficients.

The presence of electrons in cuprates is also revealed by optical spectroscopic techniques, such as optical conductivity. There is a well-known discrepancy on carrier densities calculated from doping x, the Hall coefficient and the optical effective carrier density. This is to be resolved later in this chapter.

2.2 Transport Anomalies and Non-Fermi-Liquid Behaviors

By the turn of the 21st century, the peculiar normal-state transport properties of high-T_c superconductors were rather controversial and not well understood. One of most striking of these is the alleged observation of two relaxation rates. The resistivity ρ is linear in temperature T, which seems to imply a transport (longitudinal relaxation rate $1/\tau_{tr} \propto T$ from $\rho = m_c/ne^2\tau_{tr}$ assuming, respectively, the elementary charge e, carrier (either electron or hole) mass m_c and density n of carriers are T-independent. In contrast, the Hall (transverse) relaxation rate $1/\tau_H$ from the cotangent of Hall angle $\cot\theta_H = m_c/eH\tau_H$ (SI units) is quadratic in T. These two distinct T relations have various explanations. A widely known is due to Anderson, based on the spin-charge separation of a Luttinger liquid. Other scenarios include but are not limited to the near-antiferromagnetic Fermi liquid (NAFL) theory of D. Pines et al. (Pines, 1994; Stojković and Pines, 1996, 1997) and marginal Fermi liquid (MFL) theory by Varma et al. (Varma et al., 1989).

The current author NL pointed out (Luo, 2000a) that the two distinct rates could be superficial if the effective carrier density (concentration) n actually depends on T, or $n = n(T)$. Especially if the $n(T) \propto T$ (of course optimal doping is assumed), then there is basically only one rate, for both longitudinal and transverse processes. Because $1/\tau_{tr} = n(T)e^2\rho/m_c$ (applying to both Gaussian units and SI if e interpreted as the elementary charge of the respective unit system) and as long as $n(T) \propto T$ and $\rho \propto T$ we then get $1/\tau_{tr} \propto T^2$. It has therefore the same T dependence as $1/\tau_H$ from the Hall effect. Similar explanation was given by several other groups (Alexandrov and Mott, 1994; Hirsch and Marsiglio, 1992a; Kubo and Manako, 1992, 1994; Xing and Liu, 1991) (by Hirsch, Xing, Alexandrov/Mott), but the current author was probably the first to associate it with the excitonic Coulomb phenomenon. The initial reception of such an idea was slow and the paper could not get published until after 2008 (Luo and Miley, 2009). The experimental development of recent years basically confirms the ECM explanation

(Chan et al., 2014; Mirzaei et al., 2013). Today we can say with confidence that this explanation is essentially correct, as some of the exotic accounts are now acknowledged as misled (Anderson, 2013).

Of course, cuprates are still rather different from a standard Fermi-liquid, with peculiarities such as the pseudogap, preformed pairs and so on. Hence, talks of non-Fermi liquid behaviors are legitimate. These anomalies and non-FL behavior can mostly be understood if we accept the notion that there are a significant number of electronlike quasiparticles in *p*-type cuprates. In fact we can derive most of the marginal Fermi liquid (Kuroda and Varma, 1990) results from the excitonic interaction.

As for exciton mechanisms specifically for cuprates, we need to mention some of the earlier efforts (Bardeen, 1987; Bardeen et al., 1987): The Russian school (Efetov, 1991; Gorbatsevich and Kopaev, 1991; Gorbatsevich and Kopaev, 1990); *d*-*d* excitonic (Weber et al., 1988); Jahn–Teller related excitation (Gaididei and Loktev, 1988), etc. The research by Bishop et al. was very specific to attribute the exciton to the charge transfer gap (Bishop et al., 1989). The Ching group at Missouri-Kansas (Ching et al., 1987) attempted exciton enhancement to explain the high T_c. Freeman discussed with the author the exciton mechanism of their version (Freeman et al., 1989) and thought that a key part should be the location of the exciton and electron in the cuprate. Markiewicz explicitly relates the exciton to the van Hove singularity, with which we share many common features (Markiewicz, 1990, 1991).

The evidence for the presence of electron in *p*-cuprate will be discussed in association with transports and other properties in this chapter. Some content is taken from previous published or unpublished work of the author, included here as a coherent part of a book.

2.3 Breakdown of Single Parabolic Band

The generic doping dependence of Hall coefficients in the archetypical La-214 series is shown in Fig. 2.1a, which gives the characteristic behavior of R_H near the sign reversal point x_0 for LSCO. The author used to do many Hall measurements and at the glance of the sign reversal in R_H, he realized that electronlike carriers exist in p-type cuprates in a significant proportion. In semiconductor physics, such a reversal is the sign of electron–hole compensation or partial compensation. However, for people not familiar with the Hall test, a naive explanation for such a sign reversal based on the single parabolic band model is that the carriers (quasiparticles) are holes to the left of x_0, (i.e., when $x < x_0$), transforming into electrons when $x > x_0$ and there are no electrons and holes coexisting under any specific doping x. Unfortunately, this simple yet naive explanation is not to hold. R_H is up to a constant of order unit given by $1/ne$ (Kittel, 1996; Seeger, 2013) in a single parabolic band, where e is the elementary charge and n is the mobile carrier density or concentration. Thus, R_H basically measures the inverse of n. If we accept the single parabolic band model, then n has to reach infinity while x approaches x_0 from either side as illustrated in Fig. 2.1b. Across the boundary, there would be an infinite carrier concentration jump from holelike to electronlike. This is clearly unphysical. The problem comes from 2 assumptions made in this model: "single" and "parabolic." One has to drop at least one assumption or both, in order to get the characteristic zero-crossing in R_H. We know nowadays with a fair certainty, that even in single-layer (or single plane) cuprates that have a single Cu-$d_{x^2-y^2}$-O-p band crossing the Fermi level near the nodal $(\pi/2,\pi/2)$ point, the band is split and partially gapped due to the pseudogap near the antinodal $(\pi, 0)$ region, or the alternatively called van Hove singularity point. Then a non-negligible energy dispersion along k_z is enough to dope the cuprate with both holes and electrons. Therefore, it is legitimate to treat R_H in a phenomenological two-band model.

2.4 The Phenomenological Two-Band Model

We first describe some prior efforts by other researchers. A few authors have tried to explain the doping dependent R_H from various perspectives. Their models, generally speaking, are microscopic enough to reach a characteristic R_H crossover from positive to negative near x_0. The local density approximation (LDA) of

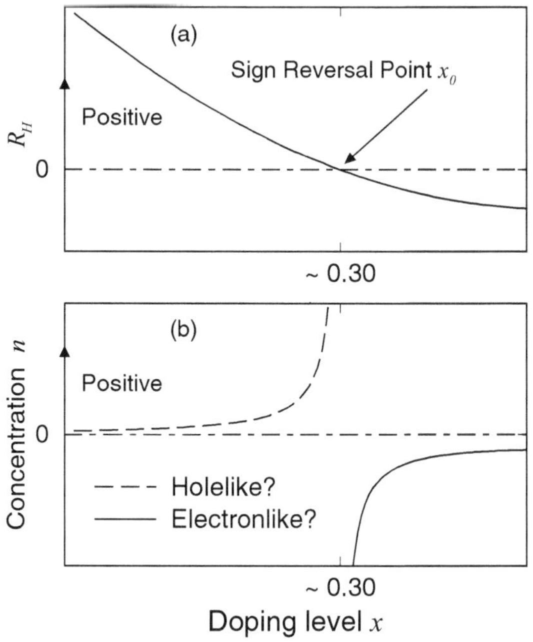

Figure 2.1 The scheme showing why the single parabolic band picture fails. (a) Characteristic R_H near sign reversal point. (b) Concentration calculated from $n = 1/R_H e$. Dot-dashed lines denote zero for R_H and n.

functional density band theory was used to calculate transport coefficients in a few cuprate superconductors (Allen et al., 1987a,b, 1988). Although the calculated R_H does show the sign reversal point in LSCO, the in-plane thermopower S_{ab} (Allen et al., 1987b) calculated with the same technique qualitatively contradicts to the experiment. (Compare the prediction (Allen et al., 1987b) with experiments (Sato et al., 1987; Sera et al., 1988); the theory basically does not get the right sign.) Moreover, LDA calculations usually describe the undoped cuprate as a metal (Pickett, 1989), which shows that the strongly correlated electron system is hard to model from conventional band theories. In general all the models solely based on LDA have difficulties in getting the strong T dependence on R_H and thermopower S in the optimally doped and underdoped regimes. This does not come as a surprise though because the strong correlation effect from the *e*-*h* interaction of the two-carrier model is neglected.

There were a few other attempts based on the Hubbard model (Fukuyama and Hasegawa, 1987; Kim et al., 1989). These calculations and those of (Kaga, 1991; Kaga and Saikawa, 1992) more or less account for the change in sign, but not without their own limitations. Slave-boson technique was used in one such attempt and surprisingly good fits to both LSCO and NCCO (Kaga and Saikawa, 1992) were obtained. However, the calculation also results into an electronlike pocket as the Fermi surface in the underdoped NCCO system, which is just opposite to what was found in ARPES experiments (Shen and Dessau, 1995), summarized by Shen et al. These discrepancies showed not only the limitation of aforementioned approaches, but in more general sense that essential physics was missing in understanding of the cuprate electronic structures. For example, the observed holelike pocket in underdoped NCCO is puzzling not only for the specific case of Hubbard model, it also seems to contradict to other experiments such as the Hall effect which is electronlike (negative) for underdoped NCCO (Uchida et al., 1989). Moreover, as already mentioned, ARPES experiments have indeed unveiled electronlike states in *p*-type cuprates near the antinode. Thus, a phenomenological two-band model is still a convenient tool in elucidating some qualitative aspects of charge dynamics, and have helped us to resolve some of the puzzling issues that existed before. This two-band picture is partly justified by various LDA band theories which show both *e*-like and *h*-like band edges crossing the FS (Pickett, 1989) for major cuprate series. Finally we need a few more words on the terminology. In our model, "two bands" actually might be better replaced by "two carriers" but because the former is more often used we still abide by it. Note that our term two-band here should not be confused with the number of bands in various Hubbard models. (There have been debates on how many Hubbard bands are necessary to high-T_c cuprates (Emery and Reiter, 1988; Zhang and Rice, 1988).) It is quite clear from ARPES and our interlayer Coulomb model that the two carriers are mainly from the antinodal $(\pi,0)$ region, besides the single-type (mainly holes) from the arc part near the nodal $(\pi/2,\pi/2)$. In single-layer cuprates, the antinodal gap is mainly due to the interlayer

(~6 Å) Coulomb instability (split) due to the van Hove singularity. In other cuprates, multilayer splitting of the conventional type helps too besides the aforementioned mechanism.

The general formula for R_H in the presence of both electrons and holes (Ziman, 1960) is

$$R_H = \frac{n_p \mu_p^2 - n_e \mu_e^2}{e(n_p \mu_p + n_e \mu_e)^2}, \tag{2.1}$$

where μ_p and μ_e are, respectively, the mobilities of holes and electrons with n_p and n_e their corresponding concentrations.

The absolute values of μ_e and μ_p are not important in Eq. (2.1). What matters is the ratio of mobilities, $b_\mu = \mu_e/\mu_p$. With this in mind, Eq. (2.1) is simplified into

$$R_H = \frac{n_p - n_e b_\mu^2}{e(n_p + n_e b_\mu)^2}. \tag{2.2}$$

Before using this formula to fit experimental data, we also want to re-investigate doping behaviors of LSCO.

2.5 Doping Model for LSCO

The doping of a Mott-insulator is in general regarded as a key issue in understanding high-T_c cuprates (Lee et al., 2006). The widely adopted yet oversimplified doping model assumes that each Sr atom contributes one hole after substituting for La. Thus, the carrier concentration n_p (only holes in the ideal doping model) is given by $2x/v_u$ where v_u is the unit cell volume, x the Sr content or doping level, and 2 indicates two formula units per unit cell. R_H would then simply be $v_u/2xe$, which is shown as the dashed curve in Fig. 2.2. Notice that the actual R_H is much smaller than $v_u/2xe$. This discrepancy is reflected in other experiments too, such as the in-plane optical conductivity $\sigma(\omega)$ obtained from optical reflectance spectroscopy (Tamasaku et al., 1994; Uchida et al., 1991). $\sigma(\omega)$ measures the frequency dependent charge dynamics in the CuO_2 plane. The corresponding integrated weight $N_{\text{eff}}(\omega)$ measures in each formula unit the effective number of carriers contributing to

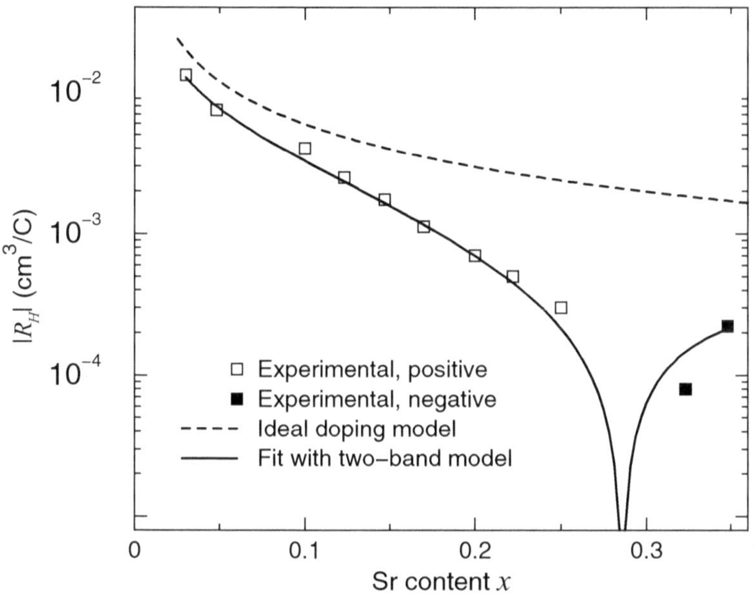

Figure 2.2 The doping-dependent Hall coefficient in LSCO. Squares are after (Takagi et al., 1989).

$\sigma(\omega)$ below ω. These two are linked by

$$N_{\text{eff}}(\omega) = \frac{2m^* v_f}{\pi e^2} \int_0^{\omega} \sigma(\omega') d\omega', \tag{2.3}$$

where m^* is the effective mass of quasiparticle and v_f is the formula unit volume.

$N_{\text{eff}}(\omega)$ is used to estimate the number of charge carries in the CuO_2 plane. In practice only low frequency (low energy) parts contribute to the ordinary transport processes such as electric conduction while the data for very high ω are usually not available and also less important to our purpose. So the total number of charge carriers that contribute to resistivity or Hall effect can be estimated below a cutoff frequency ω_c, usually between 1.2 and 1.5 eV. From the optical data (Tamasaku et al., 1994; Uchida et al., 1991) obtained by the same group (the Uchida group at University of Tokyo) which performed the Hall measurement (Takagi et al., 1989), one can see that N_{eff} is roughly 2.5 times x for low x, while it saturates and even drops after $x > 0.18$. This tells us that the idealized doping model is by no means accurate. (Ref. (Tamasaku et al., 1994) does not explicitly state whether the in-plane N_{eff} data are for a formula unit or a unit cell, but in an earlier paper (Uchida et al., 1991) by the same group similar N_{eff}-x plot for formula unit was shown and $N_{\text{eff}} \gg x$ at low x was especially mentioned and discussed. This makes us believe that data in Ref. (Tamasaku et al., 1994) are interpreted as those for formula unit.)

The reason for $N_{\text{eff}} \approx 2.5x$ at low x can be reasonably understood. The divalent Sr substitution for trivalent La results in not only one positive mobile charge (hole), but also statistically in average additional $\sim$0.5 hole and $\sim$0.5 electron or maybe more. This will make $N_{\text{eff}} \geq 2x$. There is no *a priori* reason in the doping theory against this possibility, as long as the net mobile charge is still $+1$. The latter possibility might actually be the case because the optical conductivity may be understood if one Sr gives 1.75 holes and 0.75 electrons. A bipolar self-doping effect is not unexpected the cuprates, and in later chapters we will derive this effect from interlayer Coulomb interaction.

With the coexistence of two carriers, the $N_{\text{eff}}(\omega)$ is now the sum of two terms,

$$\begin{aligned} N_{\text{eff}}(\omega) &= \frac{2m_e^* v_f}{\pi e^2} \int_0^{\omega} \sigma_e(\omega') d\omega' + \frac{2m_p^* v_f}{\pi e^2} \int_0^{\omega} \sigma_p(\omega') d\omega' \\ &= N_{\text{eff},e}(\omega) + N_{\text{eff},p}(\omega), \end{aligned} \tag{2.4}$$

where $N_{\text{eff},e}(\omega)$ and $N_{\text{eff},p}(\omega)$ are, respectively, effective numbers of electrons and holes. $N_{\text{eff},e}(\omega)$ and $N_{\text{eff},p}(\omega)$ at the cutoff frequency ω_c are related to the electron, hole concentration n_e, n_p by

$$\begin{aligned} n_e v_f &= N_{\text{eff},e}(\omega_c), \\ n_p v_f &= N_{\text{eff},p}(\omega_c). \end{aligned} \tag{2.5}$$

N_{eff}, $N_{\text{eff},e}$ and $N_{\text{eff},p}$, are functions of both w and x. At the fixed cutoff frequency ω_c, the three effective numbers become functions of x only and thus can be written explicitly as $N_{\text{eff}}(x)$, $N_{\text{eff},e}(x)$ and $N_{\text{eff},p}(x)$. n_e and n_p also depend on x. Then by using Eqs. (2.4) and (2.5), we reach

$$N_{\text{eff}}(x) = \left[n_p(x) + n_e(x)\right] v_f, \tag{2.6}$$

where the x dependence of n_p and n_e is also explicitly written.

Table 2.1 Oxygen deficiency δ for LSCO at different doping x

x	0	0.08	0.25	0.33	0.50
δ	0	0.01	0.035	0.05	0.15

From Nguyen et al. (1981).

Another notable feature in N_{eff} is the saturation when $x > 0.18$. We believe that this effect could be due to two factors: (1) the relaxation rate $1/\tau$ that increases with n; and (2) the defect chemistry of LSCO.

To understand the first factor, the generic Drude optical conductivity is related to relaxation rate $1/\tau$ by

$$\sigma(\omega) = \frac{n(T)e^2\tau}{m_c(1 + i\omega\tau)}. \tag{2.7}$$

Note that the τ in general decreases with $n(T)$, the mobile carrier density at a certain T. If the doping increases $n(T)$, then τ in general decreases: a mere fact reflecting that more crowded space brings more collisions. As the result, the mean free time (path) decreases with increasing density. Therefore, although increasing doping in general increases the optical conductivity $\sigma(\omega)$, there may exist a saturation effect from the reduction in τ. The end result then manifests as a N_{eff} saturation versus x, showing up in the optical response. The actual (static) carrier density may still increase.

The second factor relates to the defect chemistry of LSCO. On the one hand, the doping efficiency of Sr at high doping levels might be less than one net mobile positive charge per Sr and similar doping behavior of oxygen has been observed in the parent cuprate $La_2CuO_{4+\delta}$ (Li et al., 1996). One the other hand, the accurate chemical formula for LSCO is actually $La_{2-x}Sr_xCuO_{4-\delta}$ (Raveau et al., 1991) instead of $La_{2-x}Sr_xCuO_4$ especially when Sr content is high (i.e., there is deficiency δ in oxygen when x is not equal to 0). The formula is a logical result if one knows that $La_{2-x}Sr_xCuO_{4-\delta}$ is intermediate between its parent materials La_2CuO_4 (LCO) and Sr_2CuO_3 (SCO), both of which are insulators when undoped. SCO has in one formula unit only three oxygen atoms, one less than LCO. The oxygen deficiency has a superlinear dependence on x (i.e., $\delta \propto x^a$, where $a > 1$). We notice from the charge balance point of view, that oxygen deficiency δ will compensate 2δ of Sr. And because of the superlinear relation, the oxygen deficiency may outnumber the increase in Sr, contributing to a saturation or even a drop in effective doping. In the doping model that we assume in this paper, the effective doping level x' is taken as $x' = x - 2\delta$ and a doping concentration of x' results into concentrations n_e and n_p as long as $v_f(n_p - n_e) = x'$ so as to maintain charge neutrality. With the x dependence explicitly written, this relation reads

$$x'(x) = \left[n_p(x) - n_e(x)\right] v_f. \tag{2.8}$$

It is then easy to solve for $n_p(x)$ and $n_e(x)$ from Eqs. (2.6) and (2.8), and we get

$$\begin{aligned} n_p(x) &= \frac{N_{\text{eff}}(x) + x'(x)}{2v_f}, \\ n_e(x) &= \frac{N_{\text{eff}}(x) - x'(x)}{2v_f}. \end{aligned} \tag{2.9}$$

Now we need to find out $x'(x) = x - 2\delta(x)$. Ref. (Nguyen et al., 1981) gives δ versus x for a few specific x values as listed in Table 2.1. We need to be cautious on the exact values of these data because it is not clear whether samples used in Ref. (Nguyen et al., 1981) were superconducting or not, while the stoichiometry of cuprate is very much process dependent and oxygen content is vital to superconductivity. However, a survey of available data in the literature (Shafer et al., 1987; Takayama-Muromachi and Rice, 1991; Torrance et al., 1988) suggests that these are, nevertheless, qualitatively right. We then fit these data using simple power series $\delta(x) = b_0 x + b_1 x^2$. We use a modified least square fit based on the following consideration. In ordinary least square fit, one wants to minimize $\sum_i (\delta(x_i) - \delta_i)^2$, where δ_i, x_i are the experimental oxygen deficiency and Sr doping for the ith data point. This method is, however, not fair if the data to be fitted to differ by more than one order of magnitude, because it makes sure only that the absolute deviation of the fit is minimized. In our case here, δ is very small (<0.01) when x is less than 0.08 while big (~ 0.15) when $x > 0.30$; thus a least square fit in order to minimize absolute deviation results in a very large relative error for small δ. Moreover, our further derivation on n_p and n_e is more sensitive to relative errors than

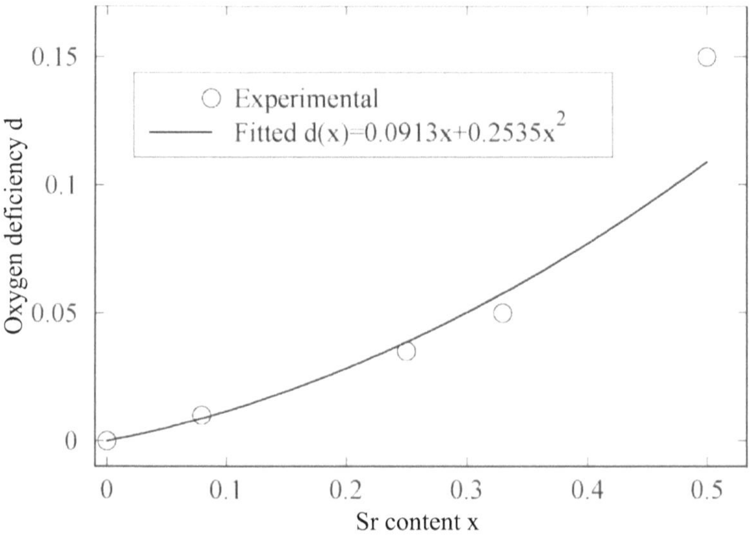

Figure 2.3 Oxygen deficiency δ, experimental and fitted curve, as a function of doping x. Experimental data are from Nguyen et al. (1981).

to absolute ones at $x < 0.2$. Thus, we change the criterion of least square fit to the minimization on $\sum_i [(\delta(x_i) - \delta_i)/\delta_i]^2$ instead. This results into a more even distribution of relative errors across the whole doping range that we are interested in. The fitting is not hard to carry out and the results are given in Fig. 2.3. Algebraically the fit reads $\delta(x) = 0.0913x + 0.253x^2$. As for N_{eff}, the difference among data is within one order of magnitude (off by a factor of 4 at most) so we still use standard least square fit and the result is given in Fig. 2.4 as the solid line. The fit is fairly good which reads $N_{\text{eff}}(x) = 0.002351 + 2.494x - 3.374x^2 - 62.79x^3 + 331.2x^4 - 625.0x^5 + 410.3x^6$.

2.6 Electrons in LSCO and Other *p*-Type Cuprates

Having found $N_{\text{eff}}(x)$ and δ, we then easily calculate $n_e(x)$ and $n_p(x)$ which are shown in Fig. 2.5a. The presence of electrons in LSCO is significant and after an average over the superconducting regime

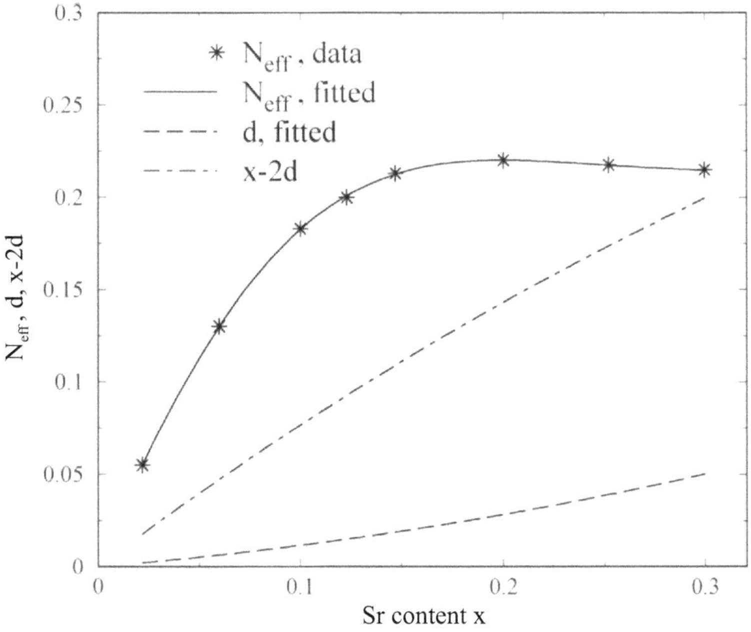

Figure 2.4 The effective carrier density N_{eff}, experimental and fitted, and the effective doping $x' = x - 2\delta$, as functions of doping x. The N_{eff} data are from Tamasaku et al. (1994).

we found $n_p/n_e \approx 3.7$. As a comparison we reproduce part of our previous result in Fig. 2.5c. One can clearly see that although $n_e(x)$ in Fig. 2.5a and 2.5c are different in their dependences on x, the average n_p/n_e are in good match. Our previous result (i.e., Fig. 2.5c) gives $n_p/n_e \approx 3.9$. At the optimal doping $x \approx 0.15$, Fig. 2.5a gives $n_p/n_e \approx 4$ while Fig. 2.5c suggests $n_p/n_e \approx 3$. These are also in rough agreement with our previous result for other p-type cuprates where $n_p/n_e \doteq 3.3$ on average (Luo, 2000b).

The mobility ratio b_μ is then readily solved for after knowing n_e, n_p and the experimental R_H. Rewrite Eq. (2.2) as

$$R_H \left[e(n_p + n_e b_\mu)^2\right] = n_p - n_e b_\mu^2. \tag{2.10}$$

Solving this simple quadratic equation for b_μ, we get,

$$b_\mu = \frac{-eR_H n_p n_e}{eR_H n_e^2 + n_e} + \frac{\sqrt{n_e n_p \left[(n_e - n_p)eR_H + 1\right]}}{eR_H n_e^2 + n_e}. \tag{2.11}$$

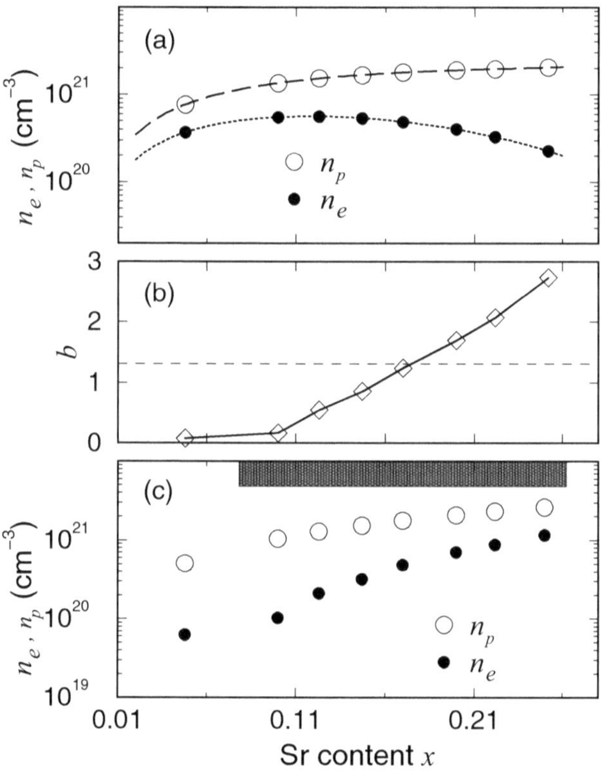

Figure 2.5 (a) The electron and hole concentration n_e and n_p versus x. The short and long dashed lines show, respectively, $n_e(x)$ and $n_p(x)$ calculated from Eq. (2.9). (b) The mobility ratios $b_\mu = \mu_e/\mu_p$ versus x, shown as the diamonds. The solid line is only a guide to the eye. The horizontal dashed line marks the average b_μ value over the superconducting regime. (c) n_e and n_p versus x from our earlier result, (Luo, 2000b). The shaded bar marks out the superconducting regime in x.

We calculate b_μ for a few dopings x where R_H data are available. The results are plotted in Fig. 2.5b. b_μ averages to 1.29 over the superconducting regime. This value is very close to what we found from average LSCO and NCCO Hall mobilities found in the literature. The close match of b_μ and the concentration ratio n_p/n_e from different methods suggests that the presence of electrons in LSCO is credible.

2.6.1 Transport Anomalies and *e-h* Coexistence

The previous section gives credible evidence to the presence of electrons in LSCO. We can generalize this to NCCO in a similar fashion and show that holes exist in this *n*-type cuprate. These two should be relatively ready to understand due to sign reversals in the R_H-x diagram. A natural question then arise: Does the coexistence of *e-h* apply to other *p*-type cuprates, especially those with the highest T_c?

In these *p*-type cuprates, doping is mostly effected with oxygen excess, and the doping is difficult to achieve beyond $x = 0.3$ and we do not see the sign reversal directly. However, $N_{\text{eff}} \gg n_H$ very similar to Fig. 1.4 of LSCO is reported (Padilla et al., 2005) in Y-123. This is a strong evidence for electrons in *p*-type cuprates with $T_c \sim 100$ K.

The existence of electrons in *p*-type cuprates can to be further understood in the context of pseudogap anomalies (Timusk and Statt, 1999). The pseudogap appears in the underdoped regime and weakens as doping level is increased, and NMR, infrared (IR), transport and photoemission (Batlogg et al., 1994; Timusk and Statt, 1999) experiments all reveal gaplike structures. We believe that the coexistence of electron and hole (EHC) should help answer this question: Electron (*e*) and hole (*h*) attract each other and form excitonic states, resulting in the loss of spectral weight in the charge (and often in spin also) excitation spectrum. This view is actually shared by a few other groups, such as the Markiewicz-Buda-Mistark-Bansil team (Markiewicz et al., 2015) and the Harshman-Dow-Fiory collaboration (Harshman et al., 2011).

There are other anomalies related to EHC, one of which is the Hall sign reversal near T_c occurring to properly doped samples. It is often explained in terms of vortex dynamics (Kopnin et al., 1993; Wang et al., 1994), which explains some phenomena. Recent experiments, however, have raised serious questions not yet answered by these theories (Nagaoka et al., 1998). We argue that vortex motion and pinning are not the whole story because there exists a similar anomaly in the thermopower S, usually measured *without* magnetic field. The S anomaly is not widely discussed but it shows up clearly in S-T plots (Cooper et al., 1987; Crommie et al., 1989; Ha et al., 1995; Kiehl et al., 1995; Martin et al., 1995; Zhou and Goodenough, 1995) like Fig. 2.6a. It occurs to major cuprate series, on samples

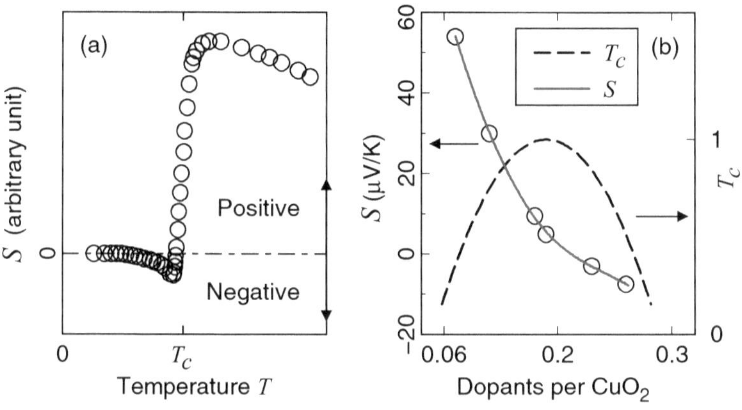

Figure 2.6 (a) Generic thermopower S versus temperature T for properly doped p-type samples near T_c (Allen et al., 1990). The dot-dashed line is $S = 0$. (b) S versus doping for p-type cuprates. S values of circles are from Ref. Allen et al. (1987a) while doping levels are inferred from a general T_c dependence on doping, shown as the dashed line. T_c are normalized to the maximal T_c, T_{cm}. The solid line is the generic behavior of S in p-types although the S values are not to be taken as exact.

from ceramics to single crystals, suggesting a nontrivial nature. This S anomaly is likely a general and doping dependent behavior of cuprate superconductors near T_c, just like that in R_H. It then poses a problem to theories based on vortices; here we have *no* magnetic field, then where do vortices come? The sign of normal-state S also depends on doping (Kaiser, 1997) as shown in Fig. 2.6b. We will see that both are explained naturally in terms of EHC.

For comparison, the doping dependent R_H of NCCO is put side-by-side with that of LSCO (Takagi et al., 1989; Uchida et al., 1989) in Fig. 2.7. The dashed line is the theoretical R_H assuming that each dopant gives one hole. The squares trace the theory relatively well when $x < 0.05$. At higher x, increasing deviation from the dashed line results in a change in sign. A calculation similar to that of LSCO is performed for NCCO. The b_μ is estimated as follows. As the zeroth order approximation, we take R_H as coming from majority carriers only and then mobility $\mu = R_H/\rho$ where ρ is the resistivity. This gives μ_e if $R_H < 0$ and μ_p if $R_H > 0$. We surveyed data in the literature (Fournier et al., 1997; Gollnik and Naito, 1998; Seng et al., 1995; Suzuki, 1989; Wang et al., 1991) from some of the best samples and got $b_\mu = 1.25$. Details of this survey will be published elsewhere. This b_μ is used for both LSCO and NCCO. Only data at 80 K are taken to reduce errors from R_H variation with T. n_e is then directly calculated from the experimental R_H. The results are in Fig. 2.7c as solid circles. n_e are then fitted by power series shown as the solid curve that reads $n_e(x) = 9.0 \times 10^{18} + 7.3 \times 10^{20}x + 6.3 \times 10^{22}x^3 + 2.3 \times 10^{25}x^{10}$ in cm^{-3}. We then plug $n_e(x)$ back in Eq. (2.2) to verify R_H, resulting in the solid line in Fig. 2.7a. Similar result for NCCO is shown in Figs. 2.7b and 2.7d with $n_p(x) = -2.1 \times 10^{20} + 1.0 \times 10^{21}x^{0.5} + 5.3 \times 10^{26}x^7$ in cm^{-3}, which holds if $x \geq 0.05$. (The triangle is a fit using $b_\mu = 0.95$. The fit cannot be carried out if $b_\mu > 0.96$. This shows b_μ dependence on x.)

The presence of electron in LSCO is significant, also as mentioned before. In the superconducting regime, $n_e \sim 0.17-0.5\ n_p$. As for NCCO, $n_p \sim 0.5-1.5\ n_e$. Thus, EHC effect is better manifested in NCCO, which might explain why it has long been suggested by various groups (Fournier et al., 1997; Wang et al., 1991).

The EHC extends into other p-type cuprates where normal-state R_H are positive regardless of doping. A sign reversal in S still occurs at high doping levels and near T_{cm}, as shown in Fig. 2.6b. This strongly suggests an EHC. Also R_H is somewhat small. The optimal doping is around 0.15 holes per CuO_2 (HPC) for most p-type cuprates, inferred from the Cu valence obtained from methods like iodometric titration (Karppinen et al., 1996). However, n_p calculated from $1/eR_H$ is nearly 2 times as large. The presence of e, as in LSCO, is thus suggested.

We now want to extend the EHC analysis to other p-type cuprates where normal-state R_H are positive regardless of doping. b_μ has to be determined first in order to find out n_e . μ_p in these cuprates is much higher than that in LSCO or NCCO, so $b < 1$ is likely. However, μ_e is not known in these materials, and we have to estimate. Because of the possible EHC, transport relaxation now involves multiple processes. There are 3 processes from the inter-carrier (IC) relaxation channel: hole–hole (h-h), electron–electron (e-e), and electron–hole (e-h) scatterings. The rate for h-h or e-e process is much less than that for e-h because of the Pauli principle. Each e-h collision involves one e and one h but because $n_p \gg n_e$,

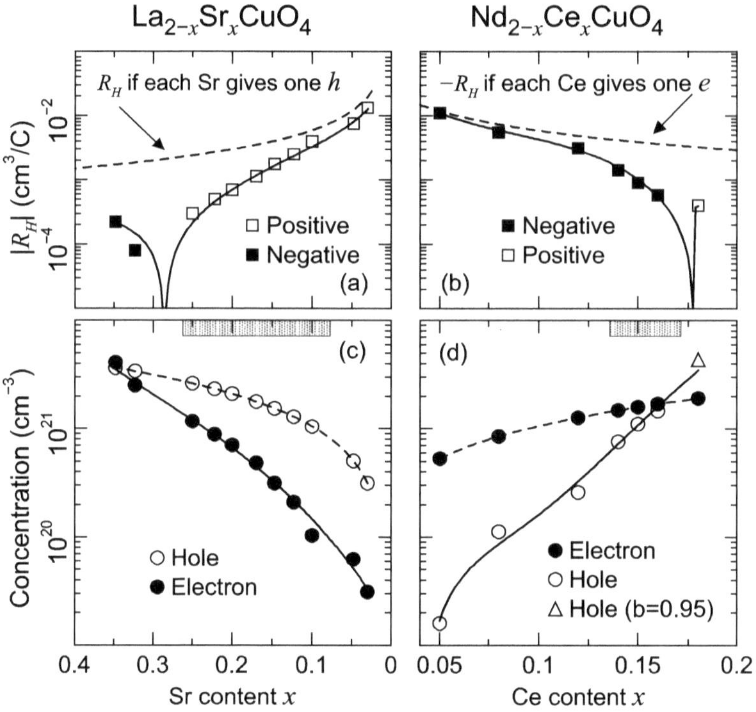

Figure 2.7 The Hall coefficients for LSCO and NCCO.

each h is scattered less frequently in average. Thus, $\mu_p \gg \mu_e$ if only IC channel is counted. However, carrier-defect (CD) scattering also contributes. Defects here are mainly ionized dopants with a density $n_d \approx n_p - n_e$ as required by charge neutrality (Martin et al., 1990). For electrons, rates from CD and IC are comparable because $n_d \approx n_p - n_e \approx n_p$, while for holes, CD dominates because $n_d \approx n_p - n_e \gg n_e$. Counting both channels, $\mu_h \approx 2\mu_e$ or $b_\mu \approx 0.5$ ($b_\mu > 1$ in LSCO may relate to the K_2NiF_4 type structure). Also, we only take data at T^* because below it many h pairs have already formed, resulting in an underestimate on n_h. T^* is selected where ρ begins to deviate below the linear ρ-T relation. The results for some of the best samples that we can find in the literature are in Table 2.2. The electron concentration is substantial: $n_e \sim n_p/3.3$ in average, partly confirming results in LSCO.

Table 2.2 The *e-h* coexistence in $YBa_2Cu_3O_{7-\delta}$ (Y-123), $Bi_2Sr_2CaCu_2O_8$ (Bi-2212), $Tl_2Ba_2CuO_{6+\delta}$ (Tl-2201) and $HgBa_2Ca_{n-1}Cu_nO_{2n+2+\delta}$ [Hg-12(n-1)n]. T_c and T^* are in K, R_H in 10^{-3} cm^3/C, the assumed n_p and calculated n_e in 10^{21} cm^{-3}

Series	T_c	T^*	R_H	Source	HPC	n_p	n_e	Notes
Y-123	93	120	2.33	Phys. Rev. B **44**, 10158 (1991)	0.15	1.73	0.64	O, SC,*
Bi-2212	82	125	2.60	Phys. Rev. B **42**, 8704 (1990)	0.15	1.33	0.68	NO, SC
Tl-2201	85	105	3.42	Physica **282–287C**, 1185 (1997)	0.15	0.87	0.57	NO, SC
Hg-1212	124	320	2.30	Phys. Rev. B **50**, 3246 (1994)	0.15	1.60	0.73	NO, PC
Hg-1223	135	320	2.38	Physica **234C**, 1;PR B **53**, 2767	0.13	1.66	0.64	NO, SC
Hg-1234	130	320	2.94	Physica **223&224B**, 512 (1996)	0.13#	1.86	0.21	NO, SC

Notes:
* Chain contribution removed.
# Assumed the same as in Hg-1223.
T^* is high in Hg-series, which are often underdoped. O: optimal. NO: near-optimal. SC: single crystal. PC: polycrystal.

The carrier densities of Table 2.2 have to be understood as those at T_c for optimal doping. If the temperature is raised, the densities also increase appreciably, up to the pseudogap T^*, above which the T dependence is weakened.

Summarizing the result of LSCO and Table 2.2, with the exception of the failed high-T_c Bi-2201 ($T_c \sim 20$ K), we can draw a generic relation between electron n_e and hole density n_p, with the fractional doping x.

$$\begin{cases} n_p = 1.5x/v_f \\ n_e = 0.5x/v_f \end{cases} ; \quad \text{when} \quad x \leq 0.2 \tag{2.12}$$

The reasonable explanation of sign reversals and pseudogap given by EHC justifies our approach this far. While more experiments are needed to establish the S anomaly, it is easily understood in terms of EHC, together with the R_H anomaly in a unified picture. Take p-type cuprates, for example. Upon cooling, holes form pairs around the pseudogap temperature T^* and then h pairs begins Bose–Einstein condensation (BEC) into supercurrent (SC) at T_c, resulting in the sharp drop in ρ (Uemura et al., 1989). Because of the large phase fluctuation in these superconductors (Emery and Kivelson, 1995), coherence across the whole sample is not yet established near T_c so that the Seebeck or Hall voltage coming from normal carriers is not yet shorted out. Because of BEC, n_p, the number of normal holes that still contribute to S or R_H, is greatly reduced. If it is so reduced that $S_p n_p \mu_p - S_e n_e \mu_e < 0$, S turns negative. Similarly $n_p \mu_p - n_e \mu_e < 0$ results into an R_H anomaly. Whether electrons pair or not does not matter much because e pairs, even if formed, have a lower density and thus enter BEC at a lower temperature $T_c^e < T_c$. Between T_c and T_c^e, e pairs are normal and thus still contribute to R_H and S negatively; anomalies are still likely. Our model depends on the delicate balance among n_e, n_p, μ_e, μ_p, S_e and S_p, which explains why it occurs only in properly doped samples. This approach thus emphasizes electronic structures, a point strongly supported by recent experiments (Nagaoka et al., 1998).

R_H anomalies in a few low-T_c superconductors such as Nb, V, MoGe/Ge multilayers (Hagen et al., 1990) are often cited as supports for the vortex mechanism. However, there exists some confusion.

The Hall effect in Nb is strongly impurity dependent. In pure Nb with residue resistivity ratio (RRR) $\gtrsim$2000 no such anomaly is reported. Reported anomalies in V were all from samples with RRR $\lesssim$200, which are likely caused by impurities, just as in Nb (Noto et al., 1976). As for MoGe/Ge, recall that Mo has both electronlike and holelike Fermi sheets (Sparlin and Marcus, 1966), and Ge is a semiconductor of either *p*- or *n*-type. Thus, even the anomalies in low-T_c superconductors seem linked more to band structures than to vortex dynamics.

2.6.1.1 EHC, a modern update

Most of preceding analyses and results given so far in Section 2.6.1 were obtained 15 years ago. The study on ratio of n_p/n_e, could have been taken as speculative by some researchers back then. Now the picture is much clearer, with notions like "electron-pocket," "two-component," and so on gaining much more acceptance. With the progress of high-precision computational tools for the electronic structure, the hole/electron ratio can now be studied with first-principles numerical methods. Our latest results from Chapter 4 gives $n_p/n_e = 4.29$, for LSCO at doping slightly higher than $x = 0.125$. This ratio is bracketed by the experimental data of Table 2.2. The near coincidence from vastly different perspectives and techniques gives the excitonic Coulomb mechanism an added evidence of support. The details of this latest research are given in Section 4.5.2.

2.6.2 Other Evidence for EHC

The EHC is already well known to *n*-type cuprates like NCCO, for example, by the R. L. Greene group (Fournier et al., 1997). The generalization to include *p*-type cuprates takes more theoretical efforts. The author is not the only one to point out the coexistence of electrons and holes in cuprates. Sometimes the same phenomenon is called two-carrier model (A. Brinkman) (Brinkman and Hilgenkamp, 2005; Gor'kov and Teitel'baum, 2014), or two-component phenomenology (Gor'kov and Teitel'baum, 2008; Gor'kov and Teitel'Baum, 2015; Haase et al., 2009; Storey and Tallon,

2012; Teitel'baum, 2017; Wulin et al., 2012), or two-fluid model (Keren et al., 2006; Sanna et al., 2008). L. P. Gor'kov and Teitel'baum, in particular, emphasize the importance of the electron contribution. K. Levin et al. interpret the pseudogap as from pairs of attractive potential, which they also think leads to the MIR. It is worth noting that the evidence for electron–hole co-presence comes not only from transports but also from optical properties and the spin sector such as NMR, μSR, etc. These are further related to phase separation effects well documented for the cuprate by K.A. Müller (Müller and Giorgio, 1993).

One confusion often encountered is the conceptual problem of having electron and hole together. In the research community, there seems to exist a conceptual barrier: It is either hole or electron. A similar misconception exists: *e-h* compensation implies that electron density equals that of hole. These oversimplified and often mistaken notions are *not* justified from either basic principles or experiments.

2.6.3 Where Is the Electron in *k* Space

Where do electrons come from? The ambipolar (self-) doping behavior can firstly be inferred from density functional theory (DFT) mapping of the FS topology. The stability of this *e-h* system against recombination should not be a big concern because of the semimetallic band structures. When ARPES first revealed electronlike (*e*-like) Fermi surface in Bi-2212 some years ago (Chuang et al., 1999), it raised a debate as to whether the claimed feature was real or spurious. Now the electronlike FS portion is firmly established (Kaminski et al., 2006; Razzoli et al., 2010; Yoshida et al., 2001), even for single CuO_2 layer Bi-2201 and LSCO. Besides, those electronlike FS sections are now seen in new techniques (Doiron-Leyraud et al., 2015; Doiron-Leyraud et al., 2007; LeBoeuf et al., 2007) such as quantum oscillations, often of the Shubnikov–de Haas effect type (Harrison and Sebastian, 2015; LeBoeuf et al., 2007; Norman, 2010; Norman and Proust, 2014; Sebastian et al., 2010), well summarized by Sebastian, Harrison, Norman, and Taillefer et al. We take the simple case of LSCO as

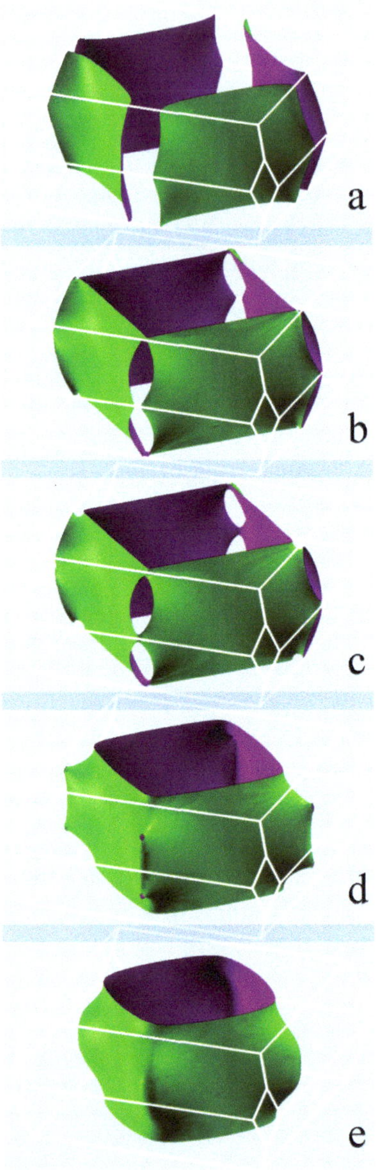

Figure 2.8 Fermi surface topology in LSCO versus doping x: (a)$\sim$0.06; (b) 0.14; (c) 0.15; (d) $\sim$0.23; (e) $\sim$0.30.

an illustrative example; once it is understood, the situation can be readily generalized to other cases.

The calculation was carried out by a full-potential code of early version with linearized augmented plane-waves, WIEN2K. The many-body correlations were handled within the general gradient approximation and the functional was that of Perdew–Burke–Ernzerhof. Diagonalization was performed over 2835 k-points in an irreducible wedge of the first Brillouin zone (BZ) to ensure a proper precision. Other important calculation parameters were: $r_{MT} \times k_{\max} = 7.0$, r_{MT} in r_B: 2.4 (La), 1.9 (Cu), 1.6 (O); $a = 3.7873$ Å, $c = 13.2883$ Å. The doping effect was handled assuming a rigid band, which had been shown to be acceptable for its purpose in the conducting phase. The Fermi-straddling band was then sampled in the reciprocal lattice to generate the three-dimensional Fermi surface.

Figure 2.8 shows the evolution of FS. At $x = 0.05$, the FS is mainly holelike centered around (π, π). Even though the FS is in 3D, notation with only 2 coordinates like (π, π) may be utilized when the feature is insensitive to k_z. In the extended reciprocal space, the FS consists of isolated holelike tubes surrounding (π, π) and the equivalent. The FS is holelike predominantly, with very tiny electronlike portion near $(\pi, 0, 0)$ in the basal plane. With increasing x, portions near $(\pi, 0)$ at planes $k_z = 0$ and $k_z = \pm\pi$ are progressively more electronlike. When x reaches $\sim$0.14, the FS slabs at $k_z = \pm\pi$ no longer intersect the BZ boundary. This represents a change of topological significance: Small tubes now connect otherwise isolated hole tubes in the extended k space. Topologically the FS changes from a 2-connected surface to a higher-order-connected one. Note that this is very near the optimal $x = 0.15$. At the optimal doping, the FS segments near $k_z = 0, \pm\pi$ are mainly electronlike, while those near $k_z = \pm\pi/2$ are still holelike.

The connectedness of the FS increases to 10-connected for FS in a single BZ. Sizeable open "necks" are formed on the BZ boundary as in Fig. 2.8c. In extended k space, the necks are actually electronlike tubes, forming a "jungle gym" structure. The coexistence of electrons and holes is best manifested near $x = 0.15$. If x is further raised, the "necks" shrink in size and finally disappear at

$x \sim 0.23$. The change at 0.23 is of topological importance too, the FS is now mainly electronlike. Beyond this point the FS is then isolated electronlike tubes in the extended k space. The highest T_c seems to coincident with: (1) the presence of both electrons and holes; (2) the topological connectedness cross-over from low to high and then back to low. When the 3D Fermi surface evolution of Fig. 2.8 is projected to the basal plane $k_z = 0$, it quite well reproduces the ARPES experiments.

The current photoemission techniques are not very sensitive in k_z. As the result, what we have seen for the supposed ARPES Fermi surface is merely its 2D projection along the k_z direction. The k_z dependence of the FS near $(\pi/2, \pi/2)$ is negligible. The ARPES FS is therefore strong and unambiguously holelike near $(\pi/2, \pi/2)$. The situation is complicated at $(\pi, 0)$, partly due to non-negligible k_z dependence ($\sim$0.1eV) near the Fermi level. As a result, the ARPES response there is smeared out, forming the "shaded" region around $(\pi, 0)$. As we know for enough confidence today, these shades corresponds to strong correlation of the excitonic Coulomb state that opens a narrow and spectral-smeared energy gap at $(\pi, 0)$. This is discussed in details in Section 3.8.

In summary, the electrons mainly reside at the extended saddle region around $(\pi, 0)$ in k-space.

2.7 Relaxation Rates in Transport

In the late 1990s, the superficial discrepancy between $\rho \propto T$ and $\cot\theta_H \propto T^2$ was treated as so abnormal in the high-T_c world, that some exotic schemes were resorted to understand it. The author was aware of similar behaviors elsewhere, typically in some semi yet traditional metals, where a single relaxation rate dominates transports. Therefore, such a discrepancy is not limited to the cuprates. One question is naturally raised, that is, do $\rho \propto T$ and $\cot\theta_H \propto T^2$ really indicate two relaxation rates? A few theories existed by then but a general consensus was not possible. Nevertheless, recent and refined experiments are increasingly indicating a single rate, as we will discuss later.

2.7.1 Relaxation Rate in Semimetals

Bismuth displays a similar discrepancy from 77 K to 300 K as shown in Figs. 2.7.1a and 2.7.1b, although with a $\cot(\theta_H) \propto T^{2.5}$ instead (Michenaud and Issi, 1972).[1] This behavior has long been understood from a single rate $1/\tau \propto T^{2.5}$ and a variable $n(T) \propto T^{1.5}$ (see Fig. 2.7.1c), which in combination give a $\rho \propto T$. It is very *unlikely* that a Luttinger liquid theory is relevant to a three-dimensional conductor, to which Bi belongs. Then taking such discrepancy as the evidence for spin-charge separation is not convincing.

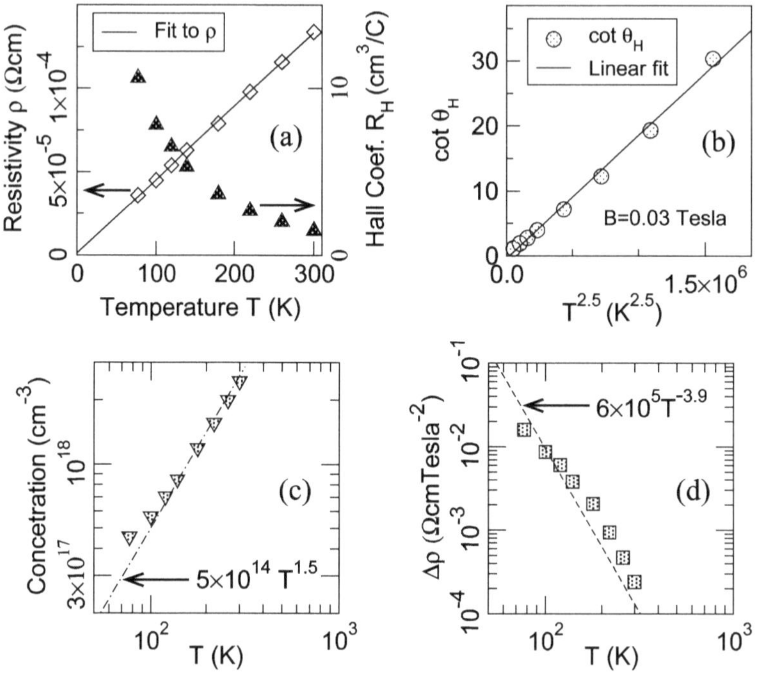

Figure 2.9 (a) Resistivity $\rho(\rho_{33})$ and Hall coefficient $R_H(R_{231})$ versus T in Bi after (Michenaud and Issi, 1972). (b) $\cot\theta_H$ versus $T^{2.5}$ calculated from ρ_{33} and $R_H(R_{231})$. (c) Concentration for both electron and hole versus T. (d) Magnetoresistance $\Delta\rho(A_{33})$ versus T.

[1] P. Michenaud confirmed to the author that tick mark labels for the vertical axis of Fig. 2 should be 10^{-7}, 10^{-8} and 10^{-9} downwards. Data are from Table 2 of their paper.

With a much lower n ($\sim 10^{17}$ cm $^{-3}$), the semimetal Bi still has a lower resistivity than most cuprates in their normal states, which are often treated as metals. This raises the question of whether the doped cuprate is a metal, a semimetal, or even a semiconductor as suggested by Alexandrov and Mott (Alexandrov, 1996; Alexandrov et al., 1996). We believe that it has characters of all of the three. It is like metal or semimetal as required by a non-zero n when approaching 0 K. On the other hand, significant electrostatic field effects on the transport (Ahn et al., 1999), the divergent 0 K resistance under a strong magnetic field (Boebinger et al., 1996; Ono et al., 2000) and the likely variable n from the Hall effect (Suzuki, 1989) put it near semiconductors and semimetals, quite far from a simple metal (Hurd, 1972; Meaden, 1965) where n is taken as fixed. n from R_H is nearly constant for group IA metals like Na, K and noble metals (100 K $< T <$ 350 K, where $\rho \propto T$). n may vary even for some metals like Al, because of its sizable R_H dependence on T. Certainly the variation in R_H cannot come from n only.

2.7.2 The Case of Cuprates

Let us go back to other evidence supporting the $n(T) \propto T$ argument. The first and foremost is actually the Hall coefficient R_H itself. Within the temperature range where $\cot\theta_H \propto T^2$, R_H can be nicely fitted by $1/(c_0 + c_1 T)$ in optimally doped (OD) p-type cuprates (Suzuki, 1989). Such a fit is even upheld in recent infrared Hall experiments at a high-energy scale of $\approx$ 0.1 eV (Shi et al., 2005). Note that $R_H = 1/nec$ or equivalently $1/ecR_H$ measures n in a simple parabolic band. However, a naive explanation of $n = 1/ecR_H$ seems not prudent because R_H may reach 0 in some cuprates. This is perhaps why the T-dependent R_H is not widely accepted as the evidence for $n(T) \propto T$. However, $R_H = 0$ can be understood with a two-carrier model as a result of compensation (Eagles, 1989; Eagles and Savvides, 1989; Hirsch and Marsiglio, 1991; Luo, 2000b, 2001a; Xing and Ting, 1988). And as long as one carrier dominates the transport, R_H is still $\propto 1/n$ although $ecR_H = 1/n$ no longer holds. To see this, suppose that the densities of two carriers are given, respectively, by $n_e = n_{e0} + n_{e1}T$ and $n_h = n_{h0} + n_{h1}T$, giving rise to a Hall coefficient

$$R_H = \frac{n_h - n_e b^2}{ec(n_h + n_e b)^2}$$
$$= \frac{(n_{h0} - n_{e0}b^2) + (n_{h1} - n_{e1}b^2)T}{ec\left[(n_{h0} + n_{e0}b) + (n_{h1} + n_{e1}b)T\right]^2},$$

where $b = \mu_e/\mu_h$ is the mobility ratio. If $n_{e1}T \gg n_{e0}$, $n_{h1}T \gg n_{h0}$, and $n_{h1} - n_{e1}b^2 \neq 0$, R_H is roughly $\propto 1/T$. Thus, the Hall effect strongly suggests $n(T) \propto T$ in the normal state of OD p-type cuprates.[2]

R_H is arguably the most widely used and quite accurate method to determine n in metals and semiconductors when compensation is not severe. It is not persuasive to treat one half of the Hall effect, say $\cot\theta_H \propto T^2$ as exact, while disregarding the physical meaning of the other half, namely $R_H \propto 1/T$, for $\cot\theta_H = m_c c/eH\tau_H$ itself holds strictly only for a simple parabolic band, just as $R_H = 1/nec$. Actually, putting these two facts at equal footing makes the physics compact and concise: There roughly exists a relaxation rate $1/\tau \propto T^2$, governing both longitudinal and transverse processes. $\rho = m_c/n(T)e^2\tau_{tr} \propto T$ is the result of $1/\tau_{tr} \approx 1/\tau \propto T^2$, and $n(T) \propto T$, while $\cot\theta_H = m_c c/eH\tau_H$ has no $n(T)$ in it thus $\cot\theta_H \propto 1/\tau_H \approx 1/\tau \propto T^2$. (We assume that m_c is relatively T-independent and this issue will be addressed later.) With a non-fixed n, we can proceed on to see that the apparent violation of Kohler's rule (Harris et al., 1995) is actually superficial.

2.7.3 Kohler's Rule

Kohler's rule in its ordinary form states that the relative magnetoresistance $\Delta\rho/\rho_0$ in a magnetic field H, can be represented in the form (Ziman, 1960)

$$\Delta\rho/\rho_0 = F(H/\rho_0), \tag{2.13}$$

where ρ_0 is the resistivity at $H = 0$ and F is a function given by the metal and its sample geometry only.

---

[2]Carrier density should increase with T in cuprates. It is better to write $n(T) = c_1 + c_2 T$ instead of $n(T) \propto T$. Unless it is necessary, we, however, in general neglect the difference of two such notations for the sake of simplicity.

In cuprates, approximately (Harris et al., 1995) $\Delta\rho/\rho_0 \propto H^2T^{-4}$ but $H/\rho_0 \propto HT^{-1}$. Because $\Delta\rho/\rho_0$ (or H^2T^{-4}) is not a function of H/ρ_0 (or HT^{-1}) only, Kohler's rule *seems* to be violated.

The argument above is, however, not valid if carrier (or alternatively referred to as quasiparticle) density n is not fixed. A scrutiny on its derivation (Ziman, 1960) shows that a more general form of Kohler's rule is

$$\Delta\rho/\rho_0 = F(H\tau), \tag{2.14}$$

where τ is the single relaxation time assumed for that conductor. For simple metals, n is roughly a constant, so that τ has the same T dependence as ρ_0, and thus there is no problem if one writes $\Delta\rho/\rho_0 = F(H/\rho_0)$ instead of $F(H\tau)$. However, for cuprates, n likely depends on T, so we should use Eq. (2.14) instead of Eq. (2.13) in order to see if Kohler's rule survives. Because $H\tau \propto HT^{-2}$ where $1/\tau$ is the single rate assumed, we have $\Delta\rho/\rho_0 \propto H^2T^{-4} = (HT^{-2})^2$ and thus $\Delta\rho/\rho_0 \propto (H\tau)^2 = F(H\tau)$ so that Kohler's rule still holds. [Similarly in Bi, $\Delta\rho/\rho_0 \propto H^2T^{-5} = (HT^{-2.5})^2$ inferred from Fig. 2.7.1d and so we get $\Delta\rho/\rho_0 \propto (H\tau)^2 = F(H\tau)$ because $1/\tau \propto T^{2.5}$ here in Bi. Kohler's rule is not violated either. A thorough derivation/analysis was given by the author elsewhere (Luo and Miley, 2002).

Recent experiments show unequivocally that the Kohler's rule is obeyed in the cuprates, at least to the regime that gives the highest T_c. The perseverance of Kohler's rule strengthens the idea of a single rate $1/\tau \propto T^2$. Thus, our previous explanation of the Hall effect was in the right direction. Angle-resolved photoemission spectroscopy (ARPES) studies also suggest the applicability of semimetal band structure to cuprates. Figure 2.10 shows some scenarios how this can be materialized in Bi-2212 with the two-dimensional Fermi surface (FS) mapping following Chuang et al. (1999). The discussion can be generalized to other p-type superconductors. Note that the discussion here represents the FS understanding of some ten years ago, but is, nevertheless, published here to serve as a historical perspective, and as part of an integral theory. More recent studies have improved our understanding and answered many puzzling questions outlined in Fig. 2.10. These will be discussed in the later part of the book.

2.7.4 Electronlike States: A Modern Update

When the shaded states in Fig. 2.10 were initially proposed, these might have been taken as a speculation rather than an educated guess. With the advent of many-body techniques to analyze the excitonic Coulomb interaction, these states are finally explained. This is a strong-correlation many-body effect that happens even in single-layer cuprates. The driving force is the interlayer Coulomb energy that can be minimized with the transfer of charge from one layer to the other. The instability is further enhanced by the van Hove singularity. The charge transfer results into hole-doping in one layer and electron-doping in the other. From this point on, the "splitting" of the single band near $(\pi, 0)$ can be even loosely understood from a traditional band picture. Of course, the problem is only tackled properly with modern techniques, such as the Bogoliubov de Gennes (BdGE) theory or the many-body Green's function formalism. The many-body solution then has some prominent features as follows: (1) The excitation spectrum (loosely the band) is gapped near the saddle point $(\pi, 0)$ (see Figs. 3.2 and 3.4); (2) the excitation spectrum is smeared and weighted down in some part of the BZ due to the many-body interaction. If described loosely in an energy-band's terminology, the split bands are blurred near $(\pi, 0)$, with portions truncated; (3) The blurred spectral density shows up as the shaded states in ARPES; (4) The antinodal gap is T and k dependent; (5) There still exist Fermi crossings near $(\pi/2, \pi/2)$; (6) In general, p-type doping reduces the Fermi level, making the partially gapped $(\pi, 0)$ "bandedge" semimetallic, which contributes electronlike quasiparticles.

2.8 Mid-Infrared

Having mentioned some evidence for the single-rate explanation, we now discuss experiments that were considered at odds with it. The apparently most serious one, in our opinion, was from the optical conductivity $\sigma(\omega)$. To account for the non-Drude behavior of $\sigma(\omega)$ in cuprates, a classical two-component model (Tanner and Timusk, 2013) is used in quite some studies,

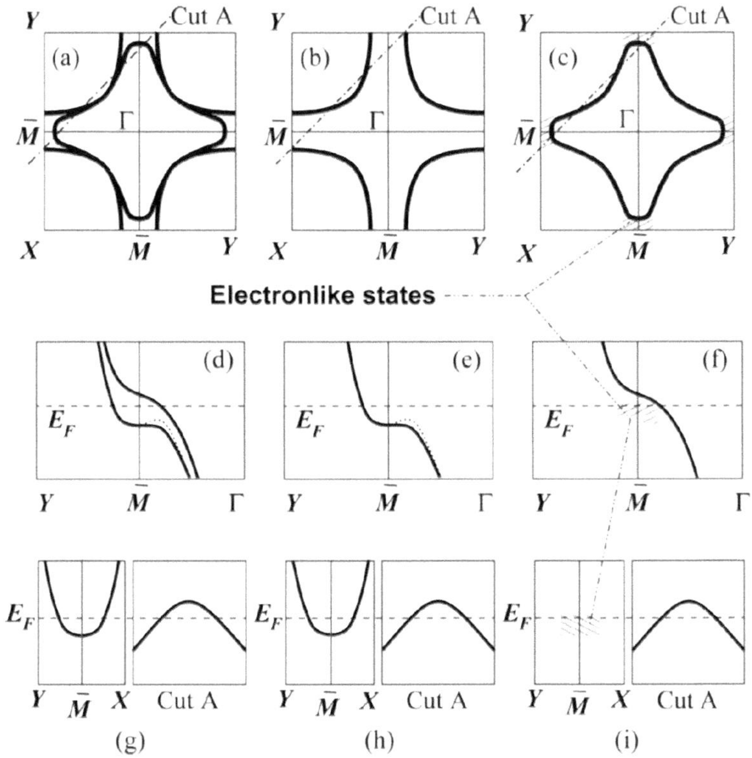

Figure 2.10 2D Fermi surface and band structure scenarios of generic cuprate superconductors from ARPES. (a) FS due to split from two CuO_2 planes. (b) Holelike FS shared by all cuprates. (c) Holelike FS with electronlike states shown as the shaded area. (d) (e) (f) Band structures near FS along Y-$\bar{M}$-Γ for cases shown in (a), (b) and (c) after Chuang et al. (1999). Dotted lines along $\bar{M}$-Γ in (d) and (e) indicate hypothesized band dispersion which stabilizes local minimum at $\bar{M}$. (g) The lower band of (d) unfolded along Y-$\bar{M}$-X on the left panel. The band is electronlike near $\bar{M}$ along Y-$\bar{M}$-X. (h) Band of (e) unfolded along Y-$\bar{M}$-X on the left panel, electronlike. (i) Electronlike states near $\bar{M}$ on the left panel. The right panels of (g), (h) and (i) show holelike band dispersion near $(\pi/2, \pi/2)$ along Cut A, indicated by dash-dotted lines in (a), (b) and (c). The possible complication in holelike bands from the bilayer splitting near $(\pi/2, \pi/2)$ as in (a) is neglected. E_F is the Fermi level in all cases. The Brillouin zone notation is that often adopted for Bi-22$(n-1)n$ systems. With slight modification, the pictures depicted here can be adopted to superconductors of space group I4/mmm, such as $La_{2-x}Sr_xCuO_{4-\delta}$.

$$\sigma(\omega, T) = \sigma_{\mathrm{D}}(\omega, T) + \sigma_{\mathrm{MIR}}(\omega, T), \tag{2.15}$$

where $\sigma_{\mathrm{D}}(\omega, T) = [n(T)e^2/m_c]\{(1/\tau)/[\omega^2 + (1/\tau)^2]\}$ is the T-dependent Drude part while $\sigma_{\mathrm{MIR}}(\omega, T)$ is the *nearly* T-independent mid-infrared (MIR) part. The spectral weight of Drude part $\int_0^\infty \sigma_{\mathrm{D}}(\omega, T)d\omega$, is found to be independent of T [the shape of $\sigma_{\mathrm{D}}(\omega, T)$ is T dependent though], suggesting a constant n if m_c is taken independent of T. Nonetheless, the typical use of such a model is oversimplified. The problem is: The MIR is treated as T-independent and modeled as classical Lorentz oscillators.

The T dependence of MIR is tied to its nature. MIR is often understood as an interband electronic transition (Collins et al., 1989; Schlesinger et al., 1990b; Timusk et al., 1988; Uchida et al., 1993). This interband transition, especially at its low ω part, is, however, not from excitation over the charge-transfer (CT) gap. Rather it perhaps comes from doping-related states and spectral weight transfers inside the CT gap as suggested by various doping-dependent photoemission, inverse photoemission, X-ray absorption, electron-energy-loss spectroscopy and IR reflectance studies (Allen et al., 1990; Chen et al., 1991a; Imada et al., 1998; Romberg et al., 1990; Uchida et al., 1993; van Veenendaal et al., 1994). Without lingering over the exact origin of these MIR-related states, we have two observations from experiments. First, MIR states are very close to the Fermi level, manifested by low-energy MIR tail approaching $\omega = 0$.[3] Such states might relate to the flat band, van Hove singularity and electronlike states (see Figs. 2.10c and 2.10f) revealed in ARPES. Second, MIR states are localized at low T as seen in $\sigma(\omega)$ at 20 K or lower (Quijada et al., 1999; Romero et al., 1992b) as shown in Fig. 2.11b, because otherwise it would behave Drudelike and there would not be an MIR at all. Then the interband transitions of MIR are likely from shallow localized states to extended ones. Later we will show how strong correlation combined with the peculiar band structure of cuprates might contribute to the localized state.

With these two properties of MIR states, σ_{MIR} should be T dependent in its low ω part (actually in far infrared, FIR),

[3] The low-energy tail of MIR often extends down to $\sim 100\ \mathrm{cm}^{-1}$ and sometimes extrapolates to $\omega \approx 0$.

because higher T should progressively set more carriers free, which corresponds to a spectral weight transfer from MIR (or FIR) to the Drude part as indicated by the arrow in Fig. 2.11d. In other words, the classical model of Lorentz oscillator with a T-independent MIR is *not* correct considering the quantum nature of interband excitations. With increasing T, there should be a reduction in the spectral weight of MIR as shown in Fig. 2.11a, and this reduction in MIR should then be made up by an increase in the Drude part (Fig. 2.11d) according to the sum rule (i.e., a spectral weight transfer). Such weight reduction (Grynberg et al., 1974) and weight transfer (Blue, 1964; Gerlach et al., 1976; Gurauskas et al., 1983) have been understood in optical studies of narrow-gap semiconductors like HgTe, $Hg_{1-x}Cd_xTe$, and semimetals such as Bi. Unlike valence-to-conduction-band type of transition there ($\Gamma_8 \rightarrow \Gamma_8$ in HgTe, $Hg_{1-x}Cd_xTe$ and possibly crossing the L-point gap in Bi), the MIR in cuprates are likely from localized to extended bands, but a similar T dependence should still hold. However, being overlapped σ_D and σ_{MIR} in cuprates have no clear-cut method to separate. (Overlap of the two parts in $Hg_{1-x}Cd_xTe$ and Bi is not as severe.) So what is often used in analysis, is assuming that the σ_{MIR} at 20 K or lower, written indiscriminately as $\sigma_{MIR}(\omega, 20\,K)$ for simplicity, be the same for all other T (Quijada et al., 1999; Romero et al., 1992b). Such a procedure without little doubt would make a constant Drude weight simply because of the conductivity sum rule.

Rewrite Eq. (2.15) as

$$\sigma_D(\omega, T) = \sigma(\omega, T) - \sigma_{MIR}(\omega, T). \tag{2.16}$$

Integrate both sides over ω from 0 to ∞, and by using sum rule for $\sigma_D(\omega, T)$, we get

$$\frac{\pi n(T)e^2}{2m_c} = \int_0^\infty \sigma(\omega, T)d\omega - \int_0^\infty \sigma_{MIR}(\omega, T)d\omega, \tag{2.17}$$

where the total weight $\int_0^\infty \sigma(\omega, T)d\omega$ of the two parts basically measures the number of states available within the CT gap.[4] It is directly integrated from experimental $\sigma(\omega, T)$ and is found independent of T. If a $\sigma_{MIR}(\omega, T) = \sigma_{MIR}(\omega, 20\,K)$ is assumed for

[4] Here $\sigma(\omega, T)$ in our notation is that from MIR and Drude only. Experimentally, this means a cutoff at ω beyond the CT gap.

all T, as is done in a typical T-dependent two-component analysis, (Quijada et al., 1999; Romero et al., 1992b) we are left with

$$n(T) = \frac{2m_c}{\pi e^2}\left[\int_0^\infty \sigma(\omega, T)d\omega - \int_0^\infty \sigma_{\mathrm{MIR}}(\omega, 20\,\mathrm{K})d\omega\right]. \quad (2.18)$$

Because $\int_0^\infty \sigma_{\mathrm{MIR}}(\omega, 20\,\mathrm{K})d\omega$ is a constant, the right hand side of Eq. (2.18) is T-invariant, and we are left with an effective carrier density $n \propto \int_0^\infty \sigma(\omega, T)d\omega - \int_0^\infty \sigma_{\mathrm{MIR}}(\omega, 20\,\mathrm{K})d\omega$ which is independent of T even if n changes with the temperature in reality. Introduction of self-consistent iteration (Quijada et al., 1999) marginally improves the result but qualitatively it would not cure the problem. There are different approaches using straightforward least-squares fit (Forro et al., 1990; Quijada et al., 1999). However, with a classical and oversimplified Drude–Lorentz model to start from, such fits are not likely to uncover the sizable T dependence of the Drude weight.

Meanwhile, the high ω part of MIR is less affected by T, as easily understood from quantum statistics and because the MIR is several times larger than Drude part, a spectral transfer on the order of one Drude weight does not contradict to the convention of nearly T-independent MIR. The transfer only occurs at low ω where the overlap with Drude peak makes its detection hard.

The analysis above explains the origin of variable n as well: As T increases, previously localized carriers are now set free and we are likely left with an $n \propto T$. Actually this variable n finds common point in the one-component model of $\sigma(\omega, T)$ (Tanner and Timusk, 2013), where renormalized relaxation rate $1/\tau^*$ and renormalized mass m_c^* both depend on T and ω. In other words,

$$\sigma(\omega, T) = \frac{ne^2}{m_c^*(\omega, T)}\frac{1/\tau^*(\omega, T)}{\omega^2 + [1/\tau^*(\omega, T)]^2}, \quad (2.19)$$

where n is, however, taken as fixed. In a typical result of this model (Romero et al., 1992a), the renormalized mass $m_c^*(\omega, T)$ at low ω say 200 cm^{-1} decreases with increasing T as $m_c^*(\omega, T) = m_{c0}c_1/(c_1 + c_2 T)$ with m_{c0} the m_c^* at low T while the renormalized rate $1/\tau^*(\omega, T)$ at low ω roughly increases as T^2. This behavior is perhaps equivalent to an alternative combination of low frequency $n(\omega, T) \propto (c_1 + c_2 T)/c_1$ and $1/\tau^*(\omega, T) \propto T^2$ if the mass m_c

is taken as fixed instead. Mathematically, by inserting $m_c^*(\omega, T) = m_{c0}c_1/(c_1 + c_2T)$ in Eq. (2.19), we have

$$\sigma(\omega, T) = \frac{ne^2}{m_{c0}c_1/(c_1 + c_2T)} \frac{1/\tau^*(\omega, T)}{\omega^2 + [1/\tau^*(\omega, T)]^2} \quad (2.20)$$

$$= \frac{n[(c_1 + c_2T)/c_1]e^2}{m_{c0}} \frac{1/\tau^*(\omega, T)}{\omega^2 + [1/\tau^*(\omega, T)]^2}. \quad (2.21)$$

If we treat mass m_{c0} as fixed, then $n(c_1 + c_2T)/c_1$ can be taken an effective carrier density which is linear in T, and we arrive at the same conclusion as we made earlier. There must exist renormalization effects to some extent, but meanwhile we cannot rule out the variation of n with T. The problem is, however, that the current optical technique is not able to differentiate one effect from the other because m_c and n are entangled together in the expression of optical conductivity.

Electronic specific heat C^{el} seems to suggest a fixed n because $\gamma = C^{el}/T$ is a constant above T_c (Loram et al., 1993). Localized states, as long as not far from the Fermi surface ($\sim kT$), nevertheless, contribute to γ because these carriers may increase their energy and jump to extended states simply by thermal excitation. Classically this is interpreted as being localized and deprived of translational degrees of freedom (DOF); they, nevertheless, have oscillatory DOF. Thus, these states still contribute a term $\propto T$ to C^{el}, such that significant variation in γ/T is not seen. Furthermore, if the states localized are the bound pairs of electron and hole, the pairs as a whole still enjoy a translational degree of freedom and therefore contribute to the specific heat. There are recent reports claiming a scattering rate linear in T shown by ARPES (Valla et al., 2000, 1999). However, we need to caution on the meaning of such a rate and its relation to the actual transport relaxation rate. Averaging to more than 0.2 eV at 300 K (Valla et al., 2000), such rate is far above $1/\tau_{tr}$ of 300–500 cm^{-1} given by current IR transport studies. On the other hand recent dynamic conductivity experiment showed that the T dependence of transport rate is much faster than T (Jenkins et al., 2010), although not really up to a T^2 relation. The THz technique adopted is in principle a much finer probe than ARPES in energy resolution (Jenkins et al., 2010), and more importantly, it directly measures the longitudinal transport rate. Similar results are obtained in dynamic optical conductivity experiments (Kusar et al., 2005) at temperature higher than T_c. The nearly T^2-rate revealed in the THz and dynamic optical experiments could well be the single rate we proposed in this Section.

2.8.1 Optical Relaxation Rate: A Modern Update

The discussion up to this point in Section 2.8 was mainly from a perspective dated 10∼15 years ago.

Recent experimental progress clearly demonstrates that the in-plane magnetoresistance obeys Kohler's rule in the pseudogap phase of cuprate superconductors (Chan et al., 2014). The transport relaxation is well explained with a single rate, and has a T^2 dependence (Mirzaei et al., 2013). More importantly, the latter experiment also directly probed the optical rate and thus has confirmed the same physics from a different experimental technique. Although more experiments with higher precision are always welcome, as of now a single relaxation rate appears to offer the most satisfying explanation.

In summary, our prediction of a single relaxation rate with T^2 dependence is now *confirmed*.

2.8.2 Charge Localization: A Modern Update

The aforementioned charge localization scenario is now buttressed by a number of solid experimental findings (Pelc et al., 2019). Although the exact microscopic mechanism is subject to further investigations, the excitonic bound state of ECM now appears to give at least a qualitative explanation: The formation of Coulomb-bound states reduces the number of charges that are available to charge transport, giving rise to an effect as charge localization. It is however different from other common localizations of the Mott or Anderson type, and can be differentiated from the latter two by charge-insensitive transports such as thermal conductivity. Another hallmark signature of this localization is its spatial inhomogeneity, which can be understood from spontaneous charge transfer of Chapter 3, and is verified by computations in a real SC cuprate in Chapter 4.

2.9 The Cause of MIR

A natural question arises as to what contributes to the MIR/midgap (and somehow localized) states. We have suggested a scenario based on excitonic interactions. We have believed that the peculiar band structure and FS topology near the optimal doping regime, combined with strong Coulomb correlation contribute to these states. First, there is strong evidence (Luo, 2000b) that electron- and holelike charge carriers coexist in a broad range of superconducting cuprates, including both the *nominally* n- and p-type. This is a fact probably best borne out of recent ARPES experiments (Kaminski et al., 2006; Nakayama et al., 2006) and is readily understood from the three-dimensional (3D) Fermi surface topology. For example, to the most rudimentary approximation, LSCO is a hole conductor with a holelike (h-like) Fermi pocket center around (π, π). However, such a notion, in the very minimum, neglects the k_z dispersion of the energy band, and the associated FS surface topological change. Although the band dispersion along k_z is weak, on the order of 0.1 eV at most, its effect on the FS topology is not trivial (Sahrakorpi et al., 2005). Aggravating the problem, is that the ARPES technique currently developed for cuprates, is not k_z sensitive. This contributes to the once much debated issue as to whether some salient features represent electronlike (e-like) FS or not (Chuang et al., 1999; Fretwell et al., 2000; Luo, 2000b, 2001a). Fortunately, the situation is much clear now due to much improved ARPES techniques and computational software. Earlier objection (Fretwell et al., 2000) to electronlike FS has now shifted to acknowledgement that such a FS topology exists in a variety of the supposed p-type superconductors (Kaminski et al., 2006; Nakayama et al., 2006), at least for proper doping levels.

In parallel to the MIR is the widely known pseudogap. There were many conflicted views on its origin, but in recent years, a clearer picture emerges. Chapter 3 will give a fairly satisfactory explanation of it with the excitonic interlayer Coulomb and the extended BdGEs. We believe that a successful theory should be general enough so as to understand all of the above phenomena in a common framework. The ECM interaction between electronlike and holelike quasiparticles is logical and general enough to accommodate these features.

In the following, let us quote a paragraph from one of our preprints (Luo, 2000b) 18 years ago:

> "Electrons and holes attract one another to form weakly bound Mott–Wannier (MW) excitons (Kittel, 1996) due to Coulomb interaction. Like that in positronium, the binding energy $E_b = e^4 m_\mu / 32\pi^2 \epsilon_0^2 \epsilon_r^2 \hbar^2$ with m_μ the reduced mass and ϵ_0 the vacuum permittivity. The mass of these quasiparticles is on the order of a few m_e, or the rest mass of a bare electron. In $YBa_2Cu_3O_7$ (YBCO or Y-123), for example, the carrier (e or h) mass $m_c \approx 2m_e$ (Schlesinger et al., 1991). The relative dielectric constant $\epsilon_r \approx 14.7$ and thus $E_b \approx 0.063$ eV $\approx$730 K in T (Samara et al., 1990). The pseudogap T^* for insulating Y-123 is not known to the author but for LSCO, $T^* \rightarrow$ 720 K when the metal-insulator boundary is approached (Batlogg et al., 1994). The good match suggests a possible connection between EHC and the pseudogap. The spectral weight loss in IR at T slightly above T_c might be similarly explained."

It is interesting to see the methodologies evolved and refined between then and now, but the basic ECM logic has always stayed the same. In the quotation above, some symbols or citation numbers are updated to this book style to be consistent.

2.9.1 Cause of MIR: A Modern Update

Sections 2.8 and 2.9 have discussed mechanisms for the spectral transfer between the Drude and MIR parts, and the origin of the MIR, etc. Figure 2.11 actually gave a phenomenological account of the spectral transfer process and has been qualitatively sound. One complication for MIR is its large variation of scale from the undoped to overdoped regime. Its magnitude could be as high as 0.4 eV near underdoped, while at the finer energy scale, it could be down to a few tens of meV. Here we focus on the low energy. At this energy range, it is readily explained by the partially gapped Fermiology outlined by single-particle excitation spectrum given in Fig. 3.2. Since the Fermi-crossing still exists, away from antinodes, the MIR spectral weight

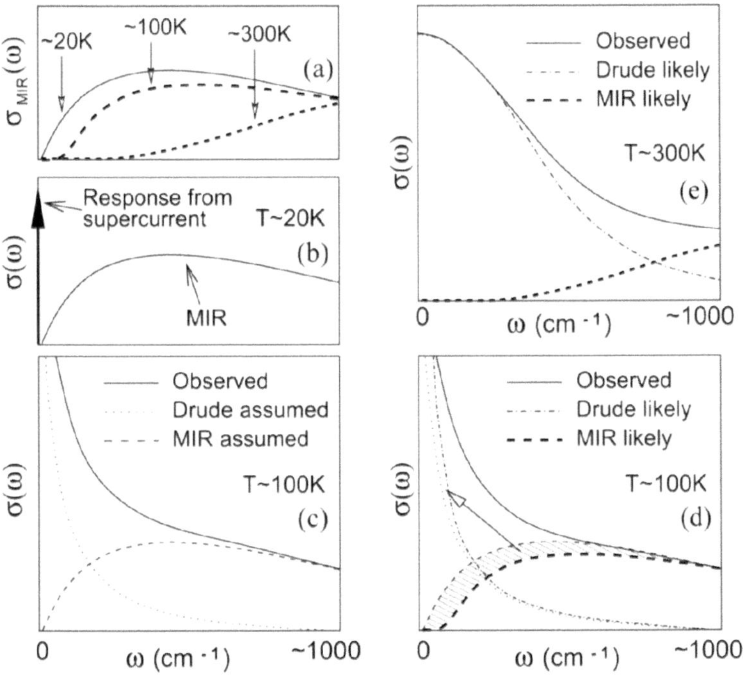

Figure 2.11 Spectral weight transfer between the Drude and MIR part of optical conductivity $\sigma(\omega, T)$ in cuprates. (a) Theoretical variation of $\sigma_{\text{MIR}}(\omega, T)$ with T in a non-Drude quantum model. (b) The $\sigma(\omega)$ as found at 20 K. (c) $\sigma(\omega)$ at $\sim$ 100 K. The dashed line is the $\sigma_{\text{MIR}}(\omega, T)$ assumed in the typical two-component analysis, which is the same as $\sigma_{\text{MIR}}(\omega)$ at 20 K. The dotted line is the calculated $\sigma_{\text{D}}(\omega)$ by using $\sigma(\omega, 100\text{ K}) - \sigma_{\text{MIR}}(\omega, 20\text{ K})$. (a), (b) and (c) share the same horizontal axis. (d) $\sigma(\omega)$ at $\sim$ 100 K. The thick dashed line is the actual $\sigma_{\text{MIR}}(\omega)$ at 100 K as shown in (a) as the long dashed line. The dash-dotted line is the actual $\sigma_{\text{D}}(\omega)$ at 100 K by using $\sigma(\omega, 100\text{ K}) - \sigma_{\text{MIR}}(\omega, 100\text{ K})$. The thin dashed and dotted lines are the same as those in (c) for comparison. Obviously the Drude weight obtained in (d) is larger than that in (c), which means an higher carrier density at 100 K compared with 20 K. The shaded area is the spectral weight transferred away from MIR as indicated by the arrow when T reaches 100 K from 20 K. (e) $\sigma(\omega)$ at $\sim$ 300 K. The spectral weight of MIR is further transferred to the Drude part.

exists down to 0 frequency (energy). The temperature dependence of MIR is well explained by the T dependence of the pseudogap, which is the same gap that shows up in Fig. 3.2.

2.10 Scattering Mechanism

All this experimental evidence suggests a promising combination of $1/\tau \propto T^2$ and $n(T) \propto T$ in explaining the peculiar transport in cuprates. Then what is the scattering mechanism behind the T^2 rate? It is likely caused by electron–electron and electron–hole scatterings. A fermion–fermion scattering results in a T^2 rate basically because of the phase space restraints from the Pauli principle. In two-dimension (2D), some nesting effect might be prominent but it has been shown that *e-e* scattering still basically follows a T^2 law (Fujimoto et al., 1991; Maebashi and Fukuyama, 1998).

Here we have, however, a variable $n(T)$, which might make us suspect a rate $1/\tau$ increasing faster than T^2. Intuitively the electrons (holes) are getting more crowded with increasing T. So we want to use a very simple argument based on power analysis to show that this worry is not needed.

Treating the scattering process using Fermi Golden rule, the probability per unit time that an electron in **k** will be scattered into another **k**′ is given by

$$w(\mathbf{k}', \mathbf{k}) = \frac{2\pi}{\hbar}\langle \mathbf{k}'|H'|\mathbf{k}\rangle^2 \delta(\varepsilon_{\mathbf{k}} - \varepsilon_{\mathbf{k}'}), \tag{2.22}$$

where $|\mathbf{k}\rangle = e^{i\mathbf{kr}}$ up to a normalization constant and $H' = e^{-\alpha r}/r$, the screened coulomb potential. In two-dimension (2D),

$$\langle \mathbf{k}'|H'|\mathbf{k}\rangle = \int \frac{e^{i(\mathbf{k}-\mathbf{k}')\mathbf{r}}e^{-\alpha r}}{r} d^2r \tag{2.23}$$

$$= \int \frac{e^{i\mathbf{qr}}e^{-\alpha r}}{r} r dr d\theta, \tag{2.24}$$

where $\mathbf{q} = \mathbf{k} - \mathbf{k}'$ and thus $q = 2k\sin(\theta/2)$ with θ the angle between **k** and **k**′.

Notice that in 2D, $e^{i\mathbf{qr}}$ can be expanded by Bessel functions of the first kind. Then

$$e^{i\mathbf{qr}} = e^{iqr\cos\theta} \tag{2.25}$$

$$= \sum_{m=-\infty}^{\infty} J_m(qr) i^m e^{im\theta}. \tag{2.26}$$

Integrate the above first over θ and $e^{im\theta}$ averages to 0 except when $m = 0$. Thus

$$\langle \mathbf{k}'|H'|\mathbf{k}\rangle = 2\pi \int_0^{\infty} J_0(qr) e^{-\alpha r} dr = \frac{2\pi}{(q^2+\alpha^2)^{\frac{1}{2}}}. \tag{2.27}$$

When the screening is not strong which should be the case for UD or OP cuprates[5] we see that $\langle \mathbf{k}'|H'|\mathbf{k}\rangle^2 \approx 4\pi^2/q^2 = \pi^2/sin^2(\theta/2)k^2$, where k will be taken as k_F because effective scatterings only occur near the FS. To be illustrative, we only consider an isotropic case (i.e., the FS is a circle), then the Fermi wave-vector $k_F = [2\pi n(T)]^{1/2}$ in 2D where $n(T)$ is taken as the area density of carriers. Without considering the complication from umklapp scattering, we conclude that the transition rate from $\mathbf{k}$ to $\mathbf{k}'$, $w(\mathbf{k}', \mathbf{k})$ is $\propto 1/T$ because $k^2 = k_F^2 \propto n(T)$ and $n(T) \propto T$. Now let us look at the phase space available for scattering. The perimeter of the FS is $2\pi k_F \propto [n(T)]^{1/2} \propto T^{1/2}$ and thus the number of states available for scattering to and from is proportional to the product of the perimeter $2\pi k_F$ and the thermal excitation width $\sim k_B T$ with k_B the Boltzmann's constant, in other words proportional $T^{3/2}$. Applied to both initial and final states this gives us a factor of T^3 from the phase space restriction. Combined with a transition probability $w(\mathbf{k}', \mathbf{k}) \propto T^{-1}$ as already seen, it results in a relaxation rate $1/\tau \propto T^2$, which is exactly what we find in experiments. Complication from screening, umklapp process and so on will be shown in a separate publication but the result is essentially the same.

A similar argument applies to *e*-*h* process where the fermion nature of *e* and *h* on their own part gives the same phase restriction we mentioned before and thus the same T dependence for charge transports.

This *e*-*h* liquid is near a Fermi liquid (FL) above the pseudogap temperature T^*. Nonetheless, it is not so under T^* because the

[5] The strong correlation nature of cuprates likely suggests a weak screening.

presence of excitonic states formed by e and h is significant at low T. These excitons are stable against recombination because of the semimetallic band structure, unlike those of fast-recombining type found in semiconductors. These bound states, as a discontinuous change from the free particle motions, manifest the loss of one-to-one correspondence from the states of free Fermi gas. This offers an explanation for various non-Fermi-liquid phenomena of cuprates. The importance of discontinuities and non-perturbative approaches in high-T_c superconductivity were well summarized in Ref. (Anderson, 2000). Some 40 years ago, Mott emphasized the interacting nature of such excitonic states (Mott, 1961). Kohn (Kohn, 1967), Halperin and Rice (Halperin and Rice, 1968a,b) pointed out the possible anomalies, like excitonic insulator, CDW, SDW, antiferromagnetic correlation, and phase separation in such a strongly interacted e-h system. This reminds us of many other correlation phenomena found in high-T_c materials, for example, the stripe phase (Emery et al., 1999), summarized by V. J. Emery, S. A. Kivelson, and J. M. Tranquada.

The excitonic Coulomb bound and quasi-bound states present at a strategic junction between semiconductor physics and superconductor physics. And in a even broader sense, these connect to other fields of condensed matter physics. This may not come as a surprise though as history had already witnessed similar situations before: The concept of an energy gap for superconductivity was extended (or borrowed) from semiconductor physics, largely thrusted forward by Bardeen. Many proximity and tunneling effects in superconductors are actually interpreted with the help of the so-called "semiconductor model."

Although Jérome, Rice and Kohn (Jérome et al., 1967), Halperin and Rice (Halperin and Rice, 1968b) also argued that the ground state of the excitonic state be an insulator, not to mention superconductivity, their conclusion is *only* true for a system with equal numbers of electrons and holes. When the densities of e and h are not equal, the ground state should be a conductor at least. The possibility of high-T_c superconductivity in systems similar to this was studied by Allender, Bray and Bardeen (Allender et al., 1973) and Ginzburg (Ginzburg, 1970). All these indicate a connection of high-T_c superconductivity to the excitonic Coulomb states of electron–hole liquid. The results of Chapters 1 and 3 of this book demonstrate with high confidence the connection of Coulomb excitonic complexes to HTSC.

2.11 Conclusion

Studies of normal (or mixed) state transport and optical processes strongly indicate the presence of electrons in *p*-type cuprates. The excitonic Coulomb coupling contributes to many "abnormal" and "puzzling" phenomena of cuprates in the normal (or mixed) state. With the introduction of electron, and the resulting excitonic Coulomb action, most of "anomalies" are either gone or now understood. Predictions of a single relaxation rate and its T^2 dependence were made based on the ECM, more than 10 years ago. Both are confirmed now by experiments. Prediction that the Kohler's rule was not violated in cuprates was also made long ago, based on the same logic. This prediction is now confirmed as well.

Chapter 3

Interlayer Coulomb Instability: The Many-Body Approach

3.1 Introduction

In previous chapters, significant evidence for the electron–hole coexistence and interaction in cuprates was discussed. Such an interaction was demonstrated to lead to high-T_c superconductivity as similarly seen in the experiment. This chapter, written by Dr. Nie Luo, proceeds to understand why such a behavior arises, at a microscopic level.

We consider the interlayer Coulomb interaction of two neighboring CuO_2 layers near the Van Hove singularity. The doping level is assumed between 0.05 and ~0.17, e.g., the underdoped (UD) regime traversing the optimal (OP) doping. We will show that due to the attractive interlayer Coulomb (like that in an exciton) lowering the potential energy, it is energetically preferable for a mobile charge to transfer from one layer to the other. The Van Hove singularity is essentially a saddle point in the energy band and gives rise to a divergent DOS. In this chapter, we start with Green's function formalism to derive the spectral function of the extended tight-binding Hamiltonian for model cuprates. Then an analytical density of states is approximated near the half-filling, or the saddle

High-T_c Superconductivity: The Excitonic Coulomb Perspective
Nie Luo and George H. Miley

ISBN 978-981-4800-57-0 (Hardcover), 978-0-429-27870-9 (eBook)
www.jennystanford.com

point. In different neighborhoods of the saddle point, quasiparticles behave differently, somewhere like an electron and elsewhere like a hole, at least to some specific directions. And it is, as we believe non-accidently, the same anti-nodal (π, 0) region where gapping phenomena (pseudogap, superconducting gap, etc.) are the most significant. By using many-body equations based on the interlayer Coulomb coupling, we will demonstrate that the pseudogap near the saddle point can be numerically accounted for.

Charge (or in more general terms, phase) segregation or inhomogeneity (Müller and Giorgio, 1993) is widely known in cuprate superconductors. Most of studies so far focus on in-plane phenomena and the static charge order, like the stripe phase. There have been many efforts to understand such phenomena. According to some theories, these charge orders may relate to Van Hove singularity and related nesting effects. Or they may be due to exciton-like interactions. So far, most studies have focused on in-plane (intra-layer) couplings. However, Leggett emphasizes in fundamental terms the importance of long-range interlayer Coulomb, and its role in saving the potential energy in cuprates. The current author has long believed the importance of attractive Coulomb interaction in cuprate superconductors and predicted the existence of electronlike quasiparticles in hole-doped cuprates. Their existence has been verified, at least in part, by recent experiments.

In this chapter, we will start with a very general Hamiltonian that includes an interlayer Coulomb interaction. When E_F is filled up to the vicinity of a peculiar band structure feature, that is, the quasi-2D saddle point also known as Van Hove singularity, an instability arises given proper conditions. This instability results in the transfer of band electrons from one layer to the other, leaving band holes in the former and extra electrons in the latter. The layers are thus *additionally* self-doped, one with electrons, while the other with holes. The electron and hole interact to form an excitonic Coulomb bound state. The exciton may also couple to *externally* doped electrons or holes, and mediate an attractive interaction, therefore pairing two like charges. There could exist other interactions. For example, two electrons are attracted to one hole, forming a trion.

This is like a pairing of two electrons in some sense, although SC is not involved in this case.

3.2 Interlayer Coulomb, Attractive and Exciton Interaction

The major limitation of prior exciton models, as believed by the current author, is that their semiconducting "glue" takes significant amount of energy to excite, and therefore these models are not much relevant to the pairing at the ground state. But there are two profound insights in the exciton model: (1) a sufficiently light thus mobile "glue" (intermediate boson) with large energy scale; (2) part of interaction that is attractive.

In the following sections, we will show how to derive the excitonic Coulomb (with attractive part) interaction from Coulomb repulsion among like charges and the Coulomb interaction with the ions. Another result is the semimetallic band-edge well explaining the stable Coulomb bound state.

We start with a very general form of the Hamiltonian for two CuO_2 planes in cuprates. The Born–Oppenheimer (BO) approximation is assumed. With the pure ionic motion neglected, the Hamiltonian is given as

$$\hat{H} = \sum_{i,\alpha}^{N,2} \left[\frac{\mathbf{p}_{i\alpha}^2}{2m} + U(\mathbf{r}_{i\alpha}) \right] + \frac{e^2}{4\pi\epsilon} \sum_{i,j}^{N} \sum_{\alpha \neq \alpha'}^{2} |\mathbf{r}_{i\alpha} - \mathbf{r}_{j\alpha'}| + \hat{H}_{intra}, \quad (3.1)$$

where, i, j are electron indices running from 1 to N. α, α' are the layer indices, either 1 or 2. The double sum describes the interlayer Coulomb among electrons. $\hat{H}_{intra}$ is the Coulomb energy term among electrons within a single layer. $U(\mathbf{r}_{i\alpha})$ is the operator for electron–ion Coulomb of $2N$ electrons and $2N$ ions. Its expectation value can be evaluated as $\langle \psi_{ei} | U | \psi_{ei} \rangle$. Here ψ_{ei} is the electron wavefunction that also implicitly depends on ion coordinates. $\langle \psi_{ei} | U | \psi_{ei} \rangle$ is *not* just a simple function of $\mathbf{r}_{i\alpha}$, but rather a functional of the total electron wavefunction ψ_{ei} of $2N$ variables $\mathbf{r}_{i\alpha}$. It is better written as $U(\mathbf{r}_{1\alpha}, \ldots, \mathbf{r}_{N\alpha}; \mathbf{r}_{1\alpha'}, \ldots, \mathbf{r}_{N\alpha'})$ and in general, $U(\mathbf{r}_{1\alpha}, \ldots, \mathbf{r}_{N\alpha}; \mathbf{r}_{1\alpha'}, \ldots, \mathbf{r}_{N\alpha'}) \neq U_1(\mathbf{r}_{1\alpha}, \ldots, \mathbf{r}_{N\alpha}) +$

$U_2(\mathbf{r}_{1\alpha'}, \ldots, \mathbf{r}_{N\alpha'})$, where the latter two were individual electron–ion energy of separate layers.

When the two layers are far apart, there is no appreciable interaction in between, and the individual Hamiltonian is then

$$\hat{H}_\alpha = \sum_i \left[\frac{\mathbf{p}_{i\alpha}^2}{2m} + U_\alpha(\mathbf{r}_{i\alpha})\right] + \hat{H}_{intra,\alpha}. \tag{3.2}$$

Suppose that we move the individual Hamiltonian system close to each other, and turning on the interlayer Coulomb interaction. Now besides the electron–electron Coulomb of

$$\frac{e^2}{4\pi\epsilon} \sum_{i,j}^{N} \sum_{\alpha\neq\alpha'}^{2} |\mathbf{r}_{i\alpha} - \mathbf{r}_{j\alpha'}|, \tag{3.3}$$

we will have additional electron–ion terms as given by

$$U(\mathbf{r}_{1\alpha}, \ldots, \mathbf{r}_{N\alpha}; \mathbf{r}_{1\alpha'}, \ldots, \mathbf{r}_{N\alpha'}) - \sum_\alpha U_\alpha(\mathbf{r}_{1\alpha}, \ldots, \mathbf{r}_{N\alpha}). \tag{3.4}$$

Equation (3.4) describes the additional (interlayer) electron–ion interaction experienced by both layers even if each layer is individually electrostatically neutral: Ions and electrons can polarize themselves to take a favorable geometry to minimize the Coulomb energy. The term of Eq. (3.4) is often negative. A similar expression when applied to neighboring (atoms, or ions plus electrons) is exactly one major part that gives rise to cohesion energy in crystals. Of course, the attractive interaction given by Eq. (3.4) often results in fixed bonding, and not the mobile electrons (charges) that we are after in SC. Still it gives an illustrative example demonstrating that for strongly correlated Hamiltonian we often cannot neglect the electron–ion interaction. The electron–ion interaction does not mean phonon descriptions though, because the $\hat{H}$ does not have dynamic degree of freedom for the ions: They are dynamically fixed. There is a subtle distinction between electron–ion energy and the standard electron–phonon interaction.

The kinetic energy (KE), electron–ion coulomb and the interlayer electron–electron, intralayer electron–electron terms in Eq. (3.1) have subtle and delicate balances in determining the ground state of interacting electron systems, and is in general very complicated matter to investigate. In the next few sections, we demonstrate that one can isolate the KE from the Coulomb part near the saddle point of cuprates, and much simplify the problem.

3.3 2D Density of States

We start from the well-known parametrized energy band in a single CO_2 plane (Markiewicz et al., 2005):

$$\begin{aligned} E(\mathbf{k}) = & -2t[\cos k_x a + \cos k_y a] - 4t'[\cos k_x a + \cos k_y a] \\ & -2t''[\cos 2k_x a + \cos 2k_y a] \\ & -2t'''[\cos 2k_x a \cos k_y a + \cos 2k_y a \cos k_x a], \end{aligned} \tag{3.5}$$

which is an expression of k-dependent eigenvalues for the Hamiltonian in Eq. (3.2) without the Coulomb interaction.

The first term in the RHS of Eq. (3.5) gives the basic saddle point at $(\pi, 0)$. The curvatures along the k_x and k_y directions are opposite, but of the same magnitude, and therefore symmetric in terms of the magnitude. The next-nearest hopping terms t' and so on, for most part will introduce additional curvatures that make the saddle point asymmetric with respect to k_x and k_y, even when only the magnitude is considered. The resulting asymmetric saddle point is often called extended, in that the DOS is enhanced.

We first study the bare (non-explicit) energy excitation and DOS properties of a single CuO_2 plane using the Green's function formalism. Green's function of such a general 2D tight-binding system is given by:

$$G(\mathbf{l}, \mathbf{m}; z) = \frac{a^2}{(2\pi)^2} \int_{1BZ} d\mathbf{k} \frac{e^{i\mathbf{k}\cdot(\mathbf{l}-\mathbf{m})}}{z - E(\mathbf{k})}, \tag{3.6}$$

where $\mathbf{l}$, $\mathbf{m}$ are lattice vectors and

$$\mathbf{k} \cdot (\mathbf{l} - \mathbf{m}) = k_1(l_1 - m_1) + k_2(l_2 - m_2), \tag{3.7}$$

and $E(\mathbf{k})$ given by Eq. (3.5).

In general, a full analytic evaluation of $G(\mathbf{l}, \mathbf{m}; z)$ is difficult, especially with so many terms in $E(\mathbf{k})$. Therefore, a proper simplification is necessary. First, the Andersen group (Pavarini et al., 2001) did extensive parametric studies on the cuprate's electronic structures and found out that t' and its ratio to t are critical in the T_c. t' is therefore to be kept with t. Second, other hopping terms like t'' and t''', for the most part, will have effects qualitatively similar to t' in making the saddle point asymmetric or extended, which results in significantly enhanced DOS. With this in mind, we rewrite the band energy as

$$E(\mathbf{k}) = \; -2t[\cos k_x a + \cos k_y a] - 4t'[\cos k_x a + \cos k_y a], \quad (3.8)$$

with t' on the order of 0.3~0.6 t, give or take. This magnitude of t' therefore has the effect of t'', t''' included in calculating DOS, given parametrization outlined in Markiewicz group's results. The value of t' are also kept in line with arguments of O. K. Andersen et al. (Pavarini et al., 2001). Further, an "exact" parametrization can be only taken as face value at this stage. Since there are so many different material families within the cuprate, which are parameterized differently, yet all give impressive T_c, a ballpark value in t' actually serves a better purpose to show the essential physics.

Equation (3.6) is in general only analytically evaluated in special cases. If $t' = 0$, the diagonal matrix element of Green's function, which can be utilized to derive the DOS, is calculated as follows:

$$\begin{aligned} G_d(z) = G(\mathbf{l}, \mathbf{m}; z)|_{\mathbf{l}=\mathbf{m}} &= \frac{a^2}{(2\pi)^2} \int_{1\mathrm{BZ}} d\mathbf{k} \frac{1}{z - E(\mathbf{k})} \\ &= \frac{a^2}{(2\pi)^2} \int_{1\mathrm{BZ}} d\mathbf{k} \frac{1}{z + 2t[\cos k_x a + \cos k_y a]}. \end{aligned} \quad (3.9)$$

Changing integration variables so that $k_x a = \phi_1$ and $k_y a = \phi_2$ and

$$-\pi/a \le k_x < \pi/a; -\pi/a \le k_y < \pi/a, \quad (3.10)$$

Equation (3.9) is rewritten as

$$\begin{aligned} G_d(z) &= \frac{1}{2\pi} \int_{-\pi}^{\pi} d\phi_1 \frac{1}{2\pi} \int_{-\pi}^{\pi} d\phi_2 \frac{1}{z + 2t[\cos\phi_1 + \cos\phi_2]} \\ &= \frac{1}{2\pi} \int_{-\pi}^{\pi} d\phi_1 \frac{1}{2\pi} \int_{-\pi}^{\pi} d\phi_2 \frac{1}{z + 4t[\cos\phi_1 \cos\phi_2]}, \end{aligned} \quad (3.11)$$

where in the last step, we used trigonometric identities and changed integral variables, which were later relabeled.

Further simplification leads to

$$\begin{aligned} G_d(z) &= \frac{1}{2\pi}\int_{-\pi}^{\pi} d\phi_1 \frac{1}{2\pi}\int_{-\pi}^{\pi} d\phi_2 \frac{1}{z + 4t\cos\phi_1\cos\phi_2} \\ &= \frac{1}{2\pi}\int_{-\pi}^{\pi} d\phi_1 \frac{1}{\sqrt{z^2 - (4t)^2\cos^2\phi_1}} \\ &= \frac{1}{\pi z}\int_0^{\pi} \frac{d\phi}{\sqrt{1-\lambda^2\cos^2\phi}} \\ &= \frac{2}{\pi z}\mathcal{K}(\lambda), \end{aligned} \tag{3.12}$$

where $\mathcal{K}$ is the complete elliptic integral of the first kind, and λ is given by

$$\lambda = \frac{4t}{z}.$$

Letting $z = E + \mathrm{i}s$, $s \to 0^{\pm}$, we may use the analytic continuation of $\mathcal{K}(\lambda)$ to obtain explicit expressions for $G_d^{\pm}(E)$.

The imaginary part of $G_d^{\pm}(E)$ is evaluated to be

$$\mathrm{Im}\{G_d^{\pm}(E)\} = \mp\frac{2}{4t\pi}\mathcal{K}\left(\sqrt{1-\frac{E^2}{16t^2}}\right), \quad \text{when} \quad |E| < 4t. \tag{3.13}$$

Hence, the density of states *per site* is given by

$$\begin{aligned} \rho(E) &= \mp\frac{1}{\pi}\mathrm{Im}\{G_d^{\pm}(E)\} \\ &= \frac{1}{2t\pi^2}\theta(4t-|E|)\mathcal{K}\left(\sqrt{1-\frac{E^2}{16t^2}}\right), \end{aligned} \tag{3.14}$$

where $\theta(4t-|E|)$ is a step function and $\mathcal{K}(x)$ is approximated by $-\log(1-x)/2$ when $x \approx 1$, which is divergent near $x = 1$. This gives the DOS the well-known van Hove singularity. The logarithmic divergency is hence a reflection of two-dimensionality at the 2D saddle point.

Since the exact analytical evaluation of the DOS is difficult when $t' \neq 0$, some approximation is needed.

Because we are mainly interested in the behavior of DOS near the saddle point $(\pi,0)$, we can approximate the energy around the saddle point in terms of k_x and k_y, measured from $(\pi,0)$, or

$$\begin{aligned} E(\mathbf{k}) &= -t[k_x^2 - k_y^2] + E_1(k), \\ &\text{with} \quad E_1(\mathbf{k}) = -4t'[\cos k_x a \times \cos k_y a], \end{aligned} \tag{3.15}$$

where k_x, k_y and a are already made dimensionless by setting a=1 and $E_1(\mathbf{k}) = -4t'[\cos k_x a \times \cos k_y a]$ is further expanded as

$$\begin{aligned} E_1(\mathbf{k}) = \; & E_1(\pi, 0) + D_k(\pi, 0) \cdot (k_x, k_y) \\ & + \frac{1}{2}(k_x, k_y) H_{kx,ky}(k_x, k_y)^t, \end{aligned} \tag{3.16}$$

where $D_k(\pi, 0)$ is the first-order partial derivative and $H_{kx,ky}$ is the Hessian of E_1 with respect to k_x, k_y at $(\pi, 0)$.

Finally the above can be readily evaluated to give

$$\begin{aligned} E(\mathbf{k}) & = -t[k_x^2 - k_y^2] + 4t' - 2t'k_y^2 - 2t'k_x^2, \\ & = 4t' + (t - 2t')k_y^2 - (t + 2t')k_x^2. \end{aligned} \tag{3.17}$$

When $t' = 0$ the saddle point is symmetric in the curvatures along k_x and k_y. On the contrary, if $t' \neq 0$ the saddle point is no longer symmetric and in general an enhanced DOS will result. Yet the Green's function formalism can still be utilized to evaluate the extended VHS.

The density of states per site is given by

$$\begin{aligned} \rho(E) & = \mp \frac{1}{\pi} \mathrm{Im}\{G_d^{\pm}(E)\} \\ & = \mp \frac{1}{\pi} \mathrm{Im} \frac{a^2}{(2\pi)^2} \int_{1\mathrm{BZ}} d\mathbf{k} \frac{1}{z - 4t' - (t - 2t')k_y^2 + (t + 2t')k_x^2} \\ & = \mp \frac{1}{\pi} \mathrm{Im} \frac{a^2}{(2\pi)^2} \int_{1\mathrm{BZ}} d\mathbf{k} \frac{1}{z - (t - 2t')k_y^2 + (t + 2t')k_x^2}, \end{aligned} \tag{3.18}$$

where in the last step a more compact form is attained by redefining the z to absorb the $4t'$ term. Now $z = E + is$ is measured from the saddle point energy.

We want to convert the integral over $\mathbf{k}$ to that over E. Thus we need to find d^2k and dE. Starting from the saddle point as the origin in the k plane, the first quadrant is divided by the asymptotic straight line given by $-(t - 2t')k_y^2 + (t + 2t')k_x^2 = 0$. Then we want to find the area between the asymptotic and an isocurve defined by E in the form $E = (t - 2t')k_y^2 - (t + 2t')k_x^2$ or $k_y = [E + (t + 2t')k_x^2)]/(t - 2t')$.

The area is then given by the integral

$$\Delta a = \int_0^{\pi/2} \left(\sqrt{\frac{E + (t+2t')k_x^2}{t-2t'}} - \sqrt{\frac{(t+2t')k_x^2}{t-2t'}} \right)$$
$$= \int_0^{\pi/2} (\sqrt{c + b^2 k_x^2} - bk_x) dk_x$$
$$= -\frac{b\pi^2}{8} + \frac{\pi}{4}\sqrt{\frac{\pi^2}{4}b^2 + c} + \frac{c}{2b}\log\left(\frac{\pi}{2} + \sqrt{\frac{\pi^2}{4} + \frac{c}{b^2}}\right)$$
$$- \frac{1}{2}\frac{c}{b}\log\left(\frac{c}{b^2}\right), \tag{3.19}$$

where substitutions $E/(t-2t') = c$ and $\sqrt{(t+2t')/(t-2t')} = b$ have been made.

Only the last term in Eq. (3.19) dominates when $E \approx 0$ and only its part proportional to $c\log(c)$ is significant in both value and variation when $E \approx 0$. Differentiate the area with respect to E and only keep the significant term one gets

$$d(\Delta a) = d\mathbf{k}$$
$$= \frac{-1}{4b(t-2t')}\log\left(\frac{E}{(t-2t')}\right) dE. \tag{3.20}$$

Rewriting the above expression with E' replacing E, and substituting into the formula for $G_d(z)$

$$G_d(z) = \int \frac{-\log(E'/(t-2t'))dE'}{4b(t-2t')}\frac{1}{z-E'}, \tag{3.21}$$
$$G_d^+(E) = \lim_{s\to 0^+} G_d(z)$$
$$= \lim_{s\to 0^+} \int \frac{-\log(E'/(t-2t'))dE'}{4b(t-2t')}\frac{1}{E-E'+is}. \tag{3.22}$$

Note that we have the Sokhotski–Plemelj identity

$$\lim_{s\to 0^+} \int_{-\infty}^{\infty} \frac{f(x)}{x \pm is} dx = \mp i\pi f(0) + \lim_{s\to 0^+} \int_{|x|>s} \frac{f(x)}{x} dx, \tag{3.23}$$

or more specifically,

$$\lim_{s\to 0^+} \int_{-\infty}^{\infty} \frac{f(x)}{x - x_0 \pm is} dx = \mp i\pi f(x_0) + P\int_{-\infty}^{\infty} \frac{f(x)}{x - x_0} dx. \tag{3.24}$$

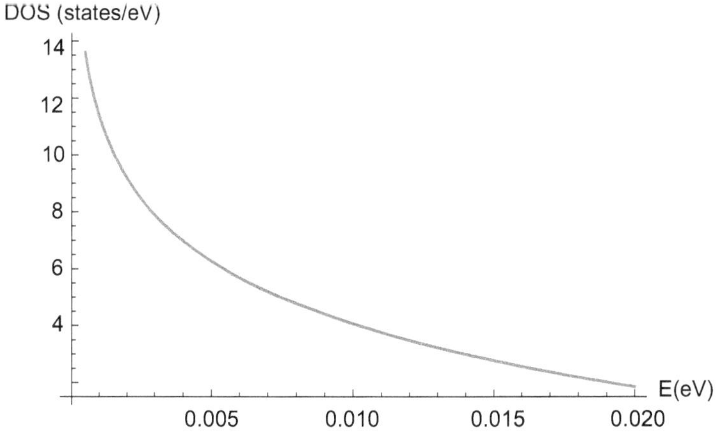

Figure 3.1 2D density of states near the extended saddle point, for a parametric cuprate band of Eq. (3.8).

Therefore, the density of state is given as follows:

$$\begin{aligned}\rho(E) &= \mp\frac{1}{\pi}\mathrm{Im}\{G_d^{\pm}(E')\}\\ &= \frac{1}{\pi}\mathrm{Im}\lim_{s\to 0^+}\int\frac{-\log(E/(t-2t'))dE}{4b(t-2t')}\frac{1}{E'-E+is}\\ &= \frac{-\log(E/(t-2t'))}{4b(t-2t')}.\end{aligned} \tag{3.25}$$

Clearly, the end result of t' is to enhance the divergent DOS near $E \sim 0$. A typical numerical result with a typical parameter set is given in Fig. 3.1.

3.4 Effects of Divergent DOS and Attractive Coulomb Interaction

Now we consider two CuO_2 planes, nearly half-filled, separated by a few angstroms as in the case of underdoped cuprates, with the Hamiltonian given in Eq. (3.1). Suppose that the mass along z (c) is heavy, which is the case for cuprates underdoped, kinetic energy change due to c-axis transport (hopping of small t_z) will be negligible. The system is hence quasi-2D.

3.4.1 Spontaneous Charge Transfer

Now suppose an electron hopping from Layer one $L1$ to Layer two $L2$ so there is a net change $-e$ in $L2$ (electron) and $+e$ in $L1$ (hole). The attractive Coulomb between the two will bind them into a Mott–Wannier-type exciton with a binding energy

$$E_b = m_h e^4/(2h^2\epsilon^2). \tag{3.26}$$

Due to the large DOS at the saddle point, the in-plane kinetic energy change is negligible. The total energy of the system is then lowered by an exciton binding energy, and thus a *spontaneous charge transfer* occurs!

The divergent DOS is equivalent to a very high in-plane mass, that reduces the KE to near zero, isolating the KE part from the potential energy, precisely what we mentioned at the end of Section 3.2.

In Eq. (3.26), the ϵ was kept vague and given as a generic dielectric permittivity. Since the exciton associated with transfer here is interplane, the effective permittivity is given by

$$\frac{1}{\epsilon_{inter}} = \frac{2}{\epsilon_1 + \epsilon_2}, \tag{3.27}$$

per derivation of Eq. (1.150). It should be noted that in Eq. (3.27) the subscripts 1 and 2 do not refer to $L1$ or $L2$; rather they refer to the dielectric and CuO_2 layer, respectively. Note that ϵ_{inter} indicates the effective dielectric constant for interlayer coupling per discussion of Section 1.18.1. Given in a numerical example as

$$\epsilon_{inter} = 3\epsilon_\perp \tag{3.28}$$

$$\epsilon_{intra} = (3 + \sqrt{6})\epsilon_\perp, \tag{3.29}$$

where ϵ_{intra} is the in-plane (or intraplane) dielectric constant of CuO_2. In this book, we do not make a strict distinction between dielectric constant and the absolute dielectric permittivity. Here "dielectric constant" should not be automatically interpreted as the relative dielectric constant, which is referred to as ϵ_r. For comparison, the dielectric constants of the dielectric and CuO_2 are

$$\begin{aligned} \epsilon_1 &= (3 - \sqrt{6})\epsilon_\perp, \quad \text{insulator;} \\ \epsilon_2 &= (3 + \sqrt{6})\epsilon_\perp, \quad CuO_2. \end{aligned} \tag{3.30}$$

The in-plane dielectric constant (more precisely permittivity), taken either as ϵ_2 or ϵ_{intra}, is large compared to the interplane

ϵ_{inter}. Note the subtle difference between the ϵ_{inter} and ϵ_{perp}. The latter is either called c-axis or sometimes out-of-plane ϵ. The difference, understandably, is due to the tensor nature of dielectrics in the layered structure: The effective value really depends on the directional angle of the field vector.

Equation (3.28) tells that an interlayer exciton is energetically more favorable to form versus an intralayer exciton. This is one major driving force for the spontaneous charge transfer across the copper oxide layers.

In ordinary materials, this type of charge transfer is hampered by penalties in kinetic energy. As already mentioned, the cuprate minimizes the KE penalty due to the large DOS associated with the saddle point singularity.

3.4.2 Factors against Spontaneous Transfer

Besides the KE, there are several additional factors impeding the spontaneous transfer: (1) The change in entropy; (2) the Coulomb repulsive energy between the transferred charge and other like charges of the destination layer; if the energy gain from the attractive Coulomb energy cannot offset the repulsive one, then the spontaneous transfer should stop; (3) the energy gain from interlayer exciton actually has to be offset by the energy loss from intralayer exciton, if the latter does exist. (This is not a major concern for the SC cuprate as already mentioned, due to the anisotropic dielectric properties.)

Let us now consider point (2) of the preceding paragraph. It turns out for cuprates, the transfer actually results in the reduction of the total Coulomb repulsion energy due to Eq. (3.28). The interplane ϵ_{inter} is smaller than the in-plane ϵ_{intra}. If both layers have N electrons, there are $N^2/2$ pairs of $e^2/4\pi\epsilon_{inter}r_{ij}$ per layer. If instead transfer happens so as to have N+1 and N-1, respectively, in the layers, there are now only $(N+1)(N-1) = N^2 - 1$ pairs left, and we end up saving the e-e interlayer Coulomb repulsion. Since the saving is in average larger than the cost in intralayer e-e coulomb the total e-e Coulomb repulsion energy is reduced (while pair number of intralayer Coulomb increases from $N^2 - N$ to $N^2 - N + 1$). The saving is still valid for interlayer distance of 6.8 Å where the thickness of the dielectric layer is twice that of CuO_2. This increases the difference between $\epsilon_2 = 7.6\epsilon_\perp$ and $\epsilon_1 = 0.7\epsilon_\perp$, and the difference between $\epsilon_{intra} = 7.6\epsilon_\perp$ and $\epsilon_{inter} = 4.15\epsilon_\perp$. The gain from interlayer exciton is still there, though at reduced magnitude. Therefore, interlayer coupling is not a phenomenon for multilayer cuprates only, it also plays a significant role in monolayer cuprates.

Meanwhile, the ion-electron energy is either unchanged or changed to a more negative value. The total numbers of interlayer and intralayer *e-i* attractive pair are still $2N^2$ before and after the transfer. (1) Suppose that the transferred charge does not perturb the original intralayer wavefunction of the receiving layer, then this means that total *e-i* Coulomb is still the same, besides the possible bonding-type gain from the transferred electron. (2) Suppose that the transferred charge does perturb the original intralayer wavefunction, we know that energetics favor bonding-type deformation. Then the potential *e-i* energy is further lowered to a more negative value.

Combining the *e-i* and *e-e* Coulomb energy, the energy saving on the order of

$$\frac{e^2}{4\pi r_{intra}}\frac{1}{\epsilon_{intra}} - \frac{e^2}{4\pi r_{inter}}\frac{1}{\epsilon_{inter}} \tag{3.31}$$

is expected, which is the energy gain of an interlayer exciton from an intralayer type.

This explains and proves strong tendency towards interlayer charge transfer or segregation from the potential energy part.

In summary, the logic for interlayer charge transfer is simple and clear: The anisotropic Coulomb repulsion is large, particularly in the interlayer part. To minimize it, electrons tend to segregate on one layer. If the DOS is high, the transfer incurs little cost in the KE part and so it becomes spontaneous. This transfer reduces overall electron–electron Coulomb and can cause further decrease in the electron–ion part, which could manifest as an excitonic bound state, or the pseudogap phase. This exciton state further modifies the inter-electron Coulomb configuration, and may give rise to Cooper pairing. The charge transfer effects a self-doping phenomenon, which is well known for 2D perovskites and some other copper compounds. The self-doping generates electronlike quasiparticles in p-type cuprates.

3.4.3 Electrostatics against Transfer

There may exist additional factors opposed to the spontaneous charge transfer. One is from electrostatics. The positive charging of one layer and the negative charging of the other will set up an interlayer electrostatic field that tends to slow down the transfer.

Take the two CuO_2 layers as the two terminals of a parallel-plate capacitor. The capacitance over the planar unit cell area is given by

$$C = \frac{\epsilon_\perp a_0^2}{d}, \tag{3.32}$$

where the dielectric permittivity $\epsilon_\perp$ is adopted: The field in a parallel-plate capacitor is perpendicular to the plate surface.

If the transferred charge per unit cell (area) is n_t, a voltage difference of the following is resulted:

$$\Delta V = \frac{n_t e}{C} = \frac{n_t e d}{\epsilon_\perp a_0^2}. \tag{3.33}$$

The potential gain of an exciton formation often involves two parts: the increase of e-i (negative) and the reduction of e-e (positive). This can be taken as twice the exciton banding energy: Note that the negative potential energy in an exciton binding is actually twice the exciton binding energy E_b. Then if the voltage difference is larger than the potential gain of an exciton formation, the transfer should stop, or

$$\Delta V > \frac{2E_b}{e}, \qquad \text{or} \qquad \frac{n_t e d}{\epsilon_\perp a_0^2} > \frac{2 m_h e^3}{2 h^2 \epsilon_{inter}^2}. \tag{3.34}$$

Take a typical ϵ_{inter} ~15, then $E_b = 0.242$ eV. Due to Eq. (3.28), $\epsilon_\perp = 5$. The corresponding n_t is then $\sim 0.484 \times 5\epsilon_0 \times a_0^2/ed$ and gives 0.028 for $d = 6.8$ Å, or 0.06 for $d = 3.2$ Å. These self-doping levels are about right for a general notion of ~0.07. Note that Eq. (3.34) outlines a kinematic limit on one layer, rather than an energetic consideration of the total system. It hence serves more of a barrier height type that limits the practical speed by which the self-doping can be achieved. In practice, such limits are possibly weakened by external dopings.

3.5 Charge Transfer, the CDW Consequence

With an attractive interaction, a transition metal oxide is susceptible to forming charge density variations or waves. Such a charge density wave is often static, and tends to compete with other orders like magnetism or superconductivity. The cuprate behaves in quite unique ways. The in-plane charge transfer does not drastically affect the charge transport: It is still a conductor, at least with doping levels above a certain threshold, say 0.05. In the out-plane transport, however, qualitatively different pictures emerge. The charge transfer tends to dimerize the alternating layers, making the *c*-axis transport more localized. In fact, if the cuprate is taken as the one-dimensional stack of many CuO_2 layers, the attractive excitonic Coulomb then tends to dimerize the charge density in the out-of-plane direction and form a gap in the *c*-axis transport. The effect is the most prominent near half filling. The phenomenon is similar to the Peierls transition and the typical CDW in this fashion tends to make a metal-insulator transition (MIT) along a specific direction.

Yet critical differences exist: In a Peierls transition, the lattice (of ions) deforms to effect the periodicity doubling. Here it is the electron degree of freedom to act as a major driving force, although, as a consequence, a slight modulation of the lattice may also result. Another difference lies in the anisotropic conductivity: An ordinary 1D metal becomes an insulator beyond the Peierls transition, yet in the cuprate, the conductivity still persists within the CuO_2-plane although the *c*-axis behaves as an insulator.

The aforementioned scenario makes it clear that charge segregation (more broadly a phase segregation) tend to occur between the CuO_2 layers in the cuprates. The exact charge transfer pattern, however, should vary with the details in the crystal structure. For single layer cuprates, there are already two sub-cases. In one case the CuO_2 planes are aligned vertically and in the other, the planes are displaced by [1/2,1/2] lattice vector in the planar direction.

In the first case, interlayer charge transfer should result in an ABAB... or a Peierls style modulation of charge density, if the cuprates are evenly doped. The latter case is more complicated as the uneven mobile charge distribution on the Cu and the O sites makes ABBAABBA... type possible. For bilayer cuprates, the two CuO_2 within a unit-cell is closely coupled, and giving therefore AB or BA within each cell. Outside a single unit cell, ABBA ... should result in the doubling of the unit cell lattice in the c-axis. For trilayer cuprates, the charge transfer is very likely ABA, which should give optimal excitonic energy and may explain why highest T_c occurs in this systems. The microscopic structure of charge density waves in underdoped $YBa_2Cu_3O_{6.54}$ has been revealed by x-ray diffraction. The work of this book predicts a generalization of it to other cuprate systems.

3.6 Many-Body Formalism of the Excitonic Interlayer Interaction

In this section, we consider two adjacent CuO_2 layers near half-filling: The electrons (holes) in the bilayer interact one another via the interlayer and intralayer Coulomb forces whose form are given by Eq. (3.1). One result of interaction is the transfer of electrons from one layer to another, say $L1$. One can think of this process as one form of "pairing": Electron 1 of k_1 leaving $L2$, leaves out Hole 1 of $-k_1$. Electron 1 and Hole 1 form a Coulomb bound state exciton X1. X1 further interacts with Electron 2 of $L1$, with k_2 not far from k_1, and binds it. In real space, this is the formation of a trion, a bound state. In the k-space, this is the "pairing" of two electrons k_1 and k_2 and can be probed by spectroscopic means such as ARPES. Since the Coulomb force involved is long range of a few lattice distance, the momentum transfer is small and k_1 should not be much different to k_2.

This "pairing" then shares some similarities to the BCS pairing, though differs in qualitative ways too. The trion is a fermion and hence totally different statistics is involved versus the Cooper pairing. Still the similarities mean that the excitonic interlayer problem can be solved with similar many-body techniques such as the Bogoliubov–de Gennes equations.

3.6.1 The Hamiltonian

The Hamiltonian of the bilayer system in Eq. (3.1) can be further written in the second-quantized form as

$$\begin{aligned}\mathcal{H} &= \mathcal{H}_{layer} + \mathcal{H}_{int} \\ &= \sum_{\alpha} \int d\mathbf{r} \Psi_{\alpha}^{\dagger}(\mathbf{r}) \hat{H}_l \Psi_{\alpha}(\mathbf{r}) \\ &\quad - \frac{1}{2} \sum_{\alpha,\beta} \int d\mathbf{r} d\mathbf{r}' V_{inter}(\mathbf{r}, \mathbf{r}') \Psi_{\alpha}^{\dagger}(\mathbf{r}) \Psi_{\beta}^{\dagger}(\mathbf{r}') \Psi_{\beta}(\mathbf{r}') \Psi_{\alpha}(\mathbf{r}), \quad (3.35)\end{aligned}$$

where α, β are actually the layer indices. In the underdoped/optimal regime, spin dynamics is in general subdued, and the Hamiltonian of Eq. (3.35) does not need to consider spin dynamics for our purpose here: Two electrons k_1 and k_2 can be paired with any spin combinations as long as $k_1 \neq k_2$. If instead $k_1 = k_2$, opposite spins for the two are assumed automatically. Therefore, α and β can be taken as widely adopted pseudo-spin indices: The transfer between the layers is like a flip of the pseudo-spin, and the pairing occurs to two electrons of opposite pseudo-spin. To a certain degree, the interlayer excitonic Coulomb problem is then mapped to the BCS formalism. This analogy is not to derive the SC gap, but rather the pseudogap due to their different quantum statistics.

V_{inter} in Eq. (3.35) can be derived from the i-e and e-e Coulomb terms. It is generally a dipole type that can be approximated by screened Coulomb potential, with a typical form given in Eq. (1.156). Its matrix elements in k-space are readily evaluated.

The single layer Hamiltonian is

$$\mathcal{H}_{layer} = \sum_{\alpha} \int d\mathbf{r} \Psi_{\alpha}^{\dagger}(\mathbf{r}) \hat{H}_l \Psi_{\alpha}(\mathbf{r}), \quad (3.36)$$

including the intralayer electron–ion and electron–electron Coulomb, and to the lowest order can be approximated by the LDA/GGA parameterized Hamiltonian.

We can further simplify the Hamiltonian of Eq. (3.35) by first dropping the subscript in V for compactness. Then by using a delta-like $V(\mathbf{r},\ \mathbf{r}') = V\delta(\mathbf{r}-\mathbf{r}')$ the derivation is made less cluttered. In other words,

$$\mathcal{H}_{int} = -\frac{V}{2}\sum_{\alpha,\beta}\int d\mathbf{r}\Psi_\alpha^\dagger(\mathbf{r})\Psi_\beta^\dagger(\mathbf{r})\Psi_\beta(\mathbf{r})\Psi_\alpha(\mathbf{r}),$$

where after tedious intermediate steps the replacement can be recovered with, $V\delta(\mathbf{r}-\mathbf{r}') \to V(\mathbf{r},\ \mathbf{r}')$ to give the final result. The Coulomb nature of V then recovers and manifests in the final steps.

The Hamiltonian is now simplified to

$$\begin{aligned}\mathcal{H} = &\sum_{\alpha}\int d\mathbf{r}\Psi_\alpha^\dagger(\mathbf{r})\hat{H}_l\Psi_\alpha(\mathbf{r}) \\ &-\frac{V}{2}\sum_{\alpha,\beta}\int d\mathbf{r}\Psi_\alpha^\dagger(\mathbf{r})\Psi_\beta^\dagger(\mathbf{r})\Psi_\beta(\mathbf{r})\Psi_\alpha(\mathbf{r}).\end{aligned} \tag{3.37}$$

The expectation value of the Hamiltonian in Eq. (3.37), as given by

$$\begin{aligned}\langle\mathcal{H}\rangle = &\sum_{\alpha}\int d\mathbf{r}\langle\Psi_\alpha^\dagger(\mathbf{r})\mathcal{H}_l\Psi_\alpha(\mathbf{r})\rangle \\ &-\frac{V}{2}\sum_{\alpha,\beta}\int d\mathbf{r}\langle\Psi_\alpha^\dagger(\mathbf{r})\Psi_\beta^\dagger(\mathbf{r})\Psi_\beta(\mathbf{r})\Psi_\alpha(\mathbf{r})\rangle,\end{aligned} \tag{3.38}$$

can be evaluated using the Wick's theorem. The 4-field product in the equation above is written as the sum of various combinatoric products of normal-order products and contractions. Only the full contractors have non-zero "vacuum" expectation values, which leave

$$\begin{aligned}\langle\Psi_\alpha^\dagger(\mathbf{r}_1)\Psi_\beta^\dagger(\mathbf{r}_2)\Psi_\gamma(\mathbf{r}_3)\Psi_\delta(\mathbf{r}_4)\rangle = &\langle\Psi_\alpha^\dagger(\mathbf{r}_1)\Psi_\beta^\dagger(\mathbf{r}_2)\rangle\langle\Psi_\gamma(\mathbf{r}_3)\Psi_\delta(\mathbf{r}_4)\rangle \\ &+\langle\Psi_\alpha^\dagger(\mathbf{r}_1)\Psi_\delta(\mathbf{r}_4)\rangle\langle\Psi_\beta^\dagger(\mathbf{r}_2)\Psi_\gamma(\mathbf{r}_3)\rangle \\ &-\langle\Psi_\alpha^\dagger(\mathbf{r}_1)\Psi_\gamma(\mathbf{r}_3)\rangle\langle\Psi_\beta^\dagger(\mathbf{r}_2)\Psi_\delta(\mathbf{r}_4)\rangle.\end{aligned} \tag{3.39}$$

Since α, β are pseudospin indices, they can be replaced with $\uparrow$ or $\downarrow$, respectively. Explicitly substituting the indices with arrows we have

$$\sum_{\alpha,\beta}\langle\Psi_\alpha^\dagger(\mathbf{r})\Psi_\beta^\dagger(\mathbf{r})\Psi_\beta(\mathbf{r})\Psi_\alpha(\mathbf{r})\rangle = \langle\Psi_\uparrow^\dagger(\mathbf{r})\Psi_\downarrow^\dagger(\mathbf{r})\rangle\langle\Psi_\downarrow(\mathbf{r})\Psi_\uparrow(\mathbf{r})\rangle$$
$$+\langle\Psi_\downarrow^\dagger(\mathbf{r})\Psi_\uparrow^\dagger(\mathbf{r})\rangle\langle\Psi_\uparrow(\mathbf{r})\Psi_\downarrow(\mathbf{r})\rangle$$
$$+\sum_{\alpha,\beta}\langle\Psi_\alpha^\dagger(\mathbf{r})\Psi_\alpha(\mathbf{r})\rangle\langle\Psi_\beta^\dagger(\mathbf{r})\Psi_\beta(\mathbf{r})\rangle$$
$$-\sum_{\alpha,\beta}\langle\Psi_\alpha^\dagger(\mathbf{r})\Psi_\beta(\mathbf{r})\rangle\langle\Psi_\beta^\dagger(\mathbf{r})\Psi_\alpha(\mathbf{r})\rangle, \tag{3.40}$$

where we have zeroed out the following terms due to anticommutator rules:

$$\langle\Psi_\uparrow^\dagger(\mathbf{r})\Psi_\uparrow^\dagger(\mathbf{r})\rangle = 0,$$
$$\langle\Psi_\downarrow(\mathbf{r})\Psi_\downarrow(\mathbf{r})\rangle = 0. \tag{3.41}$$

The first two terms of Eq. (3.40) are like the Cooper pairing ones in the BCS theory: They might not be zero at low temperature. Here, however, they are *not* the Cooper pair terms due to two reasons: (1) The effective Fermi statistics due to the presence of a third-party hole; (2) There are other Hamiltonian terms that are in competition with these two terms as to be shown later on. Nevertheless, these terms tend to pair two electrons of separate layers, to open up a (incomplete) gap which we interpret as the pseudogap above T_c, but under T^*. Hence,

$$\langle\Psi_\uparrow^\dagger(\mathbf{r})\Psi_\downarrow^\dagger(\mathbf{r})\rangle \neq 0,$$
$$\langle\Psi_\downarrow(\mathbf{r})\Psi_\uparrow(\mathbf{r})\rangle \neq 0. \tag{3.42}$$

The following two terms,

$$\langle\Psi_\uparrow^\dagger(\mathbf{r})\Psi_\downarrow(\mathrm{r})\rangle \neq 0, \tag{3.43}$$
$$\langle\Psi_\downarrow^\dagger(\mathbf{r})\Psi_\uparrow(\mathbf{r})\rangle \neq 0,$$

are further encountered in Eq. (3.40). They represents the charge transfer process between the two layers, resulting in imbalance in individual chemical potentials and giving rise to pronounced excitonic Coulomb bound states.

With the physical meanings of the terms in mind we arrive at

$$\begin{aligned}\sum_{\alpha,\beta}\langle\Psi_\alpha^\dagger(\mathbf{r})\Psi_\beta^\dagger(\mathbf{r})\Psi_\beta(\mathbf{r})\Psi_\alpha(\mathbf{r})\rangle &= 2\langle\Psi_\uparrow^\dagger(\mathbf{r})\Psi_\downarrow^\dagger(\mathbf{r})\rangle\langle\Psi_\downarrow(\mathbf{r})\Psi_\uparrow(\mathbf{r})\rangle\\ &\quad+\sum_{\alpha,\beta}\langle\Psi_\alpha^\dagger(\mathbf{r})\Psi_\alpha(\mathbf{r})\rangle\langle\Psi_\beta^\dagger(\mathbf{r})\Psi_\beta(\mathbf{r})\rangle\\ &\quad-\sum_{\alpha,\beta}\langle\Psi_\alpha^\dagger(\mathbf{r})\Psi_\beta(\mathbf{r})\rangle\langle\Psi_\beta^\dagger(\mathbf{r})\Psi_\alpha(\mathbf{r})\rangle\\ &= 2\langle\Psi_\uparrow^\dagger(\mathbf{r})\Psi_\downarrow^\dagger(\mathbf{r})\rangle\langle\Psi_\downarrow(\mathbf{r})\Psi_\uparrow(\mathbf{r})\rangle\\ &\quad+2\langle\Psi_\uparrow^\dagger(\mathbf{r})\Psi_\uparrow(\mathbf{r})\rangle\langle\Psi_\downarrow^\dagger(\mathbf{r})\Psi_\downarrow(\mathbf{r})\rangle\\ &\quad-2\langle\Psi_\uparrow^\dagger(\mathbf{r})\Psi_\downarrow(\mathbf{r})\rangle\langle\Psi_\downarrow^\dagger(\mathbf{r})\Psi_\uparrow(\mathbf{r})\rangle.\end{aligned}\tag{3.44}$$

Substituting Eq. (3.44) back into Eq. (3.38), we reach the final expression for the expectation value of $\mathcal{H}$

$$\begin{aligned}\langle\mathcal{H}\rangle = \sum_\alpha\int d\mathbf{r}\langle\Psi_\alpha^\dagger(\mathbf{r})\hat{H}_l\Psi_\alpha(\mathbf{r})\rangle\\ -V\int d\mathbf{r}[\langle\Psi_\uparrow^\dagger(\mathbf{r})\Psi_\uparrow(\mathbf{r})\rangle\langle\Psi_\downarrow^\dagger(\mathbf{r})\Psi_\downarrow(\mathbf{r})\rangle\\ +\langle\Psi_\uparrow^\dagger(\mathbf{r})\Psi_\downarrow^\dagger(\mathbf{r})\rangle\langle\Psi_\downarrow(\mathbf{r})\Psi_\uparrow(\mathbf{r})\rangle\\ -\langle\Psi_\uparrow^\dagger(\mathbf{r})\Psi_\downarrow(\mathbf{r})\rangle\langle\Psi_\downarrow^\dagger(\mathbf{r})\Psi_\uparrow(\mathbf{r})\rangle].\end{aligned}\tag{3.45}$$

The expectation value of the full Hamiltonian is thus simplified. Still the expression of Eq. (3.45) is just formal, without telling the exact value of the energy. The problem can be further simplified by the mean field approach. By introducing new auxiliary fields, we may diagonalize the Hamiltonian expectation by the Bogoliubov transformation. An effective *mean-field* Hamiltonian which is quadratic in $\hat{\Psi}$ is constructed as follows:

$$\begin{aligned}\mathcal{H}_{mf} = \sum_\alpha\int d\mathbf{r}[\Psi_\alpha^\dagger(\mathbf{r})\hat{H}_l\Psi_\alpha(\mathbf{r}) + U_\alpha(\mathbf{r})\Psi_\alpha^\dagger(\mathbf{r})\Psi_\alpha(\mathbf{r})]\\ +\int d\mathbf{r}[\Delta(\mathbf{r})\Psi_\uparrow^\dagger(\mathbf{r})\Psi_\downarrow^\dagger(\mathbf{r}) + \Delta^*(\mathbf{r})\Psi_\downarrow(\mathbf{r})\Psi_\uparrow(\mathbf{r}) + H_0(\mathbf{r})\\ +\epsilon_\delta\Psi_\uparrow^\dagger(\mathbf{r})\Psi_\downarrow(\mathbf{r}) + \epsilon_\delta^*\Psi_\downarrow^\dagger(\mathbf{r})\Psi_\uparrow(\mathbf{r})].\end{aligned}\tag{3.46}$$

The real mean fields $U_\alpha(\mathbf{r})$ and the potentially complex mean-field $\Delta(\mathbf{r})$ are unknown up to this point but will be determined later on. The subscript in $U_\alpha(\mathbf{r})$ accommodates Fermi level differences to account for charge transfers. The residue field $H_0(\mathbf{r})$ does not depend on the fields Ψ and $\Psi^\dagger$. The last two terms in Eq. (3.46) describe the charge transfer between the two layers, with the amplitude ϵ_δ, ϵ_δ^* taken as constants here for simplicity. ϵ_δ and ϵ_δ^* are in general functions of $\mathbf{r}$. They actually have the form of a hopping term (KE) in a typical tight-binding model. When taken as some constants they can be absorbed into the $\hat{H}_l$ with a proper account of different chemical potential in the two layers.

In other words one might first neglect the terms proportional to $\Delta(\mathbf{r})$ and diagonalize the $\mathcal{H}_{mf}$ to get eigenvalues to new layers $\uparrow'$ and $\downarrow'$, which are linear combinations of the original layer $\uparrow$, $\downarrow$. This gives typically different eigenvalues $\epsilon_{\delta,\uparrow'}$ and $\epsilon_{\delta,\downarrow'}$. A similar diagonalization can be done for $\langle\mathcal{H}\rangle$. Physically, this corresponds to the charge transfer that forms excitons. Under this rotation of basis, the last two terms in Eq. (3.46) disappear and the last term in Eq. (3.45) tends to zero. The other energy terms proportional to the primed Ψ (linear and quadratic) still have similar forms, though with different values. We can rewrite all primed arrows to be unprimed and the problem is now similar to a Bogoliubov–de Gennes Hamiltonian.

The expectation value of the Hamiltonian is

$$\begin{aligned}\langle\mathcal{H}\rangle \approx \sum_\alpha \int & d\langle\mathbf{r}\Psi_\alpha^\dagger(\mathbf{r})\hat{H}_l\Psi_\alpha(\mathbf{r})\rangle \\ -V\int & d\mathbf{r}[\langle\Psi_\uparrow^\dagger(\mathbf{r})\Psi_\uparrow(\mathbf{r})\rangle\langle\Psi_\downarrow^\dagger(\mathbf{r})\Psi_\downarrow(\mathbf{r})\rangle \\ & +\langle\Psi_\uparrow^\dagger(\mathbf{r})\Psi_\downarrow^\dagger(\mathbf{r})\rangle\langle\Psi_\downarrow(\mathbf{r})\Psi_\uparrow(\mathbf{r})\rangle].\end{aligned} \tag{3.47}$$

The corresponding mean-field $\mathcal{H}_{mf}$ is

$$\begin{aligned}\mathcal{H}_{mf} = \sum_\alpha \int & d\mathbf{r}[\Psi_\alpha^\dagger(\mathbf{r})\hat{H}_l\Psi_\alpha(\mathbf{r}) + U_\alpha(\mathbf{r})\Psi_\alpha^\dagger(\mathbf{r})\Psi_\alpha(\mathbf{r})] \\ & + \int d\mathbf{r}[\Delta(\mathbf{r})\Psi_\uparrow^\dagger(\mathbf{r})\Psi_\downarrow^\dagger(\mathbf{r}) + \Delta^*(\mathbf{r})\Psi_\downarrow(\mathbf{r})\Psi_\uparrow(\mathbf{r})] + H_0(\mathbf{r}).\end{aligned} \tag{3.48}$$

We want to first diagonalize the mean-field Hamiltonian with new operators $\gamma_{n,\alpha}$ and $\gamma_{n,\alpha}^\dagger$, or

$$\mathcal{H}_{mf} = E_0 + \sum_{n,\alpha}(\epsilon_n + \epsilon_{\delta,\alpha})\gamma_{n,\alpha}^\dagger\gamma_{n,\alpha}, \tag{3.49}$$

where E_0 is the ground state energy. $\gamma^\dagger_{n,\alpha}$ creates an electron of energy ϵ_n in state n at layer α. $\epsilon_{\delta,\alpha}$ account for the chemical variation in the two layers due to charge transfer: one layer anti-bonding, and the other bonding so to speak. More generally $\epsilon_{\delta,\alpha}$ can vary with space (i.e., as a function of **r**). Here for simplicity they are separately taken as a constant. We note the following identities for the commutator [,]:

$$[\mathcal{H}_{mf}, \gamma_{n,\alpha}] = -(\epsilon_n + \epsilon_{\delta,\alpha})\gamma_{n,\alpha}$$
$$[\mathcal{H}_{mf}, \gamma^\dagger_{n,\alpha}] = (\epsilon_n + \epsilon_{\delta,\alpha})\gamma^\dagger_{n,\alpha}, \tag{3.50}$$

and note that in general $\epsilon_{\delta,\uparrow} \neq \epsilon_{\delta,\downarrow}$.

We seek to find mean fields $U_\alpha(\mathbf{r})$ and $\Delta(\mathbf{r})$ in a variational approach. $\langle\mathcal{H}_{mf}\rangle$ then should have minima of the same value and functional argument as $\langle\mathcal{H}\rangle$.

$$\langle\mathcal{H}_{mf}\rangle = \mathcal{Z}^{-1}\sum_k \langle\psi^\dagger_k|\mathcal{H}_{mf}|\psi_k\rangle \exp(-E_k/k_B T),$$

where ψ_k is the set of eigenfunctions of the Hamiltonian

$$\mathcal{H}_{mf}\psi_k = E_k\psi_k.$$

The partition function is

$$\mathcal{Z} = \sum_k \exp(-E_k/k_B T).$$

And similar equations hold for the true Hamiltonian.

3.6.2 The BdG-Like Equations for Interlayer Coulomb Coupling

To diagonalize Eq. (3.46), we introduce in the following a new set of annihilation and creation operators

$$\Psi_\uparrow(\mathbf{r}) = \sum_n [\gamma_{n\uparrow}u_n(\mathbf{r}) - \gamma^\dagger_{n\downarrow}v^*_n(\mathbf{r})], \tag{3.51}$$

$$\Psi_\downarrow(\mathbf{r}) = \sum_n [\gamma_{n\downarrow}u_n(\mathbf{r}) + \gamma^\dagger_{n\uparrow}v^*_n(\mathbf{r})], \tag{3.52}$$

$$\Psi^\dagger_\uparrow(\mathbf{r}) = \sum_n [\gamma^\dagger_{n\uparrow}u^*_n(\mathbf{r}) - \gamma_{n\downarrow}v_n(\mathbf{r})], \tag{3.53}$$

$$\Psi^\dagger_\downarrow(\mathbf{r}) = \sum_n [\gamma^\dagger_{n\downarrow}u^*_n(\mathbf{r}) + \gamma_{n\uparrow}v_n(\mathbf{r})]. \tag{3.54}$$

We proceed by calculating first the commutator

$$\begin{aligned}[\mathcal{H}_{mf},\ \Psi_\uparrow(\mathbf{r})] &= \mathcal{H}_{mf}\Psi_\uparrow(\mathbf{r}) - \Psi_\uparrow(\mathbf{r})\mathcal{H}_{mf} \\ &= -\hat{H}_e\Psi_\uparrow(\mathbf{r}) - U_\uparrow\Psi_\uparrow(\mathbf{r}) \\ &\quad +\Delta(\mathbf{r})[\Psi_\uparrow^\dagger(\mathbf{r})\Psi_\downarrow^\dagger(\mathbf{r}),\ \Psi_\uparrow(\mathbf{r})] \\ &\quad +\Delta^*(\mathbf{r})[\Psi_\downarrow(\mathbf{r})\Psi_\uparrow(\mathbf{r}),\ \Psi_\uparrow(\mathbf{r})]. \end{aligned} \qquad (3.55)$$

Note that

$$\begin{aligned}[\Psi_\uparrow^\dagger\Psi_\downarrow^\dagger,\ \Psi_\uparrow] &= \Psi_\uparrow^\dagger\Psi_\downarrow^\dagger\Psi_\uparrow - \Psi_\uparrow\Psi_\uparrow^\dagger\Psi_\downarrow^\dagger \\ &= -\Psi_\uparrow^\dagger\Psi_\uparrow\Psi_\downarrow^\dagger - \Psi_\uparrow\Psi_\uparrow^\dagger\Psi_\downarrow^\dagger \\ &= -(\Psi_\uparrow^\dagger\Psi_\uparrow + \Psi_\uparrow\Psi_\uparrow^\dagger)\Psi_\downarrow \\ &= -\Psi_\downarrow^\dagger, \end{aligned} \qquad (3.56)$$

where the explicit dependence on $\mathbf{r}$ is dropped for clarity, and identities for fermion operators $\{a_k^\dagger,\ a_q\} = \delta_{kq}$ and $[a_k^\dagger,\ a_q] = 2a_k^\dagger a_q - \delta_{kq}$ were used. Similar techniques give $[\Psi_\downarrow\Psi_\uparrow,\ \Psi_\uparrow] = 0$, and putting the results back to Eq. (3.55), we have

$$\begin{aligned}[\mathcal{H}_{mf},\ \Psi_\uparrow] &= -[\hat{H}_l + U_\uparrow]\sum_n[\gamma_{n\uparrow}u_n - \gamma_{n\downarrow}^\dagger v_n^*] \\ &\quad - \Delta(\mathbf{r})\sum_n[\gamma_{n\downarrow}^\dagger u_n^* + \gamma_{n\uparrow}v_n] \\ &= \sum_n[-(\hat{H}_l + U_\uparrow)u_n - \Delta v_n]\gamma_{n\uparrow} \\ &\quad + \sum_n[(\hat{H}_l + U_\uparrow)v_n^* - \Delta u_n^*]\gamma_{n\downarrow}^\dagger. \end{aligned} \qquad (3.57)$$

On the other hand, by the virtue of Eq. (3.50),

$$\begin{aligned}[\mathcal{H}_{mf},\ \Psi_\uparrow] &= \sum_n[\mathcal{H}_{mf},\gamma_{n\uparrow}]u_n - [\mathcal{H}_{mf},\gamma_{n\downarrow}^\dagger]v_n^* \\ &= \sum_n(-\epsilon_n - \epsilon_{\delta\uparrow})\gamma_{n\uparrow}u_n - (\epsilon_n + \epsilon_{\delta,\downarrow})\gamma_{n,\downarrow}^\dagger v_n^*. \end{aligned} \qquad (3.58)$$

Comparing the terms with $\gamma_{n\uparrow}$ and $\gamma_{n,\downarrow}^\dagger$ in Eqs. (3.57) and (3.58) we find that both equations are satisfied provided that the wavefunctions u_n and v_n satisfy the following equations, which are an extension of the *Bogoliubov–de Gennes equations* (BdGE):

$$\begin{aligned}[\hat{H}_l + U(\mathbf{r})]u_n(\mathbf{r}) + \Delta(\mathbf{r})v_n(\mathbf{r}) &= (\epsilon_n + \epsilon_{\delta\uparrow})u_n(\mathbf{r}), \\ -[\hat{H}_l^* + U(\mathbf{r})]v_n(\mathbf{r}) + \Delta^*(\mathbf{r})u_n(\mathbf{r}) &= (\epsilon_n + \epsilon_{\delta\downarrow})v_n(\mathbf{r}), \end{aligned} \qquad (3.59)$$

where n is an index to a variety of fermion quasiparticle normal modes. In getting Eq. (3.59), the approximation of $U \approx U_{\uparrow}$ is made, considering that the difference can be absorbed in the re-definition of $\epsilon_{\delta\uparrow}$. The generalized BdGEs can be written in a vector form

$$\hat{\Omega}\begin{pmatrix} u_n \\ v_n \end{pmatrix} = \epsilon_n \begin{pmatrix} u_n \\ v_n \end{pmatrix},$$

where $\hat{\Omega}$ is the matrix operator defined by

$$\hat{\Omega} = \begin{pmatrix} \hat{H}_l + U - \epsilon_{\delta\uparrow} & \Delta \\ \Delta^* & -[\hat{H}_l^* + U + \epsilon_{\delta\downarrow}] \end{pmatrix}.$$

The BdGEs of Eq. (3.59) may be written without the explicit references of n. In such cases, the following form is adopted:

$$\begin{aligned} [\hat{H}_l + U(\mathbf{r})]u(\mathbf{r}) + \Delta(\mathbf{r})v(\mathbf{r}) &= (\epsilon + \epsilon_{\delta\uparrow})u(\mathbf{r}), \\ -[\hat{H}_l^* + U(\mathbf{r})]v(\mathbf{r}) + \Delta^*(\mathbf{r})u(\mathbf{r}) &= (\epsilon + \epsilon_{\delta\downarrow})v(\mathbf{r}), \end{aligned} \tag{3.60}$$

but with the understanding of many implicit sets of such two equations.

It also helps to write down orthogonal-normalization relations among u and v

$$\int d\mathbf{r}[u_m^*(\mathbf{r})u_n(\mathbf{r}) + v_m^*(\mathbf{r})v_n(\mathbf{r})] = \delta_{mn}.$$

The positive energy ϵ_n eigenstate is $\begin{pmatrix} u_n \\ v_n \end{pmatrix}$, while negative energy $-\epsilon_n$ eigenstate is $\begin{pmatrix} v_n^* \\ -u_n^* \end{pmatrix}$. The latter's negative energy tells that energy is gained by generating a hole in the interlayer Coulomb coupling system, exactly as what is discussed in Section 3.2.

3.6.3 Self-Consistency Equations

With the extended BdGEs the energy spectrum of the interacting bilayer systems can be found. To completely determine the ϵ, still we need to find the coupled equations that determined the auxiliary mean-field Δ.

We apply the variational principle for Eq. (3.47) so that the variation of the energy expectance reads

$$\begin{aligned}\delta\langle\mathcal{H}\rangle &= \sum_{\alpha}\int d\mathbf{r}\delta\langle\Psi_{\alpha}^{\dagger}(\mathbf{r})\hat{H}_{l}\Psi_{\alpha}(\mathbf{r})\rangle\\ &\quad - V\int d\mathbf{r}[(\delta\langle\Psi_{\uparrow}(\mathbf{r})\Psi_{\downarrow}(\mathbf{r})\rangle\langle\Psi_{\downarrow}(\mathbf{r})\Psi_{\uparrow}(\mathbf{r})\rangle\\ &\quad + \langle\Psi_{\uparrow}(\mathbf{r})\Psi_{\downarrow}(\mathbf{r})\rangle(\delta\langle\Psi_{\downarrow}(\mathbf{r})\Psi_{\uparrow}(\mathbf{r})\rangle]\\ &\quad - V\int d\mathbf{r}[(\delta\langle\Psi_{\uparrow}^{\dagger}(\mathbf{r})\Psi_{\uparrow}(\mathbf{r})\langle\Psi_{\downarrow}^{\dagger}(\mathbf{r})\Psi_{\downarrow}(\mathbf{r})\rangle\\ &\quad + \langle\Psi_{\uparrow}^{\dagger}(\mathbf{r})\Psi_{\uparrow}(\mathbf{r})\rangle(\delta\langle\Psi_{\downarrow}^{\dagger}(\mathbf{r})\Psi_{\downarrow}(\mathbf{r})\rangle]. \end{aligned} \tag{3.61}$$

The variation of the mean-field energy is in contrast,

$$\begin{aligned}\delta\langle\mathcal{H}_{mf}\rangle &= \sum_{\alpha}\int d\mathbf{r}\delta\langle\Psi_{\alpha}^{\dagger}(\mathbf{r})[\hat{H}_{l}+U(\mathbf{r})]\Psi_{\alpha}(\mathbf{r})\rangle\\ &\quad + \int d\mathbf{r}[\Delta(\mathbf{r})\delta\langle\Psi_{\uparrow}^{\dagger}(\mathbf{r})\Psi_{\downarrow}^{\dagger}(\mathbf{r})\rangle + \Delta^{*}(\mathbf{r})\delta\langle\Psi_{\downarrow}(\mathbf{r})\Psi_{\uparrow}(\mathbf{r})\rangle]. \end{aligned} \tag{3.62}$$

Comparing the corresponding terms in Eqs. (3.61) and (3.62) leads to the following set of relations:

$$\begin{aligned}\Delta(\mathbf{r}) &= -V\langle\Psi_{\downarrow}(\mathbf{r})\Psi_{\uparrow}(\mathbf{r})\rangle = V\langle\Psi_{\uparrow}(\mathbf{r})\Psi_{\downarrow}(\mathbf{r})\rangle,\\ \Delta^{*}(\mathbf{r}) &= -V\langle\Psi_{\uparrow}^{\dagger}(\mathbf{r})\Psi_{\downarrow}^{\dagger}(\mathbf{r})\rangle = V\langle\Psi_{\downarrow}^{\dagger}(\mathbf{r})\Psi_{\uparrow}^{\dagger}(\mathbf{r})\rangle,\\ U_{\uparrow}(\mathbf{r}) &= -V\langle\Psi_{\uparrow}^{\dagger}(\mathbf{r})\Psi_{\uparrow}(\mathbf{r})\rangle\\ U_{\downarrow}(\mathbf{r}) &= -V\langle\Psi_{\downarrow}^{\dagger}(\mathbf{r})\Psi_{\downarrow}(\mathbf{r})\rangle. \end{aligned} \tag{3.63}$$

Since $\gamma^{\dagger}$ and γ are the fermion creation and annihilation operators that diagonalize the Hamiltonian $\mathcal{H}_{mf}$, the averages in Eqs. (3.63) can be further evaluated by using the following identities:

$$\langle\gamma_{n,\alpha}^{\dagger}\gamma_{m,\beta}\rangle = \delta_{nm}\delta_{\alpha\beta}f_{n\alpha}, \tag{3.64}$$

$$\langle\gamma_{n,\alpha}\gamma_{m,\beta}\rangle = \langle\gamma_{n,\alpha}^{\dagger}\gamma_{m,\beta}^{\dagger}\rangle = 0. \tag{3.65}$$

f_n in Eq. (3.64) is governed by Fermi statistics. In equilibrium, it reduces to the Fermi function

$$f_{n\alpha} = \frac{1}{e^{(\epsilon_n+\epsilon_{\delta\alpha})/k_BT}+1}. \tag{3.66}$$

Substituting Eqs. (3.64) (3.66) into Eqs. (3.63) we find

$$\Delta(\mathbf{r}) = V \sum_n (1 - f_{n\uparrow} - f_{n\downarrow}) u_n(\mathbf{r}) v_n^*(\mathbf{r})$$
$$\Delta^*(\mathbf{r}) = V \sum_n (1 - f_{n\uparrow} - f_{n\downarrow}) u_n^*(\mathbf{r}) v_n(\mathbf{r})$$
$$U_\uparrow(\mathbf{r}) = -V \sum_n \left[|u_n(\mathbf{r})|^2 f_{n\uparrow} + |v_n(\mathbf{r})|^2 (1 - f_{n\downarrow}) \right]$$
$$U_\downarrow(\mathbf{r}) = -V \sum_n \left[|u_n(\mathbf{r})|^2 f_{n\downarrow} + |v_n(\mathbf{r})|^2 (1 - f_{n\uparrow}) \right]. \quad (3.67)$$

In general, $f_{n\uparrow} \neq f_{n\downarrow}$, but in many cases to the first-order approximation, $1 - f_{n\uparrow} - f_{n\downarrow}$ can be approximated by $1 - 2f_n$, especially at a higher temperature.

The density of particles in each layer of the system is

$$n_\uparrow = \langle \Psi_\uparrow^\dagger(\mathbf{r}) \Psi_\uparrow(\mathbf{r}) \rangle = \sum_n \left[|u_n(\mathbf{r})|^2 f_{n\uparrow} + |v_n(\mathbf{r})|^2 (1 - f_{n\downarrow}) \right],$$
$$n_\downarrow = \langle \Psi_\downarrow^\dagger(\mathbf{r}) \Psi_\downarrow(\mathbf{r}) \rangle = \sum_n \left[|u_n(\mathbf{r})|^2 f_{n\downarrow} + |v_n(\mathbf{r})|^2 (1 - f_{n\uparrow}) \right].$$
$$(3.68)$$

The total density n_t and density difference n_Δ of the two layers are

$$n_t = \sum_n \left[|u_n(\mathbf{r})|^2 (f_{n\uparrow} + f_{n\downarrow}) + |v_n(\mathbf{r})|^2 (2 - f_{n\uparrow} - f_{n\downarrow}) \right],$$
$$n_\Delta = \sum_n \left[|u_n(\mathbf{r})|^2 (f_{n\uparrow} - f_{n\downarrow}) - |v_n(\mathbf{r})|^2 (f_{n\uparrow} - f_{n\downarrow}) \right]. \quad (3.69)$$

This difference n_Δ of Eq. (3.69) can be utilized to determine $\epsilon_{\delta\uparrow}$ and $\epsilon_{\delta\downarrow}$ with the help of electrostatics in a self-consistent way.

The effective field U enters the BdGE together with the chemical potential μ. In a sense, it can be included in μ to make a new chemical potential. In what follows, we assume that it is included in μ and it is hence absent in the BdGEs of later sections.

3.6.4 Energy Spectrum of Excitonic Coulomb Interlayer Coupling

So far, we have only assumed a δ-like single-point contact interaction to simplify the derivation. In general, the interaction is a screened

Coulomb type connecting two points. Under such a realistic potential the "pairing interaction" $\Delta(\mathbf{r})$ then becomes $\Delta(\mathbf{r}, \mathbf{r}')$ as given by

$$\Delta(\mathbf{r}, \mathbf{r}') = V(\mathbf{r} - \mathbf{r}')\sum_n (1 - f_{n\uparrow} - f_{n\downarrow})u_n(\mathbf{r})v_n^*(\mathbf{r}'),$$

$$\Delta^*(\mathbf{r}, \mathbf{r}') = V(\mathbf{r} - \mathbf{r}')\sum_n (1 - f_{n\uparrow} - f_{n\downarrow})u_n^*(\mathbf{r})v_n(\mathbf{r}'). \quad (3.70)$$

The symmetry of $\Delta(\mathbf{r}, \mathbf{r}')$ is assumed under swapping of $\mathbf{r}$ and $\mathbf{r}'$ in deriving Eq. (3.70). This is the case for many real applications such as a singlet pairing and for the interlayer excitonic problem of this book.

With this extension, the corresponding generalized BdGEs are now

$$[\hat{H}_l + U(\mathbf{r})]u(\mathbf{r}) + \int d\mathbf{r}'\Delta(\mathbf{r}, \mathbf{r}')v(\mathbf{r}') = (\epsilon + \epsilon_{\delta\uparrow})u(\mathbf{r}),$$

$$-[\hat{H}_l^* + U(\mathbf{r})]v(\mathbf{r}) + \int d\mathbf{r}'\Delta^*(\mathbf{r}, \mathbf{r}')u(\mathbf{r}') = (\epsilon + \epsilon_{\delta\downarrow})v(\mathbf{r}). \quad (3.71)$$

$U(\mathbf{r})$ is a slow varying function in $\mathbf{r}$, since chemical potential changes slowly except at some boundary. In our case the two layers have quite different potentials but within each layer along the planar direction, the variation is more gradual, as indexed by $\mathbf{r}$. The assumption of slow variation $U(\mathbf{r})$ might not always be the case, but here for simplicity let us approximate $U(\mathbf{r})$ with a constant U. The equations simplify to

$$[\hat{H}_l + U]u(\mathbf{r}) + \int d\mathbf{r}'\Delta(\mathbf{r}, \mathbf{r}')v(\mathbf{r}') = (\epsilon + \epsilon_{\delta\uparrow})u(\mathbf{r}),$$

$$-[\hat{H}_l^* + U]v(\mathbf{r}) + \int d\mathbf{r}'\Delta^*(\mathbf{r}, \mathbf{r}')u(\mathbf{r}') = (\epsilon + \epsilon_{\delta\downarrow})v(\mathbf{r}). \quad (3.72)$$

Fourier transforming above equations into the momentum space $\mathbf{q}$ gives the following:

$$\left[-\frac{\hbar^2}{2m}\mathbf{q}^2 - \mu\right]u_\mathbf{q} + \Delta_\mathbf{q}v_\mathbf{q} = (\epsilon_\mathbf{q} + \epsilon_{\delta\uparrow})u_\mathbf{q},$$

$$\left[\frac{\hbar^2}{2m}\mathbf{q}^2 + \mu\right]v_\mathbf{q} + \Delta_\mathbf{q}^*u_\mathbf{q} = (\epsilon_\mathbf{q} + \epsilon_{\delta\downarrow})v_\mathbf{q}, \quad (3.73)$$

where U have been included in the chemical potential $\mu = \hbar^2k_F^2/2m$ and the energy is measured from μ.

The equation then simplifies to

$$\begin{aligned} \xi_{\mathbf{q}} u_{\mathbf{q}} + \Delta_{\mathbf{q}} v_{\mathbf{q}} &= (\epsilon_{\mathbf{q}} + \epsilon_{\delta\uparrow}) u_{\mathbf{q}}, \\ -\xi_{\mathbf{q}} v_{\mathbf{q}} + \Delta^*_{\mathbf{q}} u_{\mathbf{q}} &= (\epsilon_{\mathbf{q}} + \epsilon_{\delta\downarrow}) v_{\mathbf{q}}, \end{aligned} \tag{3.74}$$

where

$$\xi_{\mathbf{q}} = \frac{\hbar^2}{2m}[q^2 - k_F^2].$$

The generalized BdG equations differ from the ordinary BdG ones in energy shifts $\epsilon_{\delta\uparrow}$ and $\epsilon_{\delta\downarrow}$. As mentioned before these are due to the interlayer charge transfer and for the cuprates, both are non-zero. The two tend to have opposite signs but are not necessarily the same in magnitude. As discussed in Section 3.6.3, the two are related to U, $U_\uparrow$, $U_\downarrow$, and should in principle be determined with the help of self-consistent relations such as Eq.(3.69). Because $U_\uparrow$ and $U_\downarrow$ depends on $\mathbf{r}$, $\epsilon_{\delta\uparrow}$ and $\epsilon_{\delta\downarrow}$ actually vary in space. In scenarios where the variation is slow, they are taken as constant. This is the assumption taken in this book, which should already generate interesting results.

Since the charge transfer is helped with diverging DOS of the anti-nodal point in cuprates, the $\Delta_{\mathbf{q}}$ as a function of momentum is largest near the (π,0) region. And since $\epsilon_{\delta\uparrow}$ and $\epsilon_{\delta\downarrow}$ are more of an average effect, they are in general smaller than $\Delta_{\mathbf{q}}$ at the anti-nodal point where the pseudo-gap is the most prominent.

Equation (3.74) can now be connected to the long puzzling pseudogap phenomenon.

3.7 Pseudogap in Cuprates

First we focus on the anti-nodal region. Since $\epsilon_{\delta\uparrow}$ and $\epsilon_{\delta\downarrow}$ are negligible compared to $\Delta_{\mathbf{q}}$ here, Eq. (3.74) is simplified to

$$\begin{aligned} \xi_{\mathbf{q}} u_{\mathbf{q}} + \Delta_{\mathbf{q}} v_{\mathbf{q}} &= \epsilon_{\mathbf{q}} u_{\mathbf{q}}, \\ -\xi_{\mathbf{q}} v_{\mathbf{q}} + \Delta^*_{\mathbf{q}} u_{\mathbf{q}} &= \epsilon_{\mathbf{q}} v_{\mathbf{q}}. \end{aligned} \tag{3.75}$$

This is readily solved with the standard eigenvalue procedure to give

$$\epsilon_{\mathbf{q}} = \pm\sqrt{\xi_{\mathbf{q}}^2 + |\Delta_{\mathbf{q}}|^2}.$$

As a convenient notation, the following symbols are defined:

$$\epsilon_{\mathbf{q}}^{+} = \sqrt{\xi_{\mathbf{q}}^{2} + |\Delta_{\mathbf{q}}|^{2}},$$
$$\epsilon_{\mathbf{q}}^{-} = -\sqrt{\xi_{\mathbf{q}}^{2} + |\Delta_{\mathbf{q}}|^{2}}. \tag{3.76}$$

This result is similar to the solution of the BCS problem which shows a gap in the single-particle excitation spectrum. The difference though, is the explicit momentum space $\mathbf{q}$ dependence of $\Delta_{\mathbf{q}}$. In other words, the "pairing potential" (put in an analogy) here for the pseudogap could be large in the anti-nodal region, but it can be negligible or even equals zero in the nodal region.

The wavefunctions u and v for a given momentum $\mathbf{q}$ are obtained from Eqs. (3.75), once $\epsilon_{\mathbf{q}}$ is known. We have

$$u_{\mathbf{q}} = \frac{1}{\sqrt{2}}\left(1 + \frac{\xi_{\mathbf{q}}}{\epsilon_{\mathbf{q}}}\right)^{1/2}, \tag{3.77}$$
$$v_{\mathbf{q}} = \frac{1}{\sqrt{2}}\left(1 - \frac{\xi_{\mathbf{q}}}{\epsilon_{\mathbf{q}}}\right)^{1/2}.$$

The energy $|\Delta_{\mathbf{q}}|$ is the lowest single-particle excitation energy in the many-body ground-state which we take as the two-electron paired state attracted by another hole. $2|\Delta_{\mathbf{q}}|$ corresponds to an energy which is needed to break the hole pairing of two electrons, which is on par with a trion binding energy. Therefore, one can identify $2|\Delta_{\mathbf{q}}|$ as the pairing energy.

3.8 The *k*-Dependent Energy Gap

Multiplying $u_{\mathbf{q}}$ and $v_{\mathbf{q}}^{*}$ given by Eq. (3.78):

$$u_{\mathbf{q}}v_{\mathbf{q}}^{*} = \frac{1}{\sqrt{2}}\left(1 + \frac{\xi_{\mathbf{q}}}{\epsilon_{\mathbf{q}}}\right)^{1/2} \cdot \frac{1}{\sqrt{2}}\left(1 - \frac{\xi_{\mathbf{q}}}{\epsilon_{\mathbf{q}}}\right)^{1/2} = \frac{\Delta_{\mathbf{q}}}{2\epsilon_{\mathbf{q}}}. \tag{3.78}$$

Note that up to a normalization constant

$$u(\mathbf{r}) = u_{\mathbf{q}}e^{i\mathbf{r}\cdot\mathbf{q}}, \quad v(\mathbf{r}') = u_{\mathbf{q}}e^{i\mathbf{r}'\cdot\mathbf{q}}. \tag{3.79}$$

Substituting Eqs. (3.78) and (3.79) to the Fourier transformed Eq. (3.70), and relabeling the index n as $\mathbf{q}$ we reach

$$\Delta_{\mathbf{k}} = \sum_{\mathbf{q}}(1 - 2f_{\mathbf{q}})\frac{V_{\mathbf{kq}}\Delta_{\mathbf{q}}}{2\epsilon_{\mathbf{q}}}, \tag{3.80}$$

where approximation $f_{\mathbf{q}\uparrow} + f_{\mathbf{q}\downarrow} \approx 2f_{\mathbf{q}}$ is assumed. $V_{\mathbf{kq}}$ is the matrix element of the attractive screened Coulomb similar to Eq. (1.156), and is given by

$$V_{\mathbf{kq}} = \frac{V_e}{|\mathbf{k}-\mathbf{q}|^2 + q_s^2}. \tag{3.81}$$

Equation (3.81) indicates that contributions for the pseudogap are mainly from long-range part of the Coulomb. Most of the contribution to the summation of Eq. (3.80) are from $\mathbf{q}$ states near $\mathbf{k}$, within a radius $\sim q_s$.

At the superficial level, Eq. (3.80) is similar to the s-wave gap BCS self-consistency equation. However, significant differences exist since the gap function is now also $\mathbf{k}$ dependent. Both $\mathbf{k}$ and $\mathbf{q}$ are adopted to indicate the momentum- or wavevector-dependent properties. One example is already given in Eq. (3.80) to distinguish between the two different dependencies.

We proceed in the BCS fashion by replacing the sum of Eq. (3.80) with an integral,

$$\sum_{\mathbf{q}\sim\mathbf{k}} \approx \int_{\mathbf{q}\sim\mathbf{k}} \frac{d\mathbf{q}}{(2\pi)^3} \approx N(\mathbf{k}) \int d\epsilon.$$

Here $\mathbf{q} \sim \mathbf{k}$ reflects the fact that most of contribution to the integral comes from states near $\mathbf{k}$.

The fermion factor is explicitly

$$1 - 2f_{\mathbf{q}} = \tanh\left(\frac{\epsilon_{\mathbf{q}}}{2k_B T}\right).$$

Using the substitution Eq. (3.80) is now expressed as an integral in the following:

$$\begin{aligned}
\Delta_{\mathbf{k}} &= V\sum_{\mathbf{q}}(1-2f_{\mathbf{q}})\frac{V_{\mathbf{kq}}\Delta_{\mathbf{q}}}{2\epsilon_{\mathbf{q}}} \\
&\approx N(\mathbf{k})\int(1-2f_{\mathbf{q}})\frac{V_{\mathbf{kq}}\Delta_{\mathbf{q}}}{2\epsilon_{\mathbf{q}}}d\epsilon \\
&\approx \frac{N(\mathbf{k})V_e}{q_s^2}\int_{|\Delta_{\mathbf{k}}|}^{\infty}\frac{\Delta_{\mathbf{k}}}{\sqrt{\epsilon^2-|\Delta_{\mathbf{k}}|^2}}\tanh\left(\frac{\epsilon}{2k_B T}\right)d\epsilon \\
&\approx \frac{N(\mathbf{k})V_e\Delta_{\mathbf{k}}}{q_s^2}\int_{|\Delta_{\mathbf{k}}|}^{E_b}\frac{1}{\sqrt{\epsilon^2-|\Delta_{\mathbf{k}}|^2}}\tanh\left(\frac{\epsilon}{2k_B T}\right)d\epsilon,
\end{aligned} \tag{3.82}$$

where in the last step the integration upper limit is replaced by an energy cut-off on order of the exciton binding energy E_b. In the neighborhood of $\mathbf{k}$ where $V_{\mathbf{kq}}$ is appreciable, it is replaced with a simple V_e/q_s^2, and this is reflected in the third step of Eq. (3.82). Note that $\epsilon_{\mathbf{q}}$ has appreciable change from near 0 to large value on the order of eV and the fermion statistical factor $1 - 2f_{\mathbf{q}}$ can take from -1 to 1.

Meanwhile $\Delta_{\mathbf{q}}$ is a small value on the order of a few tens of meV up to a few hundreds, and it changes relatively slowly. It is also positive-definite if roughly taken as the exciton binding energy. Therefore, $\Delta_{\mathbf{q}}$ can be taken as a constant in summation or integral as in Eq. (3.82). The integral contribution is mainly from $\mathbf{q}$ states near k. Therefore, $\Delta_{\mathbf{q}}$ can be approximated by $\Delta_{\mathbf{k}}$. These attributes were adopted in the third and fourth steps of Eq. (3.82), which finally simplifies to

$$1 = \lambda_{\mathbf{k}} \int_{|\Delta_{\mathbf{k}}|}^{E_b} \frac{1}{\sqrt{\epsilon^2 - |\Delta_{\mathbf{k}}|^2}} \tanh\left(\frac{\epsilon}{2k_B T}\right) d\epsilon, \tag{3.83}$$

where the dimensionless parameter $\lambda_{\mathbf{k}} = N(\mathbf{k})V_e/q_s^2$ is the $\mathbf{k}$-dependent interaction constant. For the excitonic coupling, the effective attraction works for energies roughly on par with the excitonic binding energy E_b. This equation is utilized to derive $T = 0$ pseudogap and T dependence of the pseudogap.

Although the pseudogap equation Eq. (3.83) appears similar to the standard BCS gap equation, it has qualitative differences: (1) It applies to the normal state; (2) It has a Fermi statistics; (3) It depends on $\mathbf{k}$; (4) It has gapless section unless deeply underdoped. Therefore, Eq. (3.83) applies to pseudogap properties of the normal state, typically at T of a hundred K and above. The solution Eq. (3.83) can be discussed at different temperatures.

(1) First, at zero temperature the hyperbolic tangent function $\tanh\left(\frac{\epsilon}{2k_B T}\right) = 1$, and Eq. (3.83) simplifies to

$$\frac{1}{\lambda_{\mathbf{k}}} = \int_{|\Delta_{\mathbf{k}}|}^{E_b} \frac{d\epsilon}{\sqrt{\epsilon^2 - |\Delta_{\mathbf{k}}|^2}} = \operatorname{arcosh}\left(\frac{E_b}{|\Delta_{\mathbf{k}}|}\right) \approx \ln(2E_b/|\Delta_{\mathbf{k}}|).$$

Therefore, at $T = 0$

$$|\Delta_{\mathbf{k}}| = \Delta_{\mathbf{k}}(0) = \frac{E_b}{\sinh(1/\lambda_{\mathbf{k}})}. \tag{3.84}$$

For underdoped cuprates, $\lambda_{\mathbf{k}}$ is around unity or more when $\mathbf{k} = (\pi, 0)$ and the pseudo energy gap is given $|\Delta_{\mathbf{k}}| \sim E_b$ or more. In overdoped regime, the $\lambda_{\mathbf{k}}$ is reduced to significantly less than one due to V_e and q_s^2's decreasing with doping x. Here the pseudogap approaches "weak-coupling" limit so that

$$|\Delta_{\mathbf{k}}| \approx 2E_b \exp(-1/\lambda_{\mathbf{k}}). \tag{3.85}$$

The last approximation only holds when $\lambda_{\mathbf{k}}$ is small, which gives us an expression very similar to the BCS zero T gap at the weak coupling limit. For strong coupling which can be due to the divergent local DOS near the saddle point, i.e., the antinodal region in cuprates, the formula suggests a Δ on the order of E_b which is not a surprise at all.

The tentative conclusion from the above discussion is the antinodal region of the cuprate should be gapped near the fermi level at the immediate vicinity of $(\pi,0)$, at a temperature on the scale of the pseudogap energy. Of course, with our experience in BCS, we expect a temperature dependence of the gap energy.

(2) At finite temperature, we want to determine the pseudogap "critical" temperature T^{*c}, at which the gap Δ vanishes. We have

$$1 = \lambda_{\mathbf{k}} \int_0^{E_b} \tanh\left(\frac{\epsilon}{2k_B T^{*c}}\right) \frac{d\epsilon}{\epsilon}.$$

With a change of integration variable, this reduces to

$$\frac{1}{\lambda_{\mathbf{k}}} = \int_0^{E_b/2k_B T^{*c}} \frac{\tanh x}{x} dx.$$

Since the definite integral is given as

$$\int_0^a \frac{\tanh x}{x} dx = \ln\left(\frac{4a}{\pi} e^{\gamma}\right),$$

with $\gamma \approx 0.577$ as the Euler-Mascheroni constant. Therefore,

$$k_B T^{*c} = (2e^{\gamma}/\pi) E_b e^{-1/\lambda_{\mathbf{k}}} \approx 1.13 E_b \exp(-1/\lambda_{\mathbf{k}}). \tag{3.86}$$

There is significant difference between the pseudogap temperature $T^{*c} \times k_B$ of Eq. (3.86) and the gap energy of Eq. (3.84). The former is smaller which means in practice for OP cuprates, a significant portion of gap disappears at above T^{*c}.

The pseudo energy gap of Eq. (3.84) is derived assuming the absence of other orders like the SC in UD regime. Therefore, it might not be considered exact at very low temperature.

The pseudogap (critical) temperature T^{*c} is hence now a function of momentum $\mathbf{k}$ in the Brillouin Zone due to the $\lambda_{\mathbf{k}}$. At the nodal point near $(\pi/2, \pi/2)$, the local density of states is small from DFT calculations and ARPES, a zero or low $T^{*c}_{\mathbf{k}}$ is expected. If the $T^{*c}_{\mathbf{k}}$ here turns out to be lower than the ordinary experimental temperature of a few K or a few tens of K, then $\Delta_{\mathbf{k}}(T)$ here is zero which means we have a Fermi level crossing (Fermi surface) around here. This segment is also called a Fermi arc. Besides, there are other factors preventing the formation of gap, to be mentioned later, and in a typical T of a few tens of Kelvin in cuprates the nodal Fermi surface is well defined for optimally doped sample. However, the anti-nodal gap should be well preserved up to the pseudogap temperature.

In deriving the pseudogap equation, no assumption of a breaking of U(1) gauge symmetry is made or needed. This is a qualitative difference from the BCS type. As a result, no SC condensates are generated, and the long-range coherence factor is absent. This also explains the absence of a sharp jump in the electronic specific heat near T^*. Instead a gradual suppression of the specific heat is seen upon cooling through T^*, reflecting the $\mathbf{k}$-dependent incomplete gap nature of the pseudogap.

We proceed with the many-body Green's function formalism to obtain the spectral function $A(\mathbf{q}, \omega)$. Direct connection to spectroscopic experiments like ARPES is then made with the help of $A(\mathbf{q}, \omega)$.

The single-particle Green's function is defined in the natural units as

$$G(x, x') = -i\langle \mathcal{T}[\Psi(x)\Psi^{\dagger}(x')]\rangle, \tag{3.87}$$

where $\langle \quad \rangle$ as usual is the thermal average over a grand canonical ensemble. Moreover, x are 4-variables to be understood as $(\mathbf{r}, t)$. $\mathcal{T}$ is the time-order operator. Field operator Ψ here is related to amplitudes u, v and the fermion operators γ via Eq. (3.51) plus the dependence on time t. Since u and v are already solved by Eq. (3.77), $G(x, x')$ can be similarly worked out in the momentum and frequency domain, with the replacement $\mathbf{q} \to \mathbf{k}$ as

$$G(\mathbf{k}, i\omega) = \frac{u_{\mathbf{k}}u_{\mathbf{k}}^*}{i\omega - \epsilon_{\mathbf{k}}^+} + \frac{v_{\mathbf{k}}v_{\mathbf{k}}^*}{i\omega - \epsilon_{\mathbf{k}}^-}$$
$$= \frac{1}{i\omega - \epsilon_{\mathbf{k}}^+}\frac{\epsilon_{\mathbf{k}} + \xi_{\mathbf{k}}}{2\epsilon_{\mathbf{k}}} + \frac{1}{i\omega - \epsilon_{\mathbf{k}}^-}\frac{\epsilon_{\mathbf{k}} - \xi_{\mathbf{k}}}{2\epsilon_{\mathbf{k}}}. \quad (3.88)$$

The spectral density $A(\mathbf{q}, \omega)$ is related to $G(\mathbf{q}, i\omega)$ by

$$A(\mathbf{k}, \omega) = i[G(\mathbf{k}, \omega + is) - G(\mathbf{k}, \omega - is)]$$
$$= u_{\mathbf{k}}u_{\mathbf{k}}^*\delta(\omega - \epsilon_{\mathbf{k}}^+) + v_{\mathbf{k}}v_{\mathbf{k}}^*\delta(\omega - \epsilon_{\mathbf{k}}^-)$$
$$= \frac{\epsilon_{\mathbf{k}} + \xi_{\mathbf{k}}}{2\epsilon_{\mathbf{k}}}\delta(\omega - \epsilon_{\mathbf{k}}^+) + \frac{\epsilon_{\mathbf{k}} - \xi_{\mathbf{k}}}{2\epsilon_{\mathbf{k}}}\delta(\omega - \epsilon_{\mathbf{k}}^-), \quad (3.89)$$

where ω is real and $s \to 0^+$ and Sokhotsky's formula is utilized in the last step. This gives the one-particle excitation spectrum that can be directly probed in ARPES and the inverse photoemission. Notice that though the delta function implies long-live quasiparticle since scattering is not considered here for simplicity. $A(\mathbf{k}, \omega)$ which is essentially a function of momentum and energy can be plotted in the k-E (actually k-ω) space as in Fig. 3.2.

To get Fig. 3.2, a typical parameterized tight-binding band for a single CuO_2 layer is utilized as shown in Fig. 3.3. This is adopted to calculate $u_{\mathbf{k}}$ and $v_{\mathbf{k}}$ via relations similar to Eq. (3.77). The change is as follows. Instead of using a near-free electron kinetic energy,

$$\xi_{\mathbf{k}} = \frac{\hbar^2}{2m}[k^2 - k_F^2], \quad (3.90)$$

the TB band energy $E_{TB}(\mathbf{k})$ is adopted to give

$$\xi_{\mathbf{k}} = E_{TB}(\mathbf{k}) - E_F. \quad (3.91)$$

Equation (3.83) is also resorted to in deriving the pseudogap and hence the energy solution. $N(\mathbf{k})$ is a local density of states near $\mathbf{k}$. It is related to the $D_k(\mathbf{k})$ as in Eq. (3.16), the first-order partial derivative of the TB band at $\mathbf{k}$, given as

$$N(\mathbf{k}) \propto \frac{1}{|D_k(\mathbf{k})|} \quad (3.92)$$

where there is no singularity. Near the saddle point logarithmic DOS expression can be utilized as in Eq. (3.25).

The spectral density $A(\mathbf{k}, \omega)$ is hence obtained for the interlayer excitonic Coulomb model. For clearance, only significantly non-zero

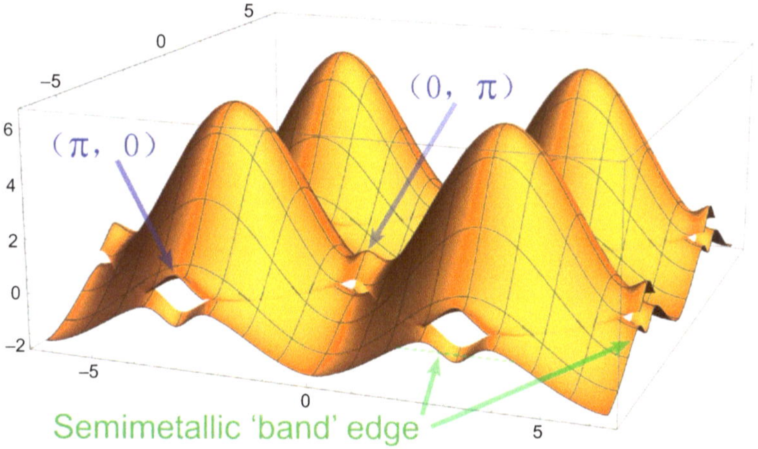

Figure 3.2 How the pseudogap is formed in the cuprates. The gapped anti-nodal case in a typical LSCO spectral density calculated from the theory in this chapter, i.e., Eq. (3.89). The blue arrows designate the anti-nodal regions like $(\pi,0)$ and $(0,\pi)$. A shift by 2π in k_x or k_y connects the equivalent k points and so there is no difference between $(\pi,0)$ and $(-\pi,0)$, and so on. The dashed green line indicates possible Fermi level variations due to interlayer charge transfer or doping, and it makes the semimetallic 'band' edge more apparent. The electron and hole portion of the 'band' are therefore obvious. This explains the puzzling electronlike quasiparticles seen in ARPES and quantum oscillations, and answers the hypotheses of Chapters 1 and 2. This feature is robust in both single-layer and multilayer cuprates, for both electron- and hole-doped HTSC copper oxides. The torn-up openings or gaps around $(\pi,0)$ and $(0,\pi)$ are the very pictorial representation of the long-mysterious pseudogap.

portion of the spectral density $A(\mathbf{k}, \omega)$ is plotted in the 3D E-$\mathbf{k}$ space (1D in energy vertically, and 2D in momentum horizontally). It is omitted altogether from the plot when small, e.g., $<10\%$.

Therefore, we have generated a k and T-dependent gap function in the cuprate energy spectrum (spectral function), with a typical result shown in Fig. 3.2. This gap can now be interpreted as the long sought-after *pseudogap*. In an independent-electron description of the electron energy state in a crystal, we have band structures. These are delta-like spectral density as a function of k, like that in Fig. 3.3. With turning-on of the strong interaction/correlation, the independent-electron band-structure often fails. Some changes

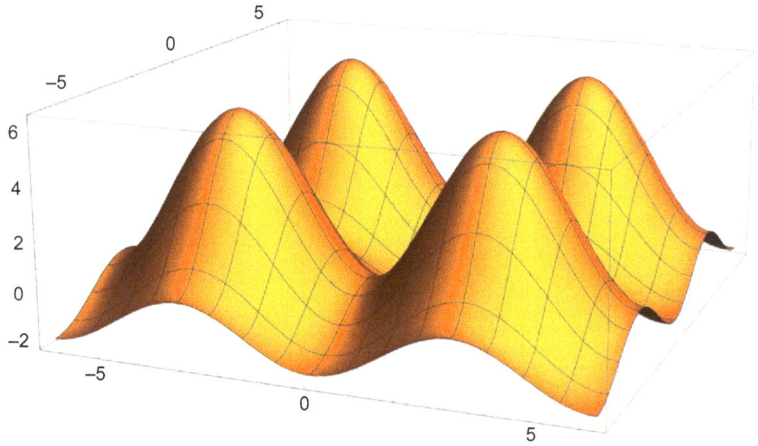

Figure 3.3 The well-known parameterized TB band structure of LSCO straddling the Fermi surface at $E = 0$.

then happen to the band state, making it a spectral density. (1) The band energy shifts, sort of renormalization effects; (2) The band "diffuses," losing spectral weight in its central locus and acquiring a spectral width (diffusiveness) that gives the quasiparticle excitation there a finite lifetime; (3) The band forms gaps, looking like tears and splits in some portion of the otherwise smooth (independent-particle) energy band in the momentum space. Some of the split-out part/branches could be so light (or faint), in the spectral weight, that they can be neglected altogether. This is the case of Fig. 3.2 near the antinodal region.

As for comparison, when the excitonic interaction is turned off, the LDA band-type energy dispersion is given in Fig. 3.3. The evolution of the LDA band to the pseudogap spectral density is evident from Fig. 3.3 to 3.2. The interlayer coupling actually doubles and splits the band near $(\pi,0)$. Away from it, the interaction effects are negligible and the "band" reverts to its LDA configuration.

If one takes a top-down look of the 3D surface as in Fig. 3.4, the characteristic Fermi arc near $(\pi/2,\pi/2)$ is clearly discernable. The spectral weight will be much reduced near the anti-nodal region at the Fermi energy because it is gapped.

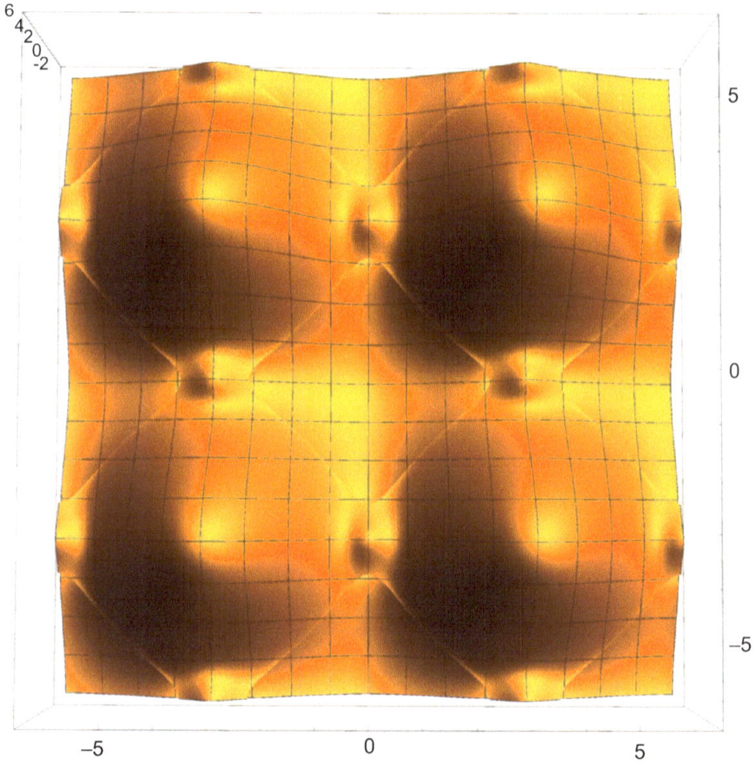

Figure 3.4 The top-down look of the 3D surface of LSCO spectral density, which was shown in an angled perspective in Fig. 3.2. The Fermi arc crossings near nodal regions are apparent.

3.9 Fermi Arc, Pocket, and Kinks in ARPES

So far we have assumed zero interlayer energy shift in Eq. (3.74), which is more relevant near the antinodes because there $\Delta_{\mathbf{k}}$ is the most significant. In more general cases $\epsilon_{\delta\uparrow}$ and $\epsilon_{\delta\downarrow}$ are not negligible versus $\Delta_{\mathbf{k}}$, and this is best represented in the nodal region.

Let us define a common energy shift $\epsilon_{\delta+}$ and deferential shift $\epsilon_{\delta-}$,

$$\begin{aligned} \epsilon_{\delta+} &= (\epsilon_{\delta\uparrow} + \epsilon_{\delta\downarrow})/2, \\ \epsilon_{\delta-} &= (\epsilon_{\delta\uparrow} - \epsilon_{\delta\downarrow})/2. \end{aligned} \tag{3.93}$$

Then Eqs. (3.74) are updated to

$$\begin{aligned} (\xi_{\mathbf{k}} - \epsilon_{\delta-})u_{\mathbf{k}} + |\Delta_{\mathbf{k}}|v_{\mathbf{k}} &= (\epsilon_{\mathbf{k}} + \epsilon_{\delta+})u_{\mathbf{k}}, \\ -(\xi_{\mathbf{k}} - \epsilon_{\delta-})v_{\mathbf{k}} + |\Delta_{\mathbf{k}}|u_{\mathbf{k}} &= (\epsilon_{\mathbf{k}} + \epsilon_{\delta+})v_{\mathbf{k}}, \end{aligned} \tag{3.94}$$

which can be routinely solved for $\epsilon_{\mathbf{k}}$ to give

$$\epsilon_{\mathbf{k}} = \pm\sqrt{(\xi_{\mathbf{k}} - \epsilon_{\delta-})^2 + |\Delta_{\mathbf{k}}|^2} - \epsilon_{\delta+}. \tag{3.95}$$

The effect of $\epsilon_{\delta+}$ is to make a global shift to the excitation spectrum. The nodal region is typically Fermi-crossing at the optimal doping if the temperature T is not too low. Since $\Delta_{\mathbf{k}}(T)$ is not zero below the **k**-local pseudogap critical temperature $T^{*c}(\mathbf{k})$, the solutions of Eq. (3.95) should still have a gap, that is, the energy solutions being separated to two branches of non-overlapping energy. However, due to the presence of $\epsilon_{\delta+}$, Fermi crossing is still possible even though the spectrum itself is gapped. This situation is illustrated in a k_x-k_y plot as in Fig. 3.5 and in a E-k plot as in Fig. 3.6.

Figure 3.5 is actually a 3D plot of energy, that is, the lower branch of solution in Eq. (3.95), versus the plane of k_x-k_y. The energy $\epsilon_{\mathbf{k}}^{-} = -\sqrt{(\xi_{\mathbf{k}} - \epsilon_{\delta-})^2 + |\Delta_{\mathbf{k}}|^2} - \epsilon_{\delta+}$ is plotted in the dark red color, versus the cyan background, which represents the zero energy Fermi level. The intersection of the two colors defines a Fermi crossing. Since Fig. 3.5 is a face-on capture normal to the k_x-k_y plane, it appears to be 2D.

When the energy solutions are plotted against along the (0, 0)-(π, π) direction, it gives Fig. 3.6. Here curves are solutions to the BdGEs to be interpreted as spectral density, neglecting line-width broadening. Solid portions are interpreted as with significant spectral density and the dotted curve insignificant, with the weight given by solutions like Eq. (3.77). The locus of spectral density with a weight near 1 might be loosely called an energy band, given as the solid segments of curves. In other words, the solid curve portions of Fig. 3.6 roughly corresponds to energy bands. Therefore, Fig. 3.6 conveys the idea that the energy band (major spectral density) can still traverse the Fermi energy twice in the ($\pi/2$, $\pi/2$) direction. All such double Fermi crossings form a crescent shaped pocket when plotted in the basal momentum space, as in Fig. 3.5. This gives a natural explanation to long puzzled phenomena known as Fermi arc versus Fermi pocket, and should help resolve the controversy as to

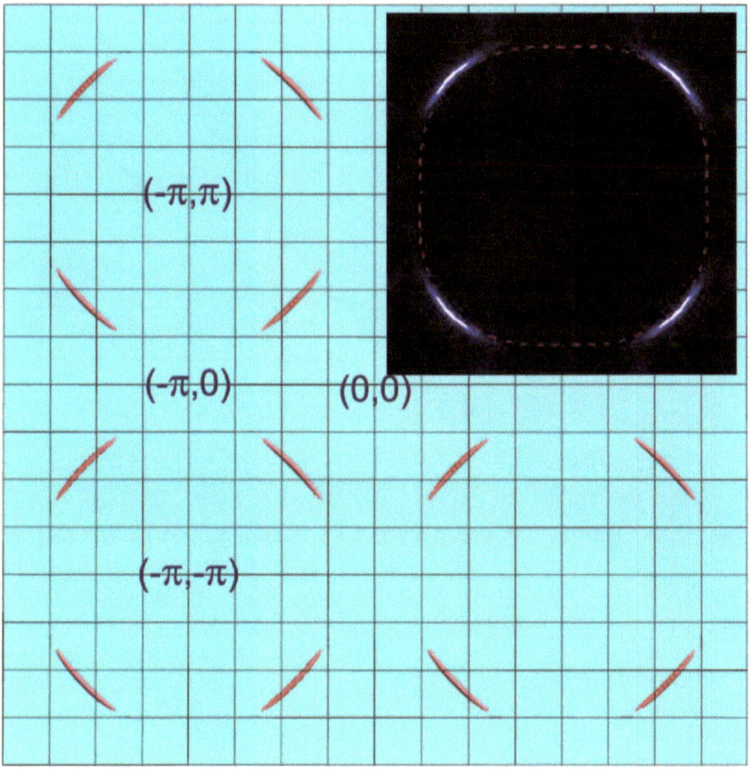

Figure 3.5 Fermi arcs or pockets of underdoped LSCO, obtained with Eq. (3.95). The upper right inset is from experiment data of the Lanzara Group at UCB and (Yoshida et al., 2003).

their origins. In the ECM picture, the arc and the pocket are two sides of the same coin.

The formalism outlined here should also help shedding a light on the widely reported kinks in ARPES near the nodal region. Going to the anti-node, the magnitude of the slope change in E vs. k is actually more prominent. However, probably since the anti-node is gapped, researchers tend not to draw too much a conclusion there. This might also be due to the fact that prior theoretical and conceptual developments were not well suited for phenomena away from the Fermi energy. As the result, a large number of investigations center

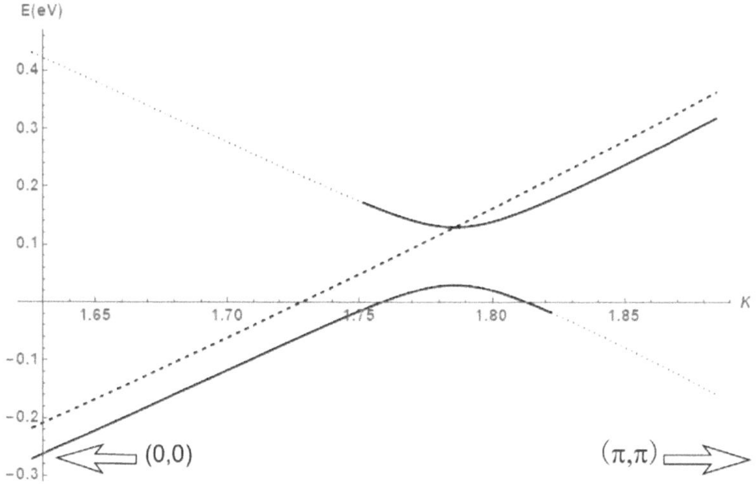

Figure 3.6 ARPES kink and pocket (arc) of underdoped LSCO along the $[\pi,\pi]$ direction. Dashed curve: tight-binding energy band; solid curve: significant spectral density assuming a zero line width; dotted curve: negligible spectral density assuming a zero line width. The energy at 0 corresponds to a Fermi crossing. The double Fermi crossing of the lower solid curve forms a Fermi arc or a Fermi pocket. The depression of slope for the solid curve near the Fermi level manifests as a kink. The kink is not very obvious in this figure due to neglecting the k-dependent chemical shift. The kink is often attributed to a bosonic mode, which we interpreted as the excitonic mode, although many other theories, like the electron–phonon interaction, exist.

around the node, due to a better defined band structure or spectral density.

But from Eq. (3.95), the kink, a relatively abrupt deviation (often a depression, near E_f) of the dE/dk slop from the DFT projection can be actually understood from the gapping behavior of $\sqrt{(\xi_{\mathbf{k}} - \epsilon_{\delta-})^2 + |\Delta_{\mathbf{k}}|^2}$. First, the presence of $|\Delta_{\mathbf{k}}|$ tend to flatten out the band near E_F. The part should then give a T-dependent kink behavior, diminishing with increasing temperature. Second, even if the $\Delta_{\mathbf{k}}$ is zero near the node (for example, at high temperature), the existence of charge transfer with the interlayer coupling still shifts the Fermi level of individual layers so that $\epsilon_{\delta+}$ and $\epsilon_{\delta-}$ accrue appreciable value. Then according to Eq. (3.95), the energy "band" is raised or dropped. As discussed in Section 3.6.4, $\epsilon_{\delta\uparrow}$ and $\epsilon_{\delta\downarrow}$ actually vary in space due to the $U_\uparrow$ and $U_\downarrow$ dependence of **r**, and hence both are k dependent as well. The rise and fall of the energy "band" are therefore k dependent, most prominent near the Fermi-level, and contribute to such a bending effect, known as the kink.

The characteristic slope depression near Fermi level in Fig. 3.6 is therefore the simulated result to model the kink. However, a quantitative match is farfetched at this point. There exist a few oversimplifications in our calculation here. (1) A constant Fermi level is assumed for each layer. In reality, it should be self-consistently derived from Eq. (3.67), this then gives rise to space-varying or dependence on **k**; (2) As a result, $\epsilon_{\delta+}$ and $\epsilon_{\delta-}$ should also be **k** dependent, and the two are related to $\Delta_{\mathbf{k}}$ via Eq. (3.67) and should also be governed by electrostatics.

On the experimental side the kink is also affected by the more-traditional bilayer splitting in Bi-2212, and therefore a clear cut understanding of what contributes to which takes more experimental and theoretical works. A thorough investigation of the problem hence is not within the scope of this book and will be deferred to a separate effort.

3.10 Conclusion

In this chapter, we stick to the Green's function formalism to investigate the DOS of layered cuprates. We further study the collective and single particle excitation spectrum of the excitonic Coulomb problem, represented by the interlayer coupling model of two-CuO_2 layers. A set of BdGEs and related self-consistency equations are derived and found to give both quantitative and qualitative explanations to the pseudogap phase. The technique naturally forms a many-body extension to the conventional band theory, and logically explains the gapped excitation spectrum. Long-puzzling phenomena such as Fermi arcs and pockets are readily understood within this model, along with the ARPES kinks, though to a lesser satisfactory level. The formalism explains the formation of semimetallic band-edge from split of the band at the saddle point, giving rise to electron-type quasiparticles, and has verified a long held conjecture in the high-T_c field. These results corroborate the theoretical and experimental results of Chapters 1 and 2.

Chapter 4

Where Is the Electron Located in the Hole-Doped Cuprate?

4.1 Introduction

The previous chapters have provided the following observations: (1) Given mobile electrons in p-doped cuprates, the model of excitonic Coulomb superconductivity accounts for major experimental phenomena in high-T_c cuprates; (2) Experimental evidence now overwhelmingly supports the notion of having mobile electrons in p-doped high-T_c cuprates. Now, a natural question may arise: Where exactly do those electrons reside?

The author of this chapter, Dr. Nie Luo, still vividly remembers the early days when he first tried to formulate a formalism of attack on this excitonic Coulomb model of pairing. It was in the mid-late 1990s and the author was seeking insights from Prof. A. J. Freeman back at Northwestern University. During a discussion one day, Arthur popped out a few words and asked the author, "Nie, I like excitons, but where are the electrons?"

On a note of history, Prof. Freeman's electronic structure calculation was (or one of) the first to help elucidate the chemical, structural (lattice constants, symmetry, etc.) and electronic proper-

High-T_c Superconductivity: The Excitonic Coulomb Perspective
Nie Luo and George H. Miley

ISBN 978-981-4800-57-0 (Hardcover), 978-0-429-27870-9 (eBook)
www.jennystanford.com

ties of La-124 and YBCO (Hwu et al., 1987), back in early 1987, at the Woodstock of physics. Their early calculations also revealed a Fermi surface for the optimal doping regime, in an early time when some exotic theories obviated the notion of FS in the high-T_c cuprates. The FS issue was finally settled a few years later in the early 1990s with the advent of precision ARPES, high quality crystals and improved surface preparation.

Back to the location of carrier electrons in p-type cuprates, now we can answer this with a fair portion of confidence: The electrons are located in the CuO_2 layer too. However, the electron-rich CuO_2 layers are generally separated in space from the hole-rich ones.

4.2 Interlayer Coupling

In Chapter 3, the consequence of interlayer Coulomb coupling was clearly outlined. Approaching half-filling, the excitonic Coulomb interaction will definitely gap a significant part of the Brillouin zone (BZ), and making an otherwise metal to become a semimetal or even an insulator. In this process, one layer becomes hole rich while the other electron rich. Mobility is very low though due to the strong Coulomb interaction. Antiferromagnetism within the layer further impedes electric conductivity, which finally leads to an insulator.

With the doping of extra holes into the system, both layers tend to be more holelike. However, one layer is always left with significant amount of electrons due to the interlayer coupling. In the real material, the imbalance of charge is further exacerbated by the fact, that statistically, some layers (or portions of a layer) are closer to the dopant than others. The proximity of a layer to the dopant significantly affects the nature of charge in it, as we will see later with LDA and GGA calculations of the electronic structure.

4.3 DFT and Band/Electronic Structures

To find out the realistic doping related electronic structure, model Hamiltonians (typically on an over-simplified mathematical lattice) like that of the Hubbard model are no longer adequate. We have to resort to computational techniques that are able to explicitly deal with the constituent chemical elements. And here come the DFT theory and the related solid-state electronic structure computational methods.

The density functional theory (DFT) was developed over the past half century or so. It has demonstrated its applicability and accuracy to condensed matter physics. With the wide adoption of the local density approximation (LDA), many complicated systems can be investigated in their electronic structures. First principles quantum mechanical calculations based on LDA and its extensions, like the generalized gradient approximations (GGA), are now indispensable for many experimentalists and theorists alike. Unit cells of one hundred atoms or more can now be computed and analyzed with high precision, on computational infrastructures accessible to a significant portion of the research community. This is largely due to great simplification of the many-body quantum mechanics from the DFT approach.

Even though the formalization of DFT was initiated by Hohenberg and Kohn in the 1960s, Slater and his collaborators started to adopt a flavor of LDA many years before in the Hartree–Fock–Slater (HFS) method. HFS was in turn predated by the Thomas–Fermi method for many-electron systems. All these methods can be categorized as LDA, in that, a complicated differential-integral form of energy that can be non-local in its exact shape, is replaced by a functional of local electron (fermion) density. This greatly simplifies the mathematics and computational efforts, though at the risk of losing important physics. A notable issue of LDA in general is the underestimate of the exchange and correlation effects. Excitonic correlations are one of the best manifestations of fermion correlation as we see in the $3e$-$1h$ case: The electrons correlate their motion in order to save the Coulomb energy. This has to be treated with a many-body wavefunction to give an accurate account of the potential and kinetic energy.

In following sections, we illustrate some features in the LDA-based electron-structure computational methods, in contrast with the more rigorous variational methods developed for the excitonic complexes. The comparison helps to understand their relative merits and deficiencies. Strategy and methodology then may be developed to combine the two to give improved results.

4.3.1 The Thomas–Fermi Method

The Fermi energy of an electron gas is given by

$$E_F = \frac{\hbar^2}{2m}\left(\frac{3\pi^2 N}{V}\right)^{2/3}. \tag{4.1}$$

This is the highest energy state of individual electron at the absolute zero T. In average each electron has $3E_F/5$ and the total energy of the N electrons is

$$E_{tot} = \frac{3N\hbar^2}{10m}\left(\frac{3\pi^2 N}{V}\right)^{2/3}, \tag{4.2}$$

and the total energy is actually the total kinetic energy of the N in a volume of V.

Equation (4.2) can be rearranged as

$$E_{tot} = \frac{3\hbar^2}{10m}V\left(\frac{3\pi^2 N}{V}\right)^{5/3}, \tag{4.3}$$

or

$$E_{tot}/V = \frac{3\hbar^2}{10m}\left(\frac{3\pi^2 N}{V}\right)^{5/3}, \tag{4.4}$$

where N/V can be further written as $\rho(\mathbf{r})$, and E_{tot}/V taken as energy density. Moreover, if the atomic units are adopted with $\hbar^2/m = 1$, then the total kinetic energy of the electron system is

$$T_{TF}[\rho] = C_F \int \rho^{5/3}(\mathbf{r})d\mathbf{r}, \tag{4.5}$$

where $C_F = 3(3\pi^2)^{2/3}/10$, and we have the total kinetic energy as a functional of local electron density.

The total energy of the electron system interacting with ions and with mutual Coulomb interaction is then

$$E_{TF}[\rho(\mathbf{r})] = C_F \int \rho^{5/3}(\mathbf{r})d\mathbf{r} - \sum_i \int \frac{Z_i \rho(\mathbf{r})}{|\mathbf{r}-\mathbf{r}_i|}d\mathbf{r} + \frac{1}{2}\iint \frac{\rho(\mathbf{r}_1)\rho(\mathbf{r}_2)}{|\mathbf{r}_1-\mathbf{r}_2|}d\mathbf{r}_1 d\mathbf{r}_2. \tag{4.6}$$

The variational calculus with the Lagrange multipliers can be applied to E_{TF} of Eq. (4.6) with respect to the density $\rho(\mathbf{r})$ to give the generalized Thomas Fermi equation

$$\mu_{TF} = \frac{5}{3} C_F \rho^{2/3}(\mathbf{r}) - \sum_i \frac{Z_i}{r} + \int \frac{\rho(\mathbf{r}_2)}{|\mathbf{r} - \mathbf{r}_2|} d\mathbf{r}_2, \tag{4.7}$$

where μ_{TF} is the Thomas–Fermi chemical potential. Overall, Eq. (4.7) is a integral equation for $\rho(\mathbf{r})$.

The TF kinetic term may underestimate the kinetic energy penalties in abrupt change of charge density but it overestimates the inter-electron Coulomb repulsion due to the exchange and correlation energy. At long distance (inter-molecule) scale the Coulomb dominates and the TF fails to give molecular bond to all atoms. This is also why the standard LDA, without some *ad hoc* inclusion of self interaction correction (SIC) terms, cannot give correct binding to negative ions. Similarly, it in general fails in the trion and tetron case.

To address this issue, one may introduce a hole function $h(\mathbf{r}_1; \mathbf{r}_2)$ and express the accurate inter-electron Coulomb energy as

$$\langle\Psi| \sum_{i<j}^{N} \frac{1}{r_{ij}} |\Psi\rangle = \iint \frac{\rho(\mathbf{r}_1)\rho(\mathbf{r}_2)}{|\mathbf{r}_1 - \mathbf{r}_2|} d\mathbf{r}_1 d\mathbf{r}_2 + \iint \frac{\rho(\mathbf{r}_1)h(\mathbf{r}_1; \mathbf{r}_2)}{|\mathbf{r}_1 - \mathbf{r}_2|} d\mathbf{r}_1 d\mathbf{r}_2. \tag{4.8}$$

The hole function $h(\mathbf{r}_1; \mathbf{r}_2)$ introduces an effective attraction to electrons in strongly correlated electron systems. And it is one of the major reasons why we emphasize EHC in cuprates. The $h(\mathbf{r}_1; \mathbf{r}_2)$ corrects for self-interaction error (SIE) and at least in formality accounts for the exchange and correlation energy in many-electron systems. The exchange term is included when spins are considered and the many-body wavefunction is written in an anti-symmetrized form.

4.3.2 The Hartree–Fock–Slater Method

This antisymmetrized (variational test) wavefunction can then be inserted into the many-body Hamiltonian to give a variety of Hartree–Fock methods at different assumptions. In the HF methods the kinetic term can be evaluated precisely nowadays with a proper selection of the basis set. Still the exchange correlation term is difficult and expensive to evaluate. Therefore, Slater made a further assumption to represent the exchange term as a functional of electron density. The total energy is given by

$$E_{HFS}[\rho(\mathbf{r})] = \sum_i^N \langle\psi_i| - \frac{\nabla_i^2}{2}|\psi_i\rangle - \sum_i^{ion} \int \frac{Z_i \rho(\mathbf{r})}{|\mathbf{r}-\mathbf{r}_i|} d\mathbf{r}$$
$$+ \frac{1}{2} \iint \frac{\rho(\mathbf{r}_1)\rho(\mathbf{r}_2)}{|\mathbf{r}_1 - \mathbf{r}_2|} d\mathbf{r}_1 d\mathbf{r}_2 + E_x[\rho(\mathbf{r})]. \quad (4.9)$$

With the Slater approximation, the last term in Eq. (4.8) is approximated by

$$E_x = -\frac{9\alpha}{8}\left(\frac{3}{\pi}\right)^{1/3} \int \rho^{4/3}(\mathbf{r}) d\mathbf{r}, \quad (4.10)$$

where $\alpha = 1$ gives the Slater exchange functional. In general, Eq. (4.10) is called the $X\alpha$ functional with α as the parameter. The related H-F calculations that utilize Eq. (4.10) are called the HFS method, which is a real forerunner of the modern DFT.

The typical Slater functional is plotted in Fig. 4.1 versus electron density.

There are now many reviews of the subject such as Koch and Holthausen (Koch and Holthausen, 2015), Parr and Yang (Parr and Weitao, 1994), Lundqvist and March (Lundqvist and March, 1983), Engel and Dreizler (Engel and Dreizler, 2011), Dreizler and da Provincia (Dreizler and da Providência, 2013), Ernzerhof, Perdew, and Burke (Ernzerhof et al., 1996), just to name a few. For a more involved exposition of DFT, the reader is directed to the aforementioned references.

4.3.3 Hohenberg and Kohn Theorem

The foundation of DFT and the LDA is the theorem (Hohenberg and Kohn, 1964) by Hohenberg and Kohn. It states that the external potential $v_{\text{ext}}(\mathbf{r})$ and thus total energy E, of a system of interacting electrons, is given exactly as a functional of the ground state electronic density, $\rho(\mathbf{r})$, or

$$E = E[\rho(\mathbf{r})] = \int \rho(\mathbf{r}) v_{\text{ext}} d\mathbf{r} + F[\rho(\mathbf{r})] \quad (4.11)$$

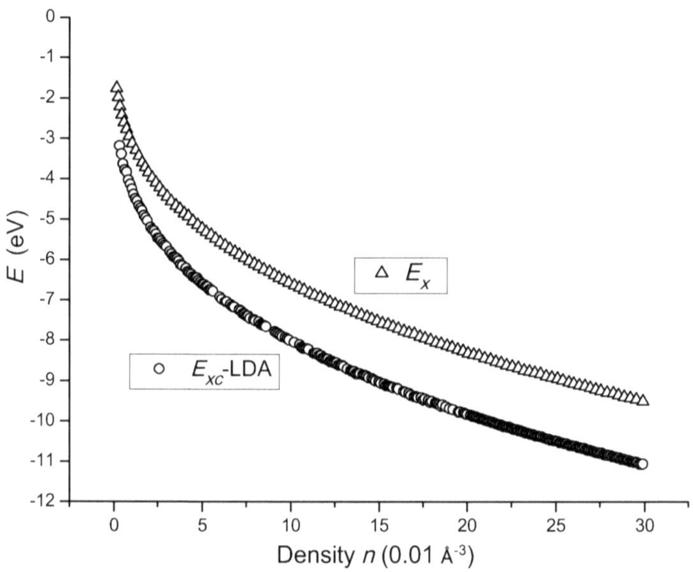

Figure 4.1 Typical exchange and correlation functionals of DFT. Triangle: the Slater exchange functional; circle: the typical LDA exchange-correlation functional.

There cannot be two different v_{ext} (thus two different wavefunctions Ψ) that generate the same ground state density $\rho(\mathbf{r})$. It therefore uniquely determines the v_{ext} and hence the system Hamiltonian. This in turn, uniquely determines all properties of the system.

Moreover, it is shown that the exact ground state density is the density that minimizes $E[\rho]$. When generalized to system where spins are explicitly included, the E and the other ground state properties become functionals of the spin density, which in a general case is given as a spin-indexed density matrix (Kubler et al., 1988; Sandratskii and Guletskii, 1986; von Barth and Hedin, 1972). This is in some aspect related to the spin-indexed Green's function in later chapters. In the often encountered collinear case, the off-diagonal terms of the matrix vanish. The densities that determine the functional are, respectively, labeled by spin-up and spin-down:

$$E = E[\rho_{\uparrow}(\mathbf{r}),\ \rho_{\downarrow}(\mathbf{r})]. \tag{4.12}$$

The Hohenberg-Kohn theorem does not, however, reveal the exact form of $E[\rho]$. What it does give us is the theoretical foundation and confidence in applying the HFS equations like Eq. (4.9). The applicability of DFT, hence, depends on the discovery of sufficiently accurate yet computationally efficient approximations.

One of the simplest formulations is then to replace the exchange functional E_x of Eqs. (4.9) and (4.10) with a more accurate exchange-correlation (xc) functional, $E_{xc}[\rho(r)]$, the result is then the DFT functional E_{DFT}

$$E_{DFT}[\rho(\mathbf{r})] = \sum_i^N \langle\psi_i| - \frac{\nabla_i^2}{2}|\psi_i\rangle - \sum_i^{ion} \int \frac{Z_i \rho(\mathbf{r})}{|\mathbf{r} - \mathbf{r}_i|} d\mathbf{r} + \frac{1}{2} \iint \frac{\rho(\mathbf{r}_1)\rho(\mathbf{r}_2)}{|\mathbf{r}_1 - \mathbf{r}_2|} d\mathbf{r}_1 d\mathbf{r}_2 + E_{xc}[\rho(\mathbf{r})]. \quad (4.13)$$

Here the first term on RHS of Eq. (4.13) denotes the single particle kinetic energy, and second term is the Coulomb interaction energy between the electrons and the ions. The third term on RHS is the Hartree component of the electron–electron energy,

$$\frac{1}{2} \iint \frac{\rho(\mathbf{r}_1)\rho(\mathbf{r}_2)}{|\mathbf{r}_1 - \mathbf{r}_2|} d\mathbf{r}_1 d\mathbf{r}_2. \quad (4.14)$$

4.3.4 LDA

The energy term $E_{xc}[\rho(\mathbf{r})]$ of Eq. (4.13) is in general an unknown functional. It can be formally expressed in terms of the pair (hole) correlation function $h(\rho, \mathbf{r}_1; \mathbf{r}_2)$, as follows:

$$E_{xc} = \iint \frac{\rho(\mathbf{r}_1)h(\rho, \mathbf{r}_1; \mathbf{r}_2)}{|\mathbf{r}_1 - \mathbf{r}_2|} d\mathbf{r}_1 d\mathbf{r}_2. \quad (4.15)$$

The pair (hole) correlation function of Eq. (4.15), $h(\rho, \mathbf{r}_1; \mathbf{r}_2)$, generalizes the hole function $h(\mathbf{r}_1; \mathbf{r}_2)$ of Section 4.3.1, and depends on external field, details of the electron degrees of freedom (e.g., band structures, elementary excitations and so on). Ironically it is in general unknown as well. However, for special cases like a homogeneous electron gas, the pair (exchange) correlation function is known to a high accuracy. In this case, Eq. (4.15) can be integrated over $\mathbf{r}_2$ to give

$$E_{xc} = \iint \frac{\rho(\mathbf{r}_1)h^{hoe}(\rho, \mathbf{r}_1; \mathbf{r}_2)}{|\mathbf{r}_1 - \mathbf{r}_2|} d\mathbf{r}_1 d\mathbf{r}_2$$
$$= \int \rho(\mathbf{r}_1) d\mathbf{r}_1 \int \frac{h^{hoe}(\rho, \mathbf{r}_1; \mathbf{r}_2)}{|\mathbf{r}_1 - \mathbf{r}_2|} d\mathbf{r}_2$$
$$= \int \rho(\mathbf{r}_1) d\mathbf{r}_1 \epsilon_{xc}^{hoe}[\rho(\mathbf{r}_1)], \quad (4.16)$$

where superscript hoe indicates the case of homogeneous electron gas. This applies in principal only to the uniform electron gas.

Here comes the local density approximation: For more general cases where density is not constant versus location, we, nevertheless, assume that the exchange-correlation energy at a point $\mathbf{r}$ is simply equal to the exchange-correlation energy of a homogeneous electron gas that has the same density at that point $\mathbf{r}$. In other words, a form of xc functional similar to Eq. (4.16)

$$E_{xc}[\rho(\mathbf{r})] = \int d\mathbf{r} \rho(\mathbf{r}) \epsilon_{xc}[\rho(\mathbf{r})], \quad (4.17)$$

gives the LDA xc-functional. In general, LDA gives good calculation on structural properties but fails in energies due to overbinding. Therefore, extensions to take into account of nonlocality were attempted.

4.3.5 GGA

A widely adopted approximation, generalized gradient approximations (GGAs), attempts to incorporate the effects of inhomogeneities by including the gradient of charge density ρ.

$$E_{xc}^{GGA}[\rho(\mathbf{r})] = \int d\mathbf{r} \rho(\mathbf{r}) \epsilon_{xc}(\rho(\mathbf{r})) F_{xc}[\rho(\mathbf{r}), \nabla \rho(\mathbf{r})], \quad (4.18)$$

where $F_{xc}[\rho(\mathbf{r}), \nabla \rho(\mathbf{r})]$ is an enhancement factor which uses local gradient as well as the density. The local gradient is adopted in order to incorporate more information about the interacting electrons.

The GGA is in general satisfactory in energetics by largely rectifying the overbinding problem. It is significantly more successful than LDA when applied to molecules. The GGA is known to greatly improve properties of 3d transition metals. For example, unlike with the L(S)DA, the correct bcc ground state of Fe is obtained with the GGA. Still it underestimates the band-gaps in semiconductors and insulators. This is due to the fact, at least in part, that Hohenberg-Kohn theorem is not valid for excited states. Additionally, strong correlation is not treated properly, in many transition metal compounds that are either Mott–Hubbard or charge-transfer insulators. This is exactly the problem of GGA calculations on cuprates. Corrective measures like LDA+U, GGA+U, GW, SIC are often utilized.

Both LDA and GGA are based on local density, where the GGA is often termed semi-local. As such in the process of integrating over $\mathbf{r}_2$ with a oversimplified homogeneous electron pair correlation in

$$\int \frac{h^{hoe}(\rho, \mathbf{r}_1; \mathbf{r}_2)}{|\mathbf{r}_1 - \mathbf{r}_2|} d\mathbf{r}_2, \tag{4.19}$$

many real correlation/exchange and kinetic energy details are lost. And this explains their general inadequacy for strongly correlated system.

4.3.6 Applicability and Limits

The formally exact expression for the exchange correlation energy in terms of the pair (hole) correlation function $h(\mathbf{r}_1; \mathbf{r}_2)$, is given by

$$E_{xc} = \iint \frac{\rho(\mathbf{r}_1) h(\rho, \mathbf{r}_1; \mathbf{r}_2)}{|\mathbf{r}_1 - \mathbf{r}_2|} d\mathbf{r}_1 d\mathbf{r}_2. \tag{4.20}$$

The exchange correlation effects digs out a hole in the electron-filled space, and the *xc*-hole has a formal charge of $+1$, which attracts the electron, and therefore E_{xc} is negative-definite thus defined. However, the exchange correlation functional includes more than just the Coulomb part, because there involves a penalty in the kinetic energy that are necessary to dig out the hole. This reduction is accounted for by using an average of running coupling constant λ

$$E_{xc} = \iint \int_0^1 \frac{\rho(\mathbf{r}_1) h_\lambda(\rho, \mathbf{r}_1; \mathbf{r}_2)}{|\mathbf{r}_1 - \mathbf{r}_2|} d\lambda d\mathbf{r}_1 d\mathbf{r}_2. \tag{4.21}$$

In LDA the function $h_\lambda(\rho, \mathbf{r}_1; \mathbf{r}_2)$ is assumed to be of the form

$$h_\lambda(\rho, \mathbf{r}_1; \mathbf{r}_2) = h_\lambda(\rho, \mathbf{r}_1, |\mathbf{r}_1 - \mathbf{r}_2|). \tag{4.22}$$

It has the dimension of a density and is a spherically symmetric function of $|\mathbf{r}_1 - \mathbf{r}_2|$. The assumption is that only the spherical average of the exchange correlation hole needs to be corrected. This is a strong assumption, however, as we have seen in the tetron part: Angular correlation may give rise to qualitative different results in strongly correlated systems.

For many cases where correlation is not strong, this assumption, nevertheless, works well and much simplifies the calculation, and it contributed to LDA's early success, in a time when the computational prowess of computers was much limited. For the significant part of superconductive regime in cuprates, especially in the overdoped case, GGA represents a good first-order understanding of physics.

4.3.7 Kohn–Sham Equations

With the energy functional given by E_{HK}, variational principle is applied: The energy functional is varied with respect to the set of KS orbitals ψ_i and to minimize the total energy under the constraints of $\langle\psi_i|\psi_j\rangle = \delta_{ij}$. Kohn and Sham (Kohn and Sham, 1965), hence, were able to derive a set of single-particle equations, or the so called Kohn–Sham (KS) equations, as given below:

$$\left\{-\frac{\nabla^2}{2} + \left[\int \frac{\rho(\mathbf{r})}{|\mathbf{r}-\mathbf{r}'|}d\mathbf{r}' + \varepsilon_{XC}(\mathbf{r}) - \sum_j^{ion} \frac{Z_j}{|\mathbf{r}-\mathbf{r}_j|}\right]\right\}\psi_i = \epsilon_i\psi_i, \tag{4.23}$$

where φ_i are the single particle Kohn–Sham orbitals. ϵ_i are the corresponding Kohn–Sham eigenvalues. The potential obviously depends on the charge density. The Kohn–Sham orbitals are related to the electron density by

$$\rho(\mathbf{r}) = \sum_{occ} \psi_i^*(\mathbf{r})\psi_i(\mathbf{r}), \tag{4.24}$$

where the sum is over occupied states, with the highest occupied orbital is governed by the total number of electron. Density ρ thus computed is then adopted to give V_H and V_{xc}, expressed, respectively, as

$$V_H(\mathbf{r}) = \mathrm{e}^2 \int \mathrm{d}^3\mathbf{r}' \frac{\rho(\mathbf{r}')}{|\mathbf{r}-\mathbf{r}|},$$

and

$$V_{xc}(\mathbf{r}) = \frac{\delta E_{xc}[\rho]}{\delta\rho(\mathbf{r})}.$$

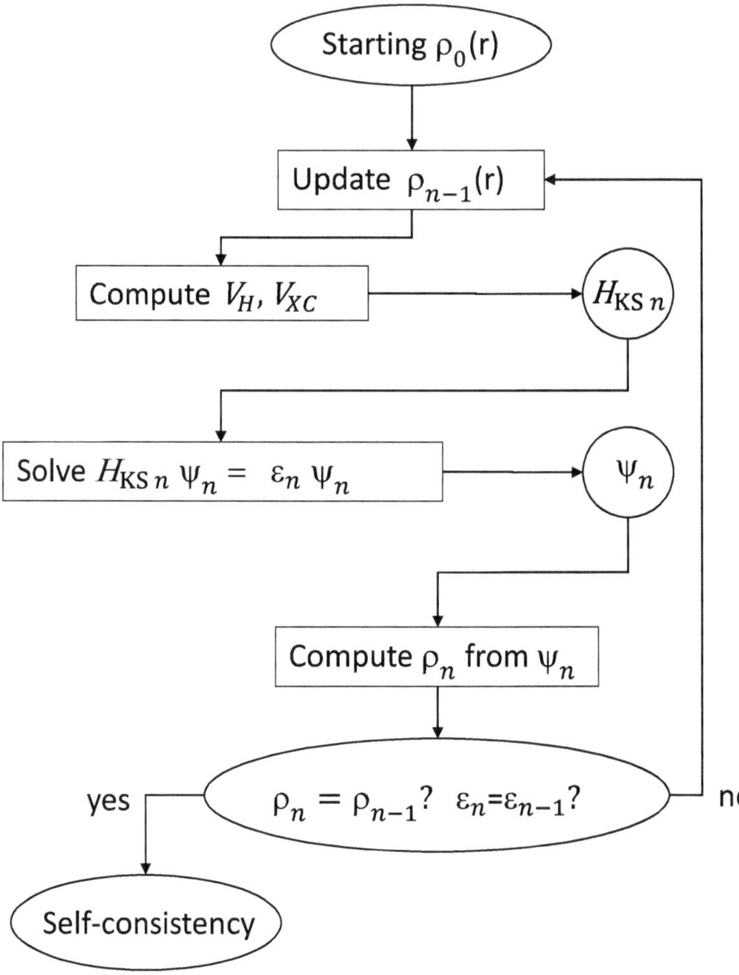

Figure 4.2 The self-consistency loop to solve the Kohn–Sham equations.

In this formalism, the self-consistent solution of Eqs. (4.23) involves an iterative procedure. That is to start from an initial electron density that can be used to calculate the Kohn–Sham potential. Such an initial charge density ρ_0, in the case of the First-principles all-electron computation for a crystal, is often a superposition of atomic charge density. Then a hamiltonian $H_{KS,1}$ is constructed with it, followed by a numerical solution of the KS equations, which gives a set of eigenvalues and wavefunctions ψ_1. Then, through these KS solutions ψ_1, a new electron density ρ_1 is obtained. Now the process can repeat so ρ_1 is used to construct $H_{KS,2}$, which will yield a ρ_2, etc. The density must converge so as to yield an effective potential that when inserted into the KS equations yields orbitals that reproduce it. Then, the convergency in both the charge and eigen-energies (with potentially other parameters like the force field and equilibrium positions) is achieved, which also implies self-consistency: This final density is then consistent with the Hamiltonian. The solution of a many-body Schrodinger equation, is then converted to a far simpler problem of solving a series of single particle equations, along with a self-consistency requirement. This iterative solution procedure is summarized in Fig. 4.2.

4.4 Basis Sets for Solving the KS

In practice the KS equations are typically solved in some specific basis set. Efficiency, simplicity and completeness are the common metrics that are used in discussing the relative merits of particular basis sets. For crystal solids, a variety of planewave basis are natural choices due to Bloch's theorem. A pure planewave basis is often combined with a set of pseudopotentials for different chemical elements to reduce the computational effort. The other often utilized basis is the augmented plane-wave (APW), first introduced by Slater (Slater, 1937). The potential was supposed to be spherically symmetrical within spheres of proper choices around the atoms, while being constant outside. This is alternatively called the muffin-tin potential. In the so-called interstitial region, far away from the nuclei, the atomic and/or crystal potential is slow-changing. Electrons are more or less "free" meaning their kinetic energy is little changed versus location, well described by an approximate fixed momentum. Quasi-free electrons as such are here best described by planewaves. Close to the nuclei, the potential is similar to an atomic type and varies strongly. The electrons behave as if they were near the nucleus of a free atom, and they are described more efficiently by atomic like functions. The crystal space is therefore divided in two regions outlined by the muffin-tin potential: Around each atom a sphere with radius R_α is given. R_α may vary for different atoms. The spheres outlined by individual radius R_α, however, do not intersect one another, with the corresponding space termed muffin tin (MT) sphere (region) or MTS. The remaining space outside the spheres is called the interstitial region, or IS. The APW bases are also implicitly designated with energy, which are treated in various approaches. In the following subsections, we basically follow the classification given by Cottenier (Cottenier, 2004).

4.4.1 APW

The augmented plane wave (APW) forms the basis to expand $\psi_{\mathbf{k}}^n$ and is defined as

$$\varphi_{\mathbf{K}}^{\mathbf{k}}(\mathbf{r},\ E) = \begin{cases} \frac{1}{\sqrt{V}} e^{i(\mathbf{k}+\mathbf{K})\cdot\mathbf{r}} & \mathbf{r} \in \mathrm{IS} \\ \sum\limits_{l,m} A_{lm}^{\alpha,\mathbf{k}+\mathbf{K}} u_l^\alpha(r',\ E_l) Y_m^l(\hat{\mathbf{r}}') & \mathbf{r} \in \mathrm{MT}_\alpha \end{cases}, \tag{4.25}$$

where $\mathbf{k}$ is the wavevector within the first Brillouin zone, and $\mathbf{K}$, being many of such, are the reciprocal lattice vectors. Here we also label the volume of the unit cell as V. Note that the APW basis set is $\mathbf{k}$- and $\mathbf{K}$ dependent, even within the MT sphere. The APW basis is further indexed by individual atom site α, with benefits to be mentioned later. $\mathbf{r}' = \mathbf{r} - \mathbf{r}_\alpha$ is the real space coordinate inside the spheres as measured with respect to the origin of each sphere MT_α. The length of $\mathbf{r}'$ is r', and the angles θ' and ϕ' specifying the direction of $\mathbf{r}'$ in spherical coordinates of atom α, are shorthanded as $\hat{\mathbf{r}}'$, for simplification. $\hat{\mathbf{r}}'$ is a direction vector of unit length. The terms Y_m^l are spherical harmonics. $A_{lm}^{\alpha,\mathbf{k}+\mathbf{K}}$ are parameters to be determined by additional conditions including the boundary ones, as is the energy parameter E_l. The summation over l,m in Eq. (4.25) is to infinity in its formality, but is in practice always up to a finite upper limit. The common practice sets $K_{\max} r_\alpha$ around 7. The physical meaning of such a value is quite evident: to give the wavefunction enough resolution power for fine details (high k, high E) while not wasting computational resources to atomic details not relevant to the crystal-scale. Therefore, as a rule of thumb,

$$K_{\max}\, r_\alpha \approx 7. \tag{4.26}$$

Each APW basis function is labeled with **k**, atomic site α, orbital quantum numbers l and m. This constitutes a major advantage of the APW-like basis: The energy band calculated can then be traced back to some specific orbitals of individual atoms. The APW basis then makes a convenient connection from the wavevector (momentum) to atomic subshells (angular momenta), or from the crystal symmetry to the orbital symmetry. This helps to sort out correlation strength of various bands for a certain energy, for example, at the Fermi level. Such a merit is utilized to its full advantage in discerning the charge type at E_F, later in this chapter.

In Eq. (4.25), u_l^α are atomic-type solutions to the radial part of the Schrödinger equations for site α, evaluated at energy E_l,

$$\left[-\frac{d^2}{dr^2}+\frac{l(l+1)}{r^2}+V(r)-E_l\right]ru_l(r)=0. \tag{4.27}$$

The solution for u_l^α is more involved than the pure atomic case. The atomic boundary conditions that the Coulomb potential $\rightarrow 0$ and $u_l^\alpha(r,\ E_l)$ vanishes for $r\rightarrow\infty$, make the energies E_l discrete and limit the number of solution u_l^α. However, the atomic boundary conditions do not apply for a crystal, since the potential approaches a negative-definite yet finite value. Further, the crystal APW wavefunction is not required to vanish at long distance. Instead, it is required to match the interstitial part at the MT sphere boundary. To achieve this, one may expand the planewave, i.e., the first line of Eq. (4.25) in spherical harmonics about the origin of the sphere of atom α:

$$\frac{1}{\sqrt{V}}e^{i(\mathbf{k}+\mathbf{K})\cdot\mathbf{r}}=\frac{4\pi}{\sqrt{V}}e^{i(\mathbf{k}+\mathbf{K})\cdot\mathbf{r}_\alpha}\sum_{l,m}i^l J_l(|\mathbf{k}+\mathbf{K}||\mathbf{r}'|)Y_m^{l*}(\hat{\mathbf{k}+\mathbf{K}})Y_m^l(\hat{\mathbf{r}}'), \tag{4.28}$$

where $J_l x)$ is the Bessel function of order l. The sphere boundary generates a boundary condition (where $\mathbf{r}'=\mathbf{R}_\alpha$, which defines $\mathbf{R}_\alpha$) that the preceding function Eq. (4.28) be equal to the lm-part of Eq. (4.25) then results in

$$A_{lm}^{\alpha,\mathbf{k}+\mathbf{K}}=\frac{4\pi i^l e^{i(\mathbf{k}+\mathbf{K})\cdot\mathbf{r}_\alpha}}{\sqrt{V}u_l^\alpha(R_\alpha,E_l)}J_l(|\mathbf{k}+\mathbf{K}|R_\alpha)Y_m^{l*}(\hat{\mathbf{k}+\mathbf{K}}).$$

$A_{lm}^{\alpha,\mathbf{k}+\mathbf{K}}$ is hereby uniquely defined, while E_l is still to be determined. The summation is in principle to an infinite numbers of l and m in Eq. (4.28), which implies an infinite number of $A_{lm}^{\alpha,\mathbf{k}+\mathbf{K}}$. In practices the summation is truncated after a maximal l_{max}. Since the l is proportion to the number of "ripples" spherically and one needs to match the spherical and plane wave, one can reason that $l_{\text{max}} > R_\alpha K_{\text{max}}$ which also makes a connection to Eq. (4.26).

A drawback of the APW basis is the uncertainty in E_l. One typically starts with a guess work of $E_l(0)$, using radial Schrödinger Eq. (4.27) to numerically solve for $u(r, E_l)$. The resulted radial functions make an APW basis. However, up to this point, whether or not this basis can be self-consistently plugged into the crystal Hamiltonian matrix and then generates new eigenstates $E_l(1)$ that match the starting guessed $E_l(0)$ is still unknown. If not, which is the typical case, then we have to try again. Due to the new $E_l(1)$, the updated APW basis is determined again, and similarly for all Hamiltonian matrix elements. With the help of root determination algorithms, this trial-and-error loop continues until a satisfactory root, say $E_l(n)$ is found. Clearly, the APW procedure involves much guess work and sometimes pure luck. A more systematic approach needs to be developed and this leads to the linearized APW basis.

4.4.2 LAPW

The often undetermined E_l makes $u_l^\alpha(r', E_l)$ computationally expensive to pin down. E_l may be written as a more general $\epsilon_{\mathbf{k}}^n$ to account for the possible implicit relation to $\mathbf{k}$. The uncertainties in both $u_l^\alpha(r', \epsilon_{\mathbf{k}}^n)$ and $\epsilon_{\mathbf{k}}^n$ can be resolved by the Taylor expansion near some likely energy E_0. That is exactly what the *Linearized Augmented Plane Wave* method enables us to do. If we have calculated u_l^α at some energy E_0, we could make a Taylor expansion to find it at energies not far away from it,

$$u_l^\alpha(r', \epsilon_{\mathbf{k}}^n) = u_l^\alpha(r', E_0) + (\epsilon_{\mathbf{k}}^n - E_0)\left.\frac{\partial u_l^\alpha(r', E)}{\partial E}\right|_{E=E_0} \tag{4.29}$$

$$= u_l^\alpha(r', E_0) + (\epsilon_{\mathbf{k}}^n - E_0)\dot{u}_l^\alpha(r', E_0), \tag{4.30}$$

where the quadratic and higher-order terms in $E_0 - \epsilon_{\mathbf{k}}^n$ are neglected.

Equation (4.29) can now be inserted into the definition of APW, that is Eq. (4.25), to give

$$\varphi_{\mathbf{K}}^{\mathbf{k}}(\mathbf{r}) = \begin{cases} \frac{1}{\sqrt{V}} e^{i(\mathbf{k}+\mathbf{K})\cdot\mathbf{r}} & \mathbf{r} \in \mathrm{IS} \\ \sum_{l,m} [A_{lm}^{\alpha,\mathbf{k}+\mathbf{K}} u_l^{\alpha}(r', E_0) + A_{lm}^{\alpha,\mathbf{k}+\mathbf{K}} (\epsilon_{\mathbf{k}}^{n} - E_0) \dot{u}_l^{\alpha}(r', E_0)] Y_m^l(\hat{\mathbf{r}}') & \mathbf{r} \in \mathrm{MT}_{\alpha} \end{cases}. \tag{4.31}$$

Since the energy $\epsilon_{\mathbf{k}}^{n}$ is still not known, one would rather just write $A_{lm}^{\alpha,\mathbf{k}+\mathbf{K}}(\epsilon_{\mathbf{k}}^{n} - E_0)$ as another unknown coefficient simply as $B_{lm}^{\alpha,\mathbf{k}+\mathbf{K}}$, which can be determined later on similar to $A_{lm}^{\alpha,\mathbf{k}+\mathbf{K}}$. Then we have arrived at the formal definition of the LAPW basis,

$$\varphi_{\mathbf{K}}^{\mathbf{k}}(\mathbf{r}) = \begin{cases} \frac{1}{\sqrt{V}} e^{i(\mathbf{k}+\mathbf{K})\cdot\mathbf{r}} & \mathbf{r} \in \mathrm{IS} \\ \sum_{l,m} [A_{lm}^{\alpha,\mathbf{k}+\mathbf{K}} u_l^{\alpha}(r', E_0) + B_{lm}^{\alpha,\mathbf{k}+\mathbf{K}} \dot{u}_l^{\alpha}(r', E_0)] Y_m^l(\hat{\mathbf{r}}') & \mathbf{r} \in \mathrm{MT}_{\alpha} \end{cases}, \tag{4.32}$$

where $\dot{u}_l^{\alpha}(r', E_0)$ is obtained by numerical differentiation of $u_l^{\alpha}(r', E_0)$. Both $A_{lm}^{\alpha,\mathbf{k}+\mathbf{K}}$ and $B_{lm}^{\alpha,\mathbf{k}+\mathbf{K}}$ can be determined by boundary conditions: The wavefunctions need to be continuous in both value and its first-order derivative versus r', at the MT spheres. Two equalities need to be utilized: The first is similar to Eq. (4.28) and the second is its derivative with respect to r', evaluated at $r' = r_{\alpha}$.

We are then only left with how to determine E_0. In practice it is often chosen near the center of the energy of a certain band-deriving orbitals. Since the orbitals are site specific and can differ much in energy even with the same l, E_0 actually depends on a particular atom and the related angular momentum. Therefore, different E_0 are selected for every physically important l (s-, p-, d- and f-states, i.e., up to $l = 3$) and for every atom, conveniently written as $E_{1,l}^{\alpha}$. The subscript label 1 is to differentiate from another energy parameter associated with the so-called local orbitals, to be introduced in Section 4.4.3. One clearly sees the difference versus APW: In LAPW, instead of a painstakingly trial-and-error search for the energy parameter, a set of well-chosen $E_{1,l}^{\alpha}$ up to $l = 3$ are already determined early in the SCF loop. For higher l, a fixed E_l can be kept. The introduction of $B_{lm}^{\alpha,\mathbf{k}+\mathbf{K}}$ also gives additional variational freedom to improve the SCF convergence and precision. In summary the LAPW basis is refined as

$$\varphi_{\mathbf{K}}^{\mathbf{k}}(\mathbf{r}) = \begin{cases} \frac{1}{\sqrt{V}} e^{i(\mathbf{k}+\mathbf{K})\cdot\mathbf{r}} & \mathbf{r} \in \mathrm{IS} \\ \sum_{l,m}[A_{lm}^{\alpha,\mathbf{k}+\mathbf{K}} u_l^{\alpha}(r', E_{1,l}^{\alpha}) + B_{lm}^{\alpha,\mathbf{k}+\mathbf{K}} \dot{u}_l^{\alpha}(r', E_{1,l}^{\alpha})] Y_m^l(\hat{\mathbf{r}}') & \mathbf{r} \in \mathrm{MT}_{\alpha} \end{cases} . \tag{4.33}$$

Since the $E_{1,l}^{\alpha}$ are fixed and the basis functions need to calculate only once, a single diagonalization of the Hamiltonian matrix will yield all band energies for a specific **k**.

The LAPW defined in Eq. (4.33) in the MT sphere is orthogonal to deep core states that are strictly confined in the MT sphere. For transitional metals, the deep-lying states like 1s, 2s, 2p, of course satisfy such a criterion. However, there are often the so-called semi-core states that are not too far from the valence band, say within 6 Ry of the Fermi level. These states often have non-negligible leakage out of the MT sphere. Therefore, the semi-core states are not strictly orthogonal to the LPAW bases of Eq. (4.33). Such states then show up as spurious energy eigenstates versus the k after diagonalization, known as ghost bands. This is exacerbated if the semi-core state is not far from the valence band; an improper choice of $E_{1,l}^{\alpha}$ actually emphasizes the semi-core energy rather than the valence, or hybridizes the two so that a proper separation of the ghost from the valence band is not possible. The early transition metals, lanthanides and actinides (or the f-block) are notable for this issue.

The so-called Local Orbitals (LO) are then introduced to complement the LAPW, which give us LAPW+LO.

4.4.3 LAPW with Local Orbitals (LAPW+LO)

Since core states are contained entirely in the muffin-tin sphere, they do not participate directly in chemical bonding with other atoms. Wavefunctions that substantially leak out of the MT sphere, are called *valence states*. Valence states participate in chemical bonds, forming energy bands, and are labeled by k, the crystal momentum. These valence states are expanded to the LAPW bases, which are labeled by k as well.

In contrast, semi-core states have some leaks, but do not form crystal energy bands. Alternatively one may say that these states only form a non-dispersive "band," despite the ghost bands that are spuriously generated in the calculation. In actuality semi-core states should not be labeled by k. Instead, these states can be accommodated with Local Orbitals, which are added as another type of basis functions to the LAPW basis set:

$$\varphi_{\alpha,LO}^{lm}(\mathbf{r}) = \begin{cases} 0 & \mathbf{r} \notin \mathrm{MT}_\alpha \\ \left[A_{lm}^{\alpha,LO} u_l^\alpha(r', E_{1,l}^\alpha) + B_{lm}^{\alpha,LO} \mathrm{u}_l^\alpha(r', E_{1,l}^\alpha) \right. & \\ \left. \quad + C_{lm}^{\alpha,LO} u_l^\alpha(r', E_{2,l}^\alpha)\right] Y_m^l(\hat{\mathbf{r}'}) & \mathbf{r} \in \mathrm{MT}_\alpha \end{cases} . \quad (4.34)$$

The LO basis function is site-specific, and notably confined to a corresponding MT sphere, being 0 in the interstitial region, and in the muffin tin spheres of other atoms. It is not labeled with the crystal momentum k, and is, hence, ideal to represent the semi-core states, which are also k independent. The addition of LO to the LAPW adds variational freedom and enforces the orthogonal relation between the valence and semi-core states. In the MT sphere of a specific atom α, the radial functions $u_l^\alpha(r', E_{1,l}^\alpha)$ and $\dot{u}_l^\alpha(r', E_{1,l}^\alpha)$ are numerically solved for, with appropriate boundary conditions, similar to the LAPW case, at the same energy parameter $E_{1,l}^\alpha$. The energy parameter $E_{1,l}^\alpha$ corresponds to the valence band. The other radial function $u_l^\alpha(r', E_{2,l}^\alpha)$ is separately solved for at the lower energy $E_{2,l}^\alpha$ that corresponds to the semi-core state. The separation of $E_{1,l}^\alpha$ and $E_{2,l}^\alpha$ removes the dilemma and ambiguity in energy selection of LAPW, giving rise to valence and semi-core states that have the same l and/or m. The three expansion coefficients (parameters) $A_{lm}^{\alpha,LO}$, $B_{lm}^{\alpha,LO}$ and $C_{lm}^{\alpha,LO}$ are to be determined. The boundary conditions require that both the value and the slope of ψ_{LO} be zero at the MT sphere. The other relation demands that ϕ_{LO} be normalized to unity. The three conditions then derives three equations, for which when solved uniquely determine the three expansion coefficients.

Since the LO enforces orthogonality between the valence and semi-core, ghost bands are largely removed, at a slight penalty of increased basis size. However, the price is rather small: a few LOs or numbers in the lower tens for small crystal unit cells versus a typical LAPW basis set of a few hundred functions. The slightly increased computational time is a small price to pay for the much improved accuracy, and therefore LOs are used in (L)APW most of time.

4.4.4 APW+lo

The success of LAPW and LO is the introduction of additional variation freedom to remove uncertainty in the energy parameter and to alleviate the tedious and time-consuming energy-searching process. Instead of taking the separate steps of linearization and Local Orbitals, there are alternative and potentially simpler approach. It turns out one can start with the APW basis and directly add an extra basis of the local orbital type. Being a LO, the only requirement is that the basis function is zero outside the MT sphere of a specific site, with an atomic numerical solution involving both the radial function and its radial derivative, or

$$\varphi_{\alpha,lo}^{lm}(\mathbf{r}) = \begin{cases} 0 & \mathbf{r} \notin \mathrm{MT}_\alpha \\ \left[A_{lm}^{\alpha,lo} u_l^\alpha(r', E_{1,l}^\alpha) + B_{lm}^{\alpha,lo} \dot{u}_l^\alpha(r', E_{1,l}^\alpha)\right] Y_m^l(\hat{\mathbf{r}}') & \mathbf{r} \in \mathrm{MT}_\alpha \end{cases}. \tag{4.35}$$

The two coefficients $A_{lm}^{\alpha,lo}$ and $B_{lm}^{\alpha,lo}$ are determined by normalization, and by requiring that φ be zero at the MT sphere. Since there are only two unknowns, there are no longer extra boundary conditions such as zero derivative. This local orbital set is different from the definition of Eq. (4.34) and for a differentiation is labeled as "lo." The lo basis then complements an APW-type basis, given by

$$\varphi_{\mathbf{K}}^{\mathbf{k}}(\mathbf{r}) = \begin{cases} \frac{1}{\sqrt{V}} e^{i(\mathbf{k}+\mathbf{K})\cdot\mathbf{r}} & \mathbf{r} \in \mathrm{IS} \\ \sum\limits_{l,m} A_{lm}^{\alpha,\mathbf{k}+\mathbf{K}} u_l^\alpha(r', E_{1,l}^\alpha) Y_m^l(\hat{\mathbf{r}}') & \mathbf{r} \in \mathrm{MT}_\alpha \end{cases}. \tag{4.36}$$

Compared to archetypical APW of Eq. (4.25), the energy parameters are both l- and α-specific. With this setting, the

E-searching of the APW iteration loop is no longer needed, augmented with the added variational freedom from the "lo"-basis.

The APW+lo is small in basis size, and as the result the SCF-loop is fast, further helped with a single diagonalization to generate the eigenvalues and eigenstates.

The optimization philosophy and techniques inherited from the APW, LAPW, LO and lo development can be further refined to give combinations like LAPW/APW+lo, APW+lo+LO, etc. These are optimized for various cases where the selection of basis gives unique benefits.

4.4.5 Correlation and LDA/GGA

In prior sections, we notice that LDA/GGA lacks the proper pair (or higher-order) correlation functions to account for strong-correlation effects at different translational and angular positions. However, the LDA/CGA still have certain aspects of correlation effects built-in, with various xc-functionals. When some variational degrees of freedom in the basis are combined with xc-functionals to solve for the eigen-problem with diagonalization of the Hamiltonian, some characteristically valuable effects of correlation are retained. In the following section, a standard APW+lo electronic-structure package will be utilized to investigate the band and doping behaviors of the archetypical high-T_c cuprate, LSCO. For the first time to the best of the author's knowledge, the calculation demonstrates that the doping of the CuO_2 planes is site-specific. More surprisingly, some CuO_2 plane can be doped with *electron* when far enough from the Sr dopant site. This counter-intuitive result is very difficult to understand from a simple picture of "throwing a jellium of electrons to the crystal by doping." Therefore, there must exist a mechanism, in the LDA/GGA (L)APW-lo calculation, to make a selective affine to some of the Cu-O_2 planes while avoiding others. This can be understood as the manifestation of a correlation effect.

4.5 The FLAPW Package WIEN2k

The program package WIEN2k (Blaha et al., 2001) allows to perform electronic structure calculations of solids using density functional theory (DFT). It is based on the full-potential (linearized) augmented plane-wave (L)APW + local orbitals (lo) method, one among the most accurate procedures that are available today for the band-structure and related calculations. Being full potential, WIEN2k is an all-electron scheme that uses all of the atomic electrons to generate the potential and charge density. It includes relativistic effects and has many additional features, to expedite the setup and execution of computation tasks. These include, for example, parallel tasking on both the coarse-grained and the fine-grained levels. Due to the new APW+lo basis set it is significantly faster (by up to an order of magnitude) than older LAPW methods. It also utilizes k-point parallelization in the most time consuming part of the SCF loop and thus is very applicable and efficient for speed-up in heterogeneous clusters. It further offers a suite of post-SCF analysis packages making possible the convenient investigation of charge density, band structure/characters and many other critical physical parameters.

A typical calculation is performed in parallel on a supercomputing cluster that comprises ~100 CPU cores. The computation cores are typically of the Intel Core type spanning from the Nahelem to the Haswell architecture, and/or of the AMD Ryzen type. For computation-demanding tasks, like the explicit example of $La_{1.875}Sr_{0.125}CuO_4$ to be given in the following subsection, a careful selection of parameters is essential for the completion of task. When optimized, this level of computation typically takes about 15 min for an SCF iteration when some tens of inequivalent k points are adopted. The convergence in both energy and charge should be achieved within 100 iterations/cycles. Such a computation should complete in about one working day, thus allowing researchers ample opportunities to adjust parameters and benchmark the results.

4.5.1 The Case of $La_{1.875}Sr_{0.125}CuO_4$

The mobile charge in a representative $La_{1.875}Sr_{0.125}CuO_4$ superconductor is studied here. The doping represents p=0.125 that is on the underdoped regime but, nevertheless, near the optimal doping of 0.15~0.16. Here the correlation in the real SC system is still strong, and therefore should form gapped structure near the antinodal region, as we have already demonstrated in Chapter 3. By working at this underdoped regime, we hope to shed lights on the strongly correlated phase. Also, the doping $p = 0.125$ gives a supercell not too large, which can still be handled by WIEN2k in a reasonable speed.

With the computation based on DFT, we do not expect to see energy gaps in the band structure: These are not our purpose to start with. Instead we take the DFT results as the indication of an electronic structure that should otherwise exist at a sufficiently high temperature, for example, at $\geq wT^*$. This is equivalent to hypothetically turning-off the strong correlation, and see if the cuprate still possesses some surprising physical features not typical of ordinary conductors. If surprises do exist then one can reason that these features may relate to strong coupling or superconductivity. Additional studies may then follow to investigate how these features may be affected if strong coupling and correlation are turned on.

The purpose of this study is to answer the question in the introductory part of this chapter: Where are electrons located in the p-type cuprate? In Chapter 3, a rigorous many-body treatment of the problem made us to believe that alternating copper layers are doped with opposite mobile charges. Still those studies were not completely material-specific, which leave gaps to a complete confidence. By using a full-potential electronic-structure package to study the mobile charge distribution, we can further answer the question with more details and clearer insights.

The unit cell that we choose for $La_{1.875}Sr_{0.125}CuO_4$ is given in Fig. 4.8 on both sides. It comprises 8 times the size of an primitive La_2CuO_4 cell (LCO). Therefore there are 56 atoms altogether, of which 34 atoms are nonequivalent. In terms of symmetry the space group is 99 or P4mm and therefore does not have centrosymmetry or an inversion center. As a result, the numericals are in complex numbers and take more computer memory and much more computation time to complete the numerical study, in comparison of a centrosymmetric case where the calculation is in real numbers.

The lattice constant adopted is 5.3435 Å in a and 26.3292 Å in c, corresponding to $a = 3.7784$ Å and $c = 13.1646$ Å for the La_2CuO_4. A number of different lattice parameters were also tried

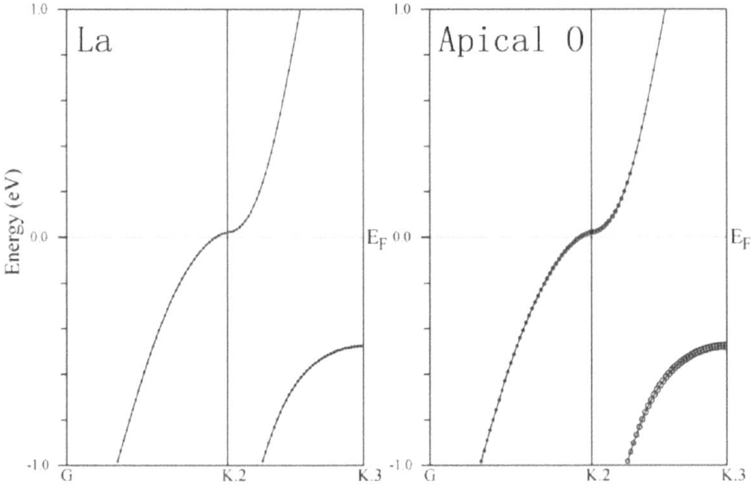

Figure 4.3 Band character plot of atoms, La and apical O near E_f, in LCO.

and the characteristic results are not sensitive to small variations of the lattice parameters chosen here.

Other important parameters that are relevant: RMT: O, 1.56; Cu, 2.0; La, 2.6; Sr, 2.6; all in bohr. Exchange-correlation potential: PBE; Energy separation of the core and valence electrons: −6 Ry; $RK_{\text{max}} = 7$; linear tetrahedral method to determine the Fermi energy; 56 nonequivalent k-points; Convergence criteria: 0.0001 Ry in energy and 0.001 e in charge. In the remaining part of the chapter, all energies are given in the Rydberg unit.

The SCF cycle typically converges around a working day, after which the band structure, electron density and differential density are investigated.

It is well known that the cuprate bands crossing the Fermi energy is mainly of the Cu-$3d_{x^2-y^2}$ type. The calculation of course confirms this. To be illustrative the band character plots for each inequivalent atoms are given side-by-side in Figs. 4.3, 4.4, and 4.5. Although these figures are calculated for the LCO, very similar elemental characters hold for the LSCO.

We can further see the band character plot of Cu d-subshell and the sub-orbital symmetry that is near the Fermi surface, as in Fig. 4.5.

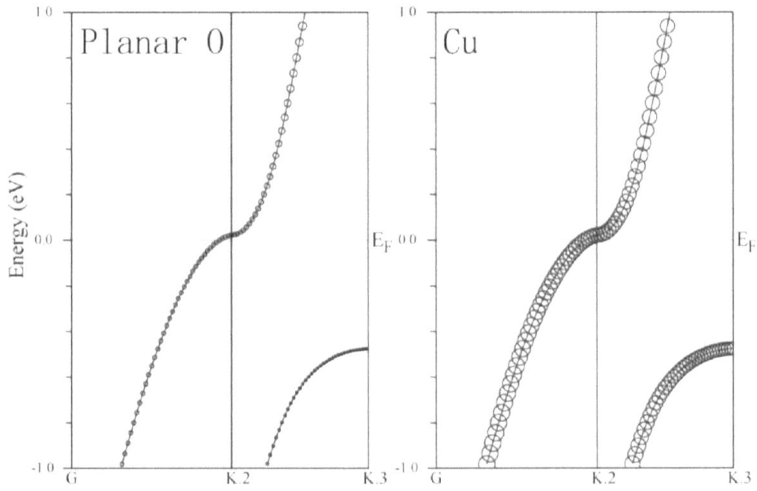

Figure 4.4 Band character plot of atoms, planar O and Cu near E_f, in LCO.

4.5.2 Difference Charge Density

WIEN2k is a full-potential package and hence offers convenience to calculate and plot either complete or partial charge density ρ within the crystal cells. The ρ can also be a valence charge density, or a difference charge density. It may include or exclude the core charge at will too. The purpose of our studies here though is to focus on the mobile charge: that is from the energy states near the Fermi surface. The relevant energy window can be reasonably set to that of thermal excitation scale, for example, 0.03 eV for T between 300 and 400 K, the typical temperature range just enough to break the pseudogap.

We computed the difference charge density at a variety of different energies below the Fermi level $\Delta\rho(E, r)$, where the E is measured from the E_F. An integration of the difference charge $\int \Delta\rho(E, r)dr$ over the relevant cell volume gives the supposed doping fraction (charge fraction) x. Here a rigid band doping behavior is assumed, in that the energy at the top of occupied electron state varies with change in the total electron number, without a characteristic or qualitative change of the band structure such as the formation or disappearance of energy gaps. Of course this is not true for LSCO at $x \ll 0.05$, but it should not matter for

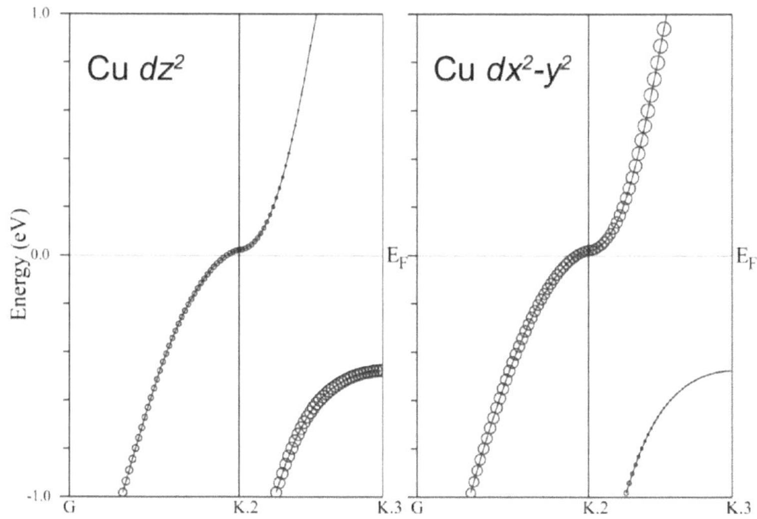

Figure 4.5 Copper character plot of atoms near E_f, in LCO.

our discussion here because we focus on $x \sim 0.125$. The procedure is performed at a couple of different E, for both La_2CuO_4 and $La_{1.875}Sr_{0.125}CuO_4$. A plot of x-E can then be generated. The plot indicates the variation of the Fermi level at different doping. The results are given in Figs. 4.6 and 4.7.

The slope of the curve in both figures is proportional to the density of states. The DOS decreases when it is doped away from the undoped Fermi energy, and this is clearly reflected in Fig. 4.6. The average slope for LCO is 44.715 Ry^{-1} while after $x = 0.125$, it is 31.114 Ry^{-1}.

The average slope for LSCO is 23.036 Ry^{-1}, evaluated at x up to 0.594. But immediately after $x = 0.125$, it is instead 27.425 Ry^{-1}, closely matching 31.114 Ry^{-1} for LCO at the same x range. This gives us added confidence that the numerical results are self-consistent.

So far, all of our discussion are based on a rigid-band picture of the doping, which assumes a change in the Fermi energy. This technique should be acceptable for average physical quantities such as the E_F and average doping x. However, it could very well neglect important correlation effect. We know that after all one has to replace an original atom, in this case, a La with a Sr atom to effect

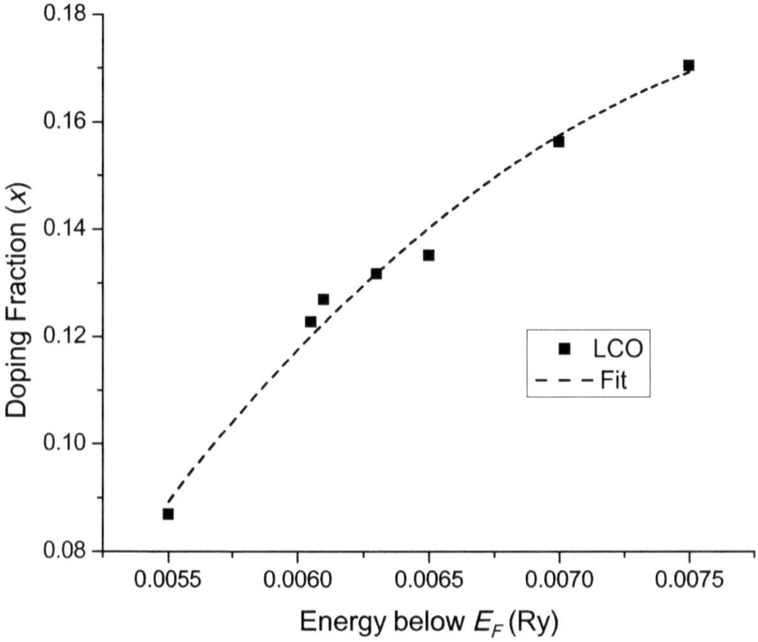

Figure 4.6 LCO doping fraction x as a function of energy below the undoped Fermi Energy.

the real doping. This replacement of atom hints at subtracting charge contribution of the parent matrix (LCO) from the doped one to see a real doping effect. In doing so, we expect to further reduce the background and average effects to make the correlation effect more prominent. Then the recipe is clear: We calculate first a difference charge density $\Delta\rho_{\text{LSCO}}(-0.002, r)$. Corresponding to this energy window, a difference density for LCO is calculated at $x \geq 0.125$ between $E_1 = 0.00605$ and $E_1 = 0.0075$ that is $\Delta\rho_{\text{LCO}}(-0.0075, r)$ $- \Delta\rho_{\text{LCO}}(-0.00605, r)$. The slightly smaller energy windows in the latter is to account for higher DOS in LCO. Then one subtract the LCO (background) charge density from the LSCO one:

$$\Delta\rho_d(r) = \Delta\rho_{\text{LSCO}}(-0.002, r) \\ - [\Delta\rho_{\text{LCO}}(-0.0075, r) - \Delta\rho_{\text{LCO}}(-0.00605, r)], \qquad (4.37)$$

where the subscript d denotes the "doped" to emphasize the doped structural and other correlation effects.

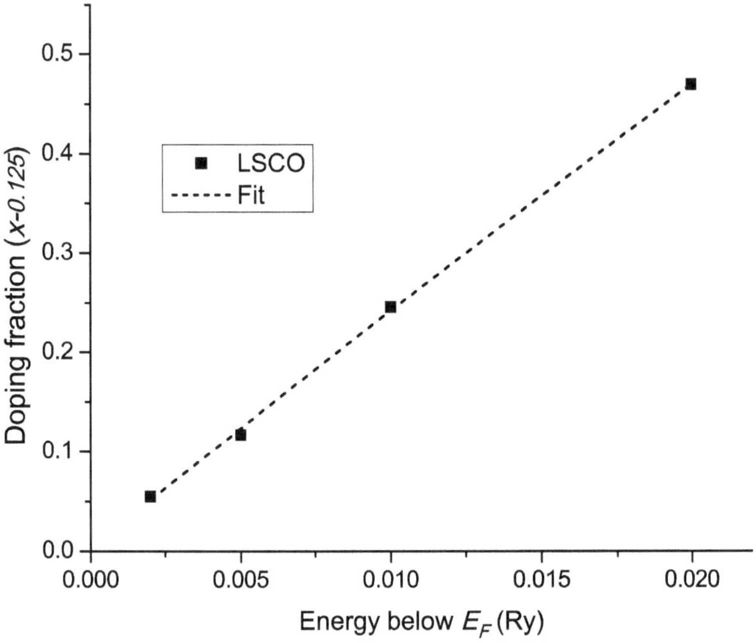

Figure 4.7 LSCO doping fraction x as a function of energy below Fermi Energy at $x = 0.125$.

$\Delta\rho_d(r)$ is a three-dimensional function of r, whose isosurface can be plotted in the crystal cell in 3D plots. The two isosurfaces at 0.0002 and −0.0002 charge/bohr^3 are plotted left and right in Fig. 4.8. Both isosurfaces are golden colored. The isosurface at a positive value indicates positive charge and thus holes. In contrast, a negative valued isosurface then indicates negative charge or (band) electrons.

The "surprising" result is that the LSCO's CuO_2 layers are alternatively doped with either electrons or holes! The presence of mobile electron in LSCO is actually predicted nearly 20 years ago based on very limited information (Luo, 2001a), in this sense it may not be considered as a surprise after all. However, it certainly should come as a surprise to the notion that still refuses to acknowledge the existence of negative charge carriers in p-type cuprates.

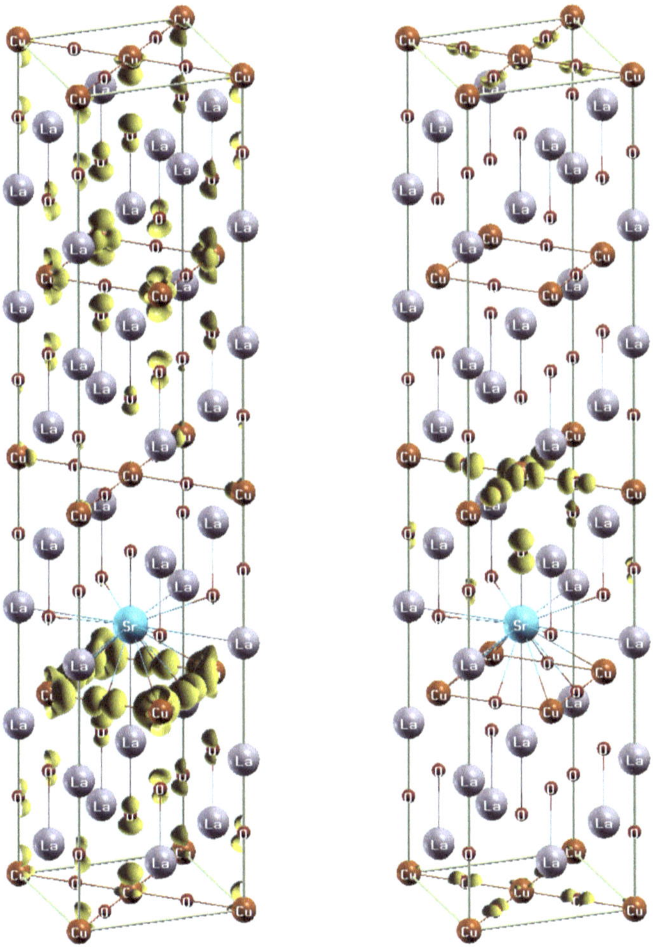

Figure 4.8 Location of electrons in LSCO. Left: Positive charge, and hence the hole density contour (of 0.0002 charge/bohr^3) in golden color. The holes are most significant in the CuO_2 plane nearest the Sr dopant. Cu-$d_{x^2-y^2}$ holes are also significantly present in the fourth CuO_2 plane nearest the Sr. Right: Negative charge, and hence the electron charge contour (of −0.0002 charge/bohr^3) in golden color. The electrons are mostly in the CuO_2 plane second nearest to the Sr dopant. There are also significant O-$p_{x,y}$ electrons in the top or bottom CuO_2 layer, which are believed to be at least mobile in-plane. The holes in the top or bottom CuO_2 layer are of Cu-d_{z^2} type. How the Cu-d_{z^2} hole may contribute to charge dynamics remains an open question as of this writing.

The characteristic alternating doping of the CuO_2 layers, featured in Fig. 4.8, is robust against reasonable variation of the energy windows across the various doping levels that have been tested so far, ranging from 0.05 to 0.20. The prominent highlights of the computational finding are summarized below.

Comparing the band-character of Figs. 4.3, 4.4, and 4.5, and the charge orientation of Fig. 4.8, the in-plane transport is dominated by Cu-$d_{x^2-y^2}$-O-$p_{x,y}$ holes and electrons. These orbitals have the right energy scale, and the proper symmetry to first contribute to normal-state conductivity and then superconductivity. There are also Cu-d_{z^2} and apical O-p contribution at E_F and should behave as a mobile charge, but these are more likely to contribute to the c-axis transport. More quantitative investigations on Cu-d_{z^2} and apical O-p are needed to clarify their possible role.

The in-plane mobile charge are centered around the Cu and planar O ions. This is verified by changing the isosurface value, making it expand or shrink: Still the in-plane holes and electrons are concentrated there. The holes are more of a Cu-$d_{x^2-y^2}$ type, while the electrons have substantial O-p fraction.

The CuO_2 plane nearest the Sr dopant is hole-doped. The next (second) nearest CuO_2 is *electron-doped*. Leaving the Sr further in the direction of the first e-doped CuO_2, the next copper-oxide plane is hole doped, and vice versa. This (nearly) alternating doping behavior is *profound*. It removes any doubt that the electron doping might just be an accident purely because it is further away from the alleged "hole"-only dopant, the Sr site. It corroborates the results in Chapters 1 and 3 from classical electrostatics, electrodynamics and the many-body strong-correlation theory.

The hole concentration in different p-doped layers varies, which should not come as much of a surprise considering the different distance from the dopant. Similar behaviors exist for electron concentrations in the n-doped layers too.

$\Delta\rho_d(r)$ of Eq. (4.37) is numerically integrated to give a total difference charge of 0.010524. When just integrating the positive part of $\Delta\rho_d(r)$, the positive charge is 0.013721, which corresponds to the difference hole charge. Similarly the difference electron charge is integrate to −0.003197. This gives a hole:electron ratio of 4.29. This ratio is still close to the optimal doping ratio of 3:1 that we predicted a long time ago (Luo, 2000b, 2001a,b). And the off-ideal difference may actually explain why the T_c in LSCO is not as high as 90 K.

Ambipolar doping of metal oxides is often encountered in the solid state. The CuO_2 plane's being doped as both *n*- and *p*-type should not come as a surprise after all, although the detailed mechanism is well worth investigation. The physics is rich. For example, internal electrostatic field makes the chemical potential (Fermi level) different with respect to local band states, and therefore different doping behaviors exist across the layers. Further, the mixed valence of transition metal ions, Cu included, makes them susceptible to losing electron (hole) again even if once doped with electron (hole). This can be further coupled with the fact that the oxygen ion should have a mixed valence as well. This is proven by spectroscopic evidence upon the MIT transition in cuprates, where the holes are actually first introduced on the oxygen sites rather than the copper ones. This is logical to understand from an ion physics perspective. Formal valences of 2+ and 2− are, respectively, assigned to Cu and O in cuprates when undoped, according to the conventional yet oversimplified chemical designation. Some early models assume upon doping, Cu^{2+} is further ionized to $Cu^{(2+x)+}$. This is simply not plausible. First, unlike in vacuum and with simple ionics, it is rather difficult and actually not right to assign a formal integer valence to an ionlike species in the solid state: The actual charges are shared across boundaries. Without definition of a "sphere" and its size, it is too vague to define a true valence, at least for atoms in solids with significant co-valence. Second, the ionizaiton energy of Cu^{2+} is so much higher than that of a O^{2-}, that O^{2-} will actually be first ionized, if it indeed exists in vacuum, in the first place. But we know it does not exist, since the second electron affinity of oxygen is positive, in vacuum. In solid states, this means hole will preferably first dope to O^{2-} site. And not surprisingly this is just what has been seen in experiments. Note that we are still borrowing notation in the conventional sense, but number like 2+ should not be taken too literally. With the hole added, O^{2-} becomes O^-. O^- in a solid environment, especially when surrounded by positive ions, may still have appreciable electron affinity. This then explains the electron doping of the O-*p* orbitals as seen in the LSCO charge mapping of Fig. 4.8.

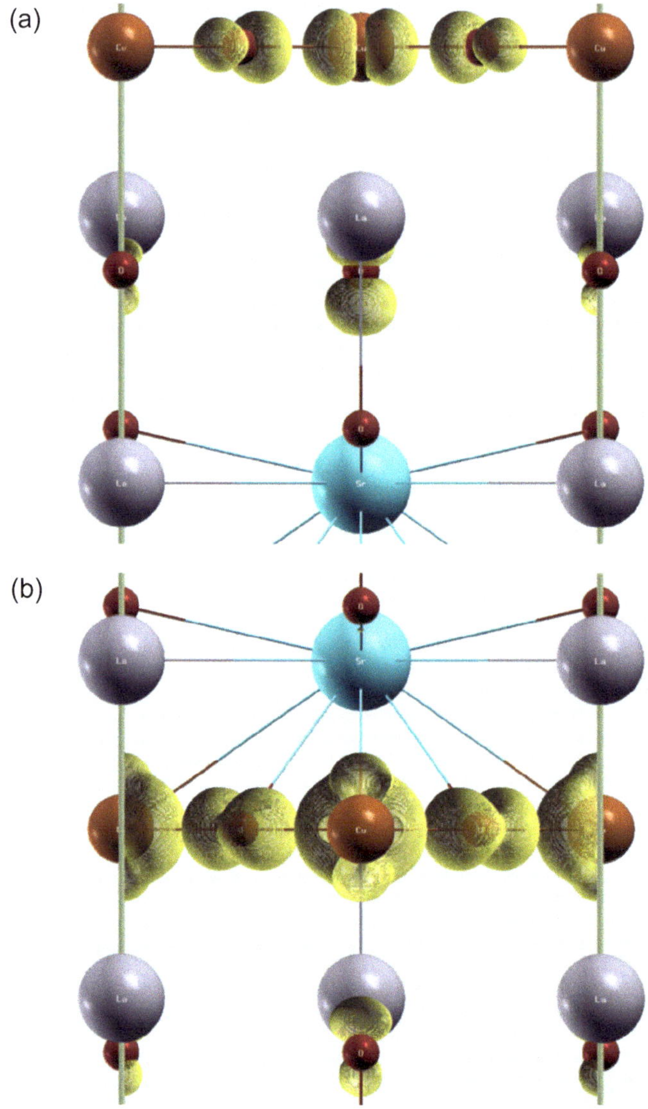

Figure 4.9 Close-up of mobile electrons and holes in LSCO. (a) Upper plate: electron charge contour in golden color. The electrons exist most significantly in the CuO_2 plane second nearest the Sr dopant. (b) Lower plate: hole charge contour in golden color. The holes are most significantly present in the CuO_2 plane nearest the Sr dopant. Brown: copper; grey: lanthanum; red: oxygen; cyan: strontium.

Some cautions need to be mentioned here. The ABAB ... alternating *e*-*h* doping configuration is based on an ideal doping model where dopant atoms form a perfect lattice. In reality, there must be a certain degree of randomness. In that case one may expect variations like ABABBA... and so on. Still charge segregation is expected as a general trait of the superconducting cuprates, and this can be interpreted as the so-called heterogeneous charge localization (Pelc et al., 2019).

If we consider the realistic random distribution of dopant atoms in the cuprate, it becomes apparent that even for a single CuO_2 layer, it can be doped into electron-rich and hole-rich domains of a few to a few tens angstroms across. In this case, even if there is no interlayer coupling, the excitonic Coulomb interaction should still exist among electrons and holes, which likely generates SC with non-trivial T_c, though at much reduced magnitudes.

4.6 Conclusion

The aim of the chapter is to locate the "alleged" mobile electrons in *p*-type cuprates. We started with a state-of-the-art computational condensed matter physics tool, the WIEN2K package. Although the length of this chapter is not very long, at some 20 pages, the search started about 17 years ago, when the author was first introduced to the FLAWP computational techniques. Between 2007 and 2008, the first indications of electron-doping of the CuO_2 layers emerged. Accurate and quantitative studies were still not possible by then though.

Over the years, our understanding of correlation deepens at all levels: at the few-body-level of 3 or 4 electrons/holes, at the many-body BdGE/Green's function level, and at the LDA/GGA level. We carefully studied the internal working mechanism of the WIEN2k package, the functionals that it employs, the strength and weakness of each functionals in coping with correlation effects, and how to avoid their weakness while playing to their strength. Some of the research and finds were reflected in the early sections of the chapter. Then appropriate procedures and programming ensued, which developed into the difference charge technique adopted in the latter part of the chapter.

By early 2017, computational hardware available to the author has matured to the stage that high-precision numerical studies on underdoped LSCO can be systematically carried out. In the following time of about one year, precise differential charge densities were mapped. And then finally in early Spring of 2018, the charge ratio of hole to electron was finally obtained with high confidence.

In summary, mobile electrons exist in CuO_2 planes too, phase-separated from the hole-doped ones. This result verifies our prior theory that charge-transfer happens between neighboring layers and this gives rise to the excitonic Coulomb pairing interaction.

Chapter 5

Superconducting Properties

5.1 Introduction

The previous chapters have focused on (1) the few-body approach of tetron to solve for the excitonic Coulomb pairing in the real coordinate space; (2) transports and electron–hole phenomenologies; (3) many-body field theoretical approach to the interlayer Coulomb interaction and related pseudogap phenomena; and (4) the DFT-based electronic-structure calculation to corroborate, at least at the weak correlation limit, the many-body theoretical results. The match of our theory to many experiments gives us confidence that the ECM pairing model is in the right direction. In this chapter, written by Dr. Nie Luo, the focus is on the superconductive state.

5.2 Penetration Depth, Uemura Relations, and the Supercarrier Mass

The real-space pairing model (pre-formed pairs or tetrons) has been shown to explain many experiments in the normal states. It is a worthy test to investigate how the ECM applies to the

High-T_c Superconductivity: The Excitonic Coulomb Perspective
Nie Luo and George H. Miley

ISBN 978-981-4800-57-0 (Hardcover), 978-0-429-27870-9 (eBook)
www.jennystanford.com

superconductive phase. A critical question to any serious theory is: Does it have some prediction power?

We start with the penetration depth, which is important for both theory and applications. For one, it is where the Uemura relation was discovered, which gives a strong support to the concept of preformed pairs and real-space pairing, for a range of non-conventional superconductors.

The *ab*-plane (planar) London penetration depth is given by

$$\lambda_{L,ab} = \left(\frac{m^*_{s,ab}}{\mu_0 n_s q^2}\right)^{1/2} = \left(\frac{m^*_{s,ab} d}{\mu_0 n_{s2D} q^2}\right)^{1/2}, \tag{5.1}$$

where $m^*_{s,ab}$ is the planar effective mass of the supercarrier, with subscript s indicating the supercarrier. q is the *charge* of the *supercarrier*, rather than the elementary charge. Note that this is very important, and can be checked from the derivation of the λ from classical electrodynamics (Tinkham, 2004). In the expression of λ_L, the three variables n_{2d}, q^2 and m^* are entangled together as in $n_{2d}q^2/m^*$, and many works in the literature are refrained from giving a definite answer to m^* due to such a complication. However, for the ECM, the tetron pairing model dictates an effective supercarrier mass that is at least 4 times the band mass of hole (electron). Therefore, we need a detailed investigation.

A survey on the current status of understanding on supercarrier mass turns out to be not very conclusive. A few authors may have found significantly heavy mass in the supercarrier (Kristoffel et al., 2006), but they typically refrained from giving a definite answer. For example, a mass of $\sim 10m_e$ could have been discovered in their data, but this is not easy to understand from a traditional Cooper pair which only doubles the band mass of a hole (or an electron), with the latter typically found to be $2m_e$ for optimally doped cuprates. Therefore, the significant extra mass was often attributed to other effects, such as the mass renormalization.

A potential pitfall in utilizing Eq. (5.1) is from the q. A survey of the literature reveals that different q are employed inconsistently across the field. Some assumed $q = 2e$ (Hirsch and Marsiglio, 1992b), while others $q = e$ (Puźniak et al., 1996; Puźniak, 1997), or it is kept vague (Orlando et al., 2018). This is a *huge* difference. All other things equal, the two different interpretations should give m_{ab} differing by a factor of 4. The average supercarrier mass as given in Ref. (Puźniak et al., 1996; Puźniak, 1997) for the Hg-1201, 1212, 1223 is 1.625 m_e. This is a value way too small due to neglecting the $q = 2e$ factor in Eq. (5.1). If multiplied with 4, this gives 6.5 m_e, which is a more reasonable value matching other constraints.

Now we proceed to clarify on this issue. We notice that even though there are some un-certainties in n_s, at $T \approx 0$ and for OP, the n_s is typically half of the free carrier density. The standard doping model puts $0.15 \sim 0.16$ holes per basel unit cell at OP and this gives $n_{2d} \sim 0.16/a^2$ at the highest T_c. Note that n_{2d} overestimates n_{s2d} since $n_{s2d} = n_{2d}/2$. For our doping model, $n_h = 1.5n_{2d}$, $n_e = 0.5n_{2d}$ and still we get the same n_{s2d}, which is half of $0.16/a^2$ for n_{s2d} in OP cuprates. We adopt $a = 3.787$ Å, $d = 6.64$ Å and these parameters are also similar to our model cuprate. For λ in LSCO, we use 260 nm from (Bennemann and Ketterson, 2008). Then we obtain

$$m^*_{s,ab} = 8.03m_e. \tag{5.2}$$

Notice that this serves at a rule-of-thumb SC mass for cuprates at OP. Some of the highest critical-temperature Tl-12(n-1)n or Hg-based compound should have reduced penetration depths, nearly 200 nm. However, the reduction in λ_L is also compensated by the fact that the average interplane distance is too reduced in Tl-12(n-1)n and most Hg-based cuprates, to the order of 5 Å. This tends to restore the effective SC mass $m^*_{s,ab}$ back to $\sim 8m_e$. For Y-123, its Cu-chain gives higher carrier density and this should explain the small λ_L, which suggests that the in-plane mass is similar to that of LSCO.

The chapter by Carlson et al. (Carlson et al., 2004) in (Bennemann and Ketterson, 2004) gives a table summarizing λ_L in many cuprates. The Bi-series have λ_L fairly scattered from 170 up to 300 nm. These average to $\sim$240 nm but for the most OP sample, $\lambda_L = 270$ nm (Anukool et al., 2009; Nideröst et al., 1998; Prozorov et al., 2000; Weber et al., 1991). The Tl-22(n-1)n series typically have λ_L above 200 nm even when overdoped and reach near 250 nm when optimally doped (Kim et al., 2002; Niedermayer et al., 1993; Uemura et al., 1993). From such a penetration depth, a $m^*_{s,ab}$ of near or above $8m_e$ is obtained for the Bi-22(n-1)n and Tl-22(n-1)n series.

Therefore, $m^*_{s,ab} \approx 8m_e$ should be a good estimate for the supercarrier mass across board for the cuprates. One probable exception is the Hg-12(n-1)n series, which seem to indicate $m^*_{s,ab} \approx 6.5m_e$ for some of the highest T_c samples under pressure. But this is understandable from a reduced band mass at pressure. It is well known that hydrostatic pressure tends to increase band energy dispersion and thus decreases band mass. If, for example, the bare band mass of the Hg-12(n-1)n series is reduced to 1.625 m_e then the total SC mass of $6.5m_e$ is well understood within the tetron model.

There is yet another reason why a supercarrier mass between 1.5 and $\sim$2 m_e does not make much sense: the single particle band mass in cuprates. The chapter by Tanner and Timusk (Tanner and Timusk, 2013) in the Series edited by Ginsberg (Ginsberg, 1992) summarizes optical mass in many cuprates. The mass of a single hole is found from the optical reflectance to be between 1.6 $\sim$2 m_e, for OP samples at the low-frequency (DC) and low-T limit. These low-frequency and low-T limits comprise what are relevant and critical to SC, and hence reveal the band mass range critical to superconductivity (Jehl et al., 1992; Romero et al., 1992a; Schlesinger et al., 1990a; Uchida et al., 1991). These are mostly from optical conductivity experiments. If a single hole has this mass (1.6 $\sim$2 m_e), then it is very unlikely that the pair mass will still have a similar value, let along to say the four-body tetron. Therefore, a careful review of λ_L actually removes prior inconsistency among theoretical interpretations of its relation to the effective mass of the SC condensate. It further provides direct evidence to the heavy mass of the supercarrier. The λ_L experiments, hence, strongly support the four-body tetron as the supercarrier of the cuprate superconductivity.

5.3 Pairing Symmetry

The phase-sensitive probes of cuprate pairing symmetry were reviewed by Van Harlingen (Van Harlingen, 1995), and by Tsuei and Kirtley (Tsuei and Kirtley, 2000). While it is widely accepted that the SC order parameter has a predominantly $d_{x^2-y^2}$ symmetry

(Annett et al., 1996), there are some very disheartening results that are difficult to understand within the current theoretical framework based on the BCS and Ginzburg–Landau theory. The *c*-axis twist junction-type experiments pose particular puzzling results to understand (Klemm, 2005; Latyshev et al., 2004; Li et al., 1999).

We interpret this apparent contradiction actually as a manifestation of four-body local pairing. Within the conventional wisdom and under the assumption that the cuprate Cooper pair is still a two-body entity, the two-particle pairing wavefunction or the superconducting order parameter can be expanded into eigenfunctions that form the irreducible representation of the Hamiltonian symmetry group. Therefore, experimental evidence such as the absence of thermodynamic transitions likely excludes symmetry mixing like the $d_{x^2-y^2}+is$ type. However, we should point out that this preceding argument is but only an approximation, *if and when* additional types of spontaneous symmetry breaking occur besides the U(1). Spontaneous symmetry breaking of the latter is associated with the emergency of superconductivity in the BCS framework.

Let us understand this from a very simple example. In a space free of external potential, one throws in protons and electrons of equal number. The system's total Hamiltonian has continuous translational and rotational symmetry. But we know that at low temperature and a not-very-low density, the ground state of such a (protons+electrons) system is a hydrogen ice, definitely breaking the continuous translational symmetry, at least in the inter-particle coordinates $\mathbf{r}_{ij}$. The culprit is the spontaneous symmetry breaking caused by the strong correlation of electrons and protons that forms bound states. Therefore, the inter-particle wavefunction is no longer a faithful irreducible representation of the original overall Hamiltonian.

The lesson learned from the example of hydrogen ice is that: Due to the strong electron correlation, the pair wavefunction of a cuprate SC is not necessarily the representation of the C_{4v} crystal point group. Now if we consider the complication from the four-particle tetron the symmetry breaking effect is even more pronounced. For example, an exciton on one side of two electrons (holes) definitely causes orthorhombic-type distortion to the crystal field, and then d-wave and s-wave mixing can be a natural result.

5.3.1 ODLRO and Other Orders

In previous chapters, we have seen strong evidence for local pairing, pre-formed pairs, and competing orders (Ekino et al., 2011). Here we need to cast the description in a many-body formalism and make a connection to the few-body physics via the four-body (8-point) Green's function, defined in the natural units as

$$i^4 G(x_1, x_2, x_3, x_4; x_1', x_2', x_3', x_4') \\ = \langle\psi|\mathcal{T}[c(x_1)c(x_2)c(x_3)d(x_4)c^\dagger(x_1')c^\dagger(x_2')c^\dagger(x_3')d^\dagger(x_4')]|\psi\rangle, \tag{5.3}$$

where $c(x)$ and $c^\dagger(x')$ are annihilation and creation operators of a hole (electron) at x and x'; and $d(x)$ and $d^\dagger(x')$ are those for an electron (hole). $\mathcal{T}$ is the time ordering operator. The subscripted x and x' are four-vectors of space and time. The four-body Green's function is related to the four-particle reduced density matrix $\rho(1, 2, 3, 4; 1', 2', 3', 4')$ in the coordinate representation (Yang, 1962) by proper time arrangements such as $t_1' \to t_1 + 0^+ = t_1^+$ and so on. Here we have also adopted abbreviation: For example, 1 is a shorthand for x_1 and $1'$ for x_1'.

The 8-point Green's function of Eq. (5.3) can be evaluated to include the following non-zero terms by using Wick's theorem:

$$iG(x_1, x_2, x_3, x_4; x_1', x_2', x_3', x_4') \\ = \cdots + \langle\psi|\mathcal{T}[c(x_1)c^\dagger(x_1')]|\psi\rangle\langle\psi|\mathcal{T}[c(x_2)c^\dagger(x_2')]|\psi\rangle \\ \times \langle\psi|\mathcal{T}[c(x_3)c^\dagger(x_3')]|\psi\rangle\langle\psi|\mathcal{T}[d(x_4)d^\dagger(x_4')]|\psi\rangle + \cdots \\ = \cdots + G(x_1, x_1')G(x_2, x_2')G(x_3, x_3')G(x_4, x_4') + \cdots \tag{5.4}$$

In Eq. (5.4), the four-particle Green's function is evaluated with the strong-correlation interaction turned on, at T_c or lower temperature. The single particle Green's function $G(x_1, x_1')$ is defined by $iG(x_1, x_1') = \langle\psi|\mathcal{T}[c(x_1)c^\dagger(x_1')]|\psi\rangle$. Then $G(x_1, x_1')$ or similarly $G(x_2, x_2')$ can be shown as the well-known Nambu–Abrikosov–Gor'kov matrix propagator, to be better rewritten as $\mathbb{G}(x_1, x_1')$ or

$\mathbb{G}(x_2, x_2')$. Its G_{11}, G_{22} components are given by

$$G_{11}(x_1, x_1') = \int\int \frac{u(\mathbf{k})u^*(\mathbf{k})}{\omega - \epsilon(\mathbf{k}) - i\delta} + \frac{v(\mathbf{k})v^*(\mathbf{k})}{\omega + \epsilon(\mathbf{k}) - i\delta}$$
$$\times \exp[i\mathbf{k} \cdot (\mathbf{r}_1 - \mathbf{r}_1') - i\omega(t_1 - t_1')]d\mathbf{k}d\omega,$$
$$G_{22}(x_1, x_1') = \int\int \frac{u(\mathbf{k})u^*(\mathbf{k})}{\omega + \epsilon(\mathbf{k}) - i\delta} + \frac{v(\mathbf{k})v^*(\mathbf{k})}{\omega - \epsilon(\mathbf{k}) - i\delta}$$
$$\times \exp[i\mathbf{k} \cdot (\mathbf{r}_1 - \mathbf{r}_1') - i\omega(t_1 - t_1')]dkd\omega, \tag{5.5}$$

and the off-diagonal terms $F(x_1, x_1') = G_{12}(x_1, x_1')$ and $F'(x_1, x_1') = G_{21}(x_1, x_1')$ are similarly evaluated.

The formation of bound states requires that the Green's function like $G(x_1, x_1')$ actually be a matrix propagator of $\mathbb{G}(x_1, x_1')$, e.g., 2×2 matrix counting the Nambu space, or 4×4 with additional spin indices. The product of $G(x_1, x_1')G(x_2, x_2')$ in Eq. (5.4) is actually a direct product of matrices, or $\mathbb{G}(x_1, x_1') \otimes \mathbb{G}(x_2, x_2')$. This matrix direct product describes the two electrons in forming a bound state of a Cooper pair. Given the lowest-energy trial wavefunction of Eq. (1.105) from Chapter 1, a pairing gap symmetry of $\Delta_k = \Delta\cos(2\theta_k)$ is derived for the two-particle Green's function $\mathbb{G}(x_1, x_1')\otimes\mathbb{G}(x_2, x_2')$. k here refers to the relevant wavevector defined by coordinates r_1 and r_2. This gives a d-wave pairing, especially with additional consideration of the cuprate crystal field.

Meanwhile, the product $G(x_3, x_3') \otimes G(x_4, x_4')$ describes the electron–hole part of the tetron which should give an s-wave in the *inter-particle* coordinate (i.e., r_{34}), again from variational trial wavefunction of Chapter 1. The center-of-mass gap though is largely determined by the crystal field symmetry as we have seen in Chapter 3, i.e., the pseudogap solved for by using the generalized BdGEs. The generalized BdGE solution of Eq. (3.88) describes one electron (or hole) in the correlated motion of two electrons (or holes) mediated by one hole (electron), which gives the pseudogap. The hole is largely dragged by two electrons since the trion is quite stable and thus the center-of-mass (COM) motion of the two electrons are pretty much the COM motion of the electron–hole pair too, which is also approximated by the single-particle Green's function of Eq. (3.88). In the relative coordinate though the inner motion is an s-wave due to the strong attraction of the e-h pair. Therefore, the Fourier transform of Eq. (3.88) and similar expressions give the matrix elements of $\mathbb{G}(x_3, x_3')$. $\mathbb{G}(x_4, x_4')$ is obtained alike, and the direct product $\mathbb{G}(x_3, x_3') \otimes \mathbb{G}(x_4, x_4')$ has substantial s-symmetry in r_{34}.

Therefore, when the wavefunction of the tetron few-body system is generalized to the many-body formalism, the mixed symmetry in the Green's function is manifested via terms like $\mathbb{G}(x_1, x_1') \otimes \mathbb{G}(x_2, x_2') \otimes \mathbb{G}(x_3, x_3') \otimes \mathbb{G}(x_4, x_4')$.

Now the four-particle Green's function with terms like $\mathbb{G}(x_1, x_1') \otimes \mathbb{G}(x_2, x_2') \otimes \mathbb{G}(x_3, x_3') \otimes \mathbb{G}(x_4, x_4')$ can be reduced into density matrix $\rho_(1, 2, 3, 4; 1', 2', 3', 4')$ by proper alignment of time indices such as $t_1' \rightarrow t_1^+ \ldots$ Since $\rho_(1, 2; 1', 2')$ as given by the time limiting case of $G(x_1, x_2; x_1', x_2')$ has ODLRO, by using the theorems of (Yang, 1962), we know for sure that $\rho_(1, 2, 3, 4; 1', 2', 3', 4')$ has ODLRO too.

We have thus proved that the many-body system of tetrons as a whole will become superconductive if its charged constituent has acquired ODLRO.

The symmetry built in $\rho_(1, 2, 3, 4; 1', 2', 3', 4')$ and $\mathbb{G}(x_1, x_1') \otimes \mathbb{G}(x_2, x_2') \otimes \mathbb{G}(x_3, x_3') \otimes \mathbb{G}(x_4, x_4')$ also explains the otherwise puzzling c-axis twist-junction tunneling experiments. Because of identical particles in indices x_1,x_2,x_3 (x_1',x_2',x_3' too), any electron in a Cooper pair, say x_1, has a significant amount of probability to be in the excitonic state of $\mathbb{G}(x_3, x_3') \otimes \mathbb{G}(x_4, x_4')$ as well. Since the latter is in an s-state, x_1 can couple first to x_3 of next layer and then couple back to x_2. Therefore, the Cooper pair of different layers, although in a nominal $d_{x^2-y^2}$ state, still can pair-tunnel to give an critical current that is in-sensitive to the twist-angle. The results in Ref. (Klemm, 2005; Li et al., 1999) are therefore accounted for.

A. A. Abrikosov attempted to reconcile on conflicting reports of the order parameter in the cuprate by adopting mixed interactions: long-range attraction and short-ranged repulsion (Abrikosov, 1995). Although phonon was still resorted in that approach, it, nevertheless, demonstrates why balancing attraction and repulsion could resolve the puzzling issue of order parameter symmetry in cuprates. The mixing of attraction and repulsion is exactly a built-in feature of the tetron and the associated excitonic Coulomb pairing mechanism.

There is yet another way to prove the existence of ODLRO and thus SC in the ECM. The two coupled CuO_2 layers can be consider as a bipartite lattice. And we have already proved that the effective attractive force of Chapter 1 exists for charges between the layers. Then according to C. N. Yang per (Yang, 1989), ODLRO should exist in such a bipartite lattice, i.e., the two coupled CuO_2 layers, given proper controlling parameters, and suitable "pairing" configurations.

5.4 The Matthias Rules

In the 1950s, B. T. Matthias developed a set of empirical rules outlining the superconductivity in the periodic table of elements, which are later conveniently referred to as Matthias rules (shorthanded as MR hereafter).

We should emphasize that Matthias rules are only approximate and some are already obsolete with the advent of high-pressure fixtures and transition metal oxides. Still some MRs continue to offer good insights when the average valence electrons of the SC compounds are counted, and are still being utilized and studied today (Conder, 2016; Slocombe et al., 2015).

MR1 states that for chemical elements, alloys or compounds to be superconductive, it takes $2 \le n \le 8$, or $2 \le n \le 9$ as given in (Matthias et al., 1963), where n is the number of valence electrons per atom. MR1 actually applies only to the d-block transition metals and alloys. When generalized to the full periodic table, it becomes MR3, with the T_c proportional to the empirical $T(n)$ function given by Fig. 5.1.

MR3 is conveniently understood from the excitonic pairing mechanism. Non-equal numbers of electron and hole in the system helps to reduce the Coulomb repulsion. There are cases where the excitonic Coulomb interaction is not strong enough to lead to pairing by itself, still it helps other pairings like the phonon mechanism.

When the valence is less than 2, which is the case for Group 1A or 2A, the conduction electrons are of s-orbital. They are most

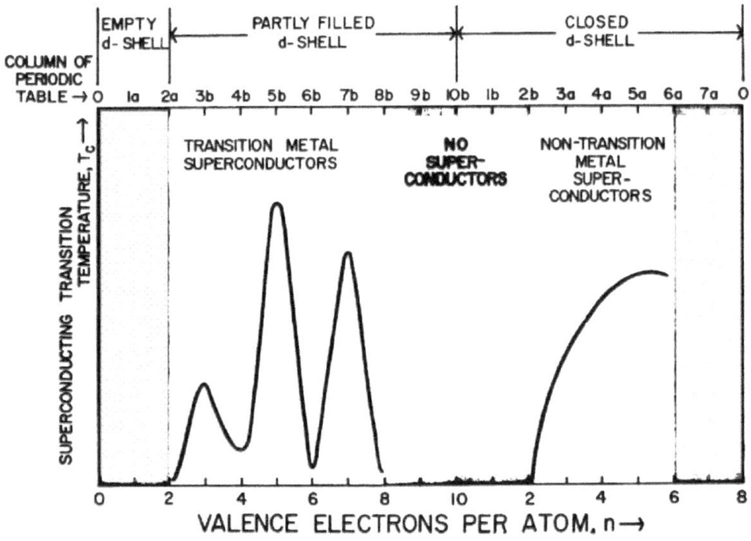

Figure 5.1 The empirical function $T(n)$ in the Matthias rule.

likely of the electron type only. There lacks attractive force to reduce the Coulomb repulsion, and therefore no (or negligible) superconductivity exists for $n \leq 2$ (at ambient pressure with the exception of Li and Be, which give mK scale T_c).

When $n \geq 8$ or 9, late transition metals are encountered, often with magnetic orders. Or it may be a platinum-group metal of filled d-band, or a simple s-like metal. Either way superconductivity is suppressed. Still the rule here is less rigorous, with notable exceptions like Ru, Os, and Ir. These are actually explained by the ECM: The three have 6 to 7 electrons in the d-subshell, favoring a 3 to 1 e/h ratio within this subshell.

When both d and s subshells are filled up like in the case of Group 2B, the availability of the higher p-orbitals make sp hybridization possible. A WIEN2k calculation of Zn crystal shows significant amount of mixed s-p characters for energy bands near the Fermi level. Besides, $sp3$ hybridization is well known for the cyanide complexes of the Group 2B elements. The s-p hybridization tends to make a hole–electron ratio of 3:1 in these elements and should, at least in part, explains the relatively high transition-temperature in Group 2B.

With the p-subshell starts to fill in Group 3A and 4A, additional coupling between the filled s-band and the partial-filled p-band should then explain the pronounced superconductivity in the sp-block. Again, this should give rise to the empirical function $T(n)$ to its right-hand-side portion as in Fig 5.1.

In Matthias' original paper (Matthias, 1955), $T(6) = 0$ when $n = 6$. This can be understood as the following reasoning. For example, the $3d$ and $4s$ subshells altogether supply 12 orbitals if counting the spin. When the valence is filled up to 6, there tend to be equal number of electron and hole that quenches superconductivity. Now if we count the d-orbitals only, d-subshells filled up to 2.5 tend to give an optimal ratio of hole/electron $=$ 3:1, and this puts the total valence at 4.5, well explaining the $T(n)$ max at $n \approx 4.6$. The secondary maximal at $n \approx 7$ is more involved, but it seems to associate with Hund's rule. Half filled d-band, preferred by Hund's rule, is likely to be holelike and supplies the majority carrier. It then couples to the s-band partially hybridized with p-states, which is likely to be electronlike. The hole/electron ratio is hence formally 5 : 2, not far from the ideal 3 : 1 that explains the high T_c.

It is impressive to see that the Matthias rules, despite their age and crude nature, still apply to new superconductors being currently investigated (Conder, 2016; Slocombe et al., 2015). There the relation of $n \approx 4.5$ still holds, which should not come as a surprise after all once we understand the optimal e/h ratio to minimize the Coulomb repulsion. Of course, the refinement of the Matthias rules in front of new experimental developments has also helped to distill their essence while discarding outdated statements.

Before the BCS, Born and Cheng (Born and Cheng, 1948) proposed a theory based on the interband interaction of electrons from separate bands folded into the BZ. Their theory, although phenomenological, actually explains MR1 of Matthias' rules. The real physics behind is due to the two band edges having opposite effective charges, minimizing the Coulomb repulsion, as can be accommodated by the excitonic Coulomb model. Benard Matthias had been skeptical of the (phonon) BCS theory and now the ECM likely reconciles the superficially conflicting views: On one hand, the excitonic action is important in reducing Coulomb repulsion, even in conventional phonon superconductors. On the other hand, BCS, as a general formalism, needs not be limited to phonon-superconductivity, as long as the boson field leads to enough attraction. In a similar fashion, Hirsch's hole superconductivity (Hirsch and Marsiglio, 1991) can now also be understood: the holes are effective in reducing electrostatic repulsion among electrons.

5.5 Other SC Material Systems

The T_c of "old" high-T_c A15 alloys like Nb_3Ge can be similarly explained: Notice a ratio of 3:1 there. In $Ba_{1-x}K_xBiO_3$ (BKBO)-type materials, electrons are likely from Ba and K, while holes are from the BiO_3 portion, and it is well known to have both electron and hole bands straddling the Fermi surface (Ali et al., 2014). Similar observations are expected in BPBO.

$LiTi_2O_4$ is not an exception either, with multiband crossing of the Fermi surface. The list goes on with MgB_2 and similar borides, iron pnictides and oxypnictides. Although we can always calculate these materials' DFT band-structure to find the exact atomic orbital origins of carriers near the Fermi surface, a simple rule of thumb can be derived from the notion of excitonic Coulomb pairing, combined with an inside-understanding of the Matthias rules. For example, for the oxypnictides, holes are expected from the $FeAsO_x$ portion, while electrons are released from the rare earth or the alkaline earth atom.

The discovery of high-T_c in the LaFeAsO systems demonstrates that an electron-doped superconductor can also reach critical temperature above 50 K (Ren et al., 2008). There are surprising similarities between the cuprates and the iron pnictogenides, such as antiferromagnetism (Nakai et al., 2008) when undoped, pseudogap, preformed pairs, quasi-2D conduction, low charge density, strong-correlations, to name a few. More importantly, the coexistence of electrons and holes are well documented (Ma and Lu, 2008; Yin et al., 2008), as the consequence of multiple bands crossing the FS. Also confirmed is the semimetal band-structure. This gives added confidence as for the validity of the ECM.

The author made a bold prediction (Greene, 2008; Luo, 2008; Zhu, 2008) about iron pnictides' Hall effect upon the cooling of temperature: The Hall coefficient, R_H, will change its sign for a certain range of doping levels. R_H is negative in the normal state, but will change to positive in the mixed state, for a certain temperature and doping range. Again, this was verified by experiment (Padilla, 2010).

For MgB_2, we notice the coexistence of electron and hole once again: Here both the hole and the electron band cross the Fermi level. Although the phonon mechanism is often considered to be responsible for pairing in MgB_2, we, nevertheless, think that the ECM at least plays an important role. At the minimum, it helps to increase the critical temperature by effectively reducing the Coulomb repulsion among the holes. Magnesium diboride with the highest T_c is predominately a hole conductor. Again, not-surprisingly the latest experiments start to demonstrate the pseudogap state for MgB_2 (Patil et al., 2017).

As for the Chevrel phase (for example, $PbMo_6S_8$) and other chalcogenides (for example, the iron-chalcogenide), the Matthias rules and 3-to-1 rule apply too. Here holes likely form out of the chalcogen, or plus late-TMs, where TM means transition metal. Meanwhile, the early-TM donates electrons. Along this line of chalcogenide SC, the recently reported very high T_c in the H_2S system may actually fit into this category.

The recent experiment on possibly very high T_c is worth noting (Drozdov et al., 2015). At the face value, the notion of a sulfur hydrogen compound does not contradict to the Matthias rules: The sulfur supplies holes, and electrons could be from the hydrogen atoms. However, the aforementioned simple argument may need a second thought considering other factors. We know that hydrogen is not the best electron donor, and superconductivity that only happens at extreme conditions always warrant more scrutiny, being less repeated by labs in other parts of the world. There are many complications for experiments at high pressure, and especially for carrying out electrical tests with a highly compressed gas. For example, metal electrodes used in the transport probe may easily form compounds with sulfur, many of which are known SC (Nagata and Atake, 1999), making an unambiguous interpretation of data difficult. Keep in mind that sulfur is often a transition metal corroder more potent than oxygen. The diamond anvil cell (DAC) is typically operated with a metal gasket, often made of rhenium and tungsten, which forms well-known sulfides. And both sulfides should be good candidates for SC according to the Matthias rules. This then makes the real "culprit" of SC really difficult to nail down.

Ref. Drozdov et al. (2015) interpreted the experiment as the conventional (phonon) SC. Many DFT calculations (Quan and Pickett, 2016; Sano et al., 2016) are based on the assumption of an H_3S compound. It is believed by many that the high Debye frequency (temperature) of the hydrogen phonon mode should give high T_c. However, nothing concrete of SC has come out of the metallic hydrogen research in the past 50 years, and if the Coulomb repulsion is not suppressed enough in the first place, SC will not emerge no matter how high the Debye temperature is. This is especially significant noting that sulfur itself has one of the highest T_c among all chemical elements, albeit only under a substantial pressure.

Some helpful information is still gained, nevertheless, from these studies: The Van Hove singularity near FS in this category of sulfides. From Chapter 3, the VHS favors the formation of excitonic interaction.

Carbon-based SC is robustly established since early investigation in alkali-metal intercalated fullerenes (Hebard et al., 1991; Nomura et al., 2015; Tanigaki et al., 1991). More recent studies extended this into graphite, fused benzene rings and graphene (Emery et al., 2005; Ludbrook et al., 2015; Mitsuhashi et al., 2010; Xue et al., 2012). Since these three species are closely related, it is not a surprise to see that SC discovered in one can predict that in others. Helped by the Matthias rules we understand that the carbon group has a propensity to donating holes. Although graphite is ambipolar, band structure calculations show that the hole tends to dominate. This is also evident from experiments, for example, the Hall effect. This tendency is augmented if graphite (or graphene) is edge oxidized. Then when these positive charge carriers encounter electrons doped from the alkali-metal (and possibly rare-earth as well), SC may arise.

5.6 Some Predictions

The formalism developed in the previous chapters has demonstrated its applicability to the cuprates. It self-consistently explains a majority of experiments. This line of thought had already made predictions before. Some date back to nearly 20 years ago, and some have been experimentally verified.

In this section, we make a few new predictions based on the ECM pairing model, and discuss future directions in achieving higher T_c in a broad range of potentially HTSC materials.

The EHC requires further and broader range of experimental verification, possibly with techniques beyond what are so far utilized for the cuprates.

One such experiment is the carrier drifting test schematically shown in Fig. 5.2.

Suppose that a laser pulse is fed through the fiber probe of a near-field scanning optical microscope (NSOM) as shown in Fig. 5.2. The pulse excites the cuprate sample in a small spot, say a few hundreds of nm across, generating non-equilibrium electrons and holes, which drift under E-field and upon reaching the metal probe induce a voltage pulse there. By probing the direction at which carriers move relative to E, their signs are found. The distance between two probes needs to be small, say ≤ 1 μm to cope with the

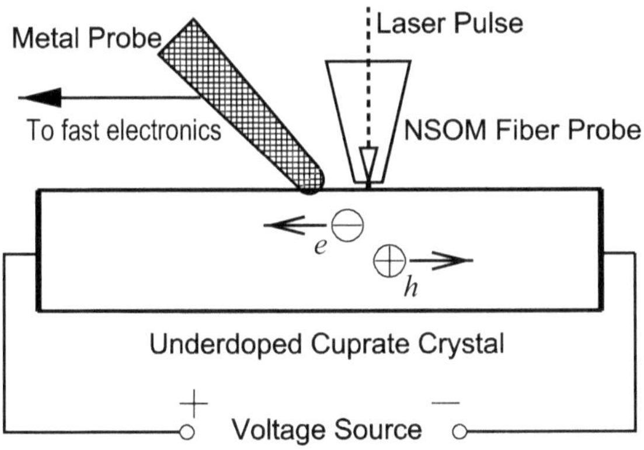

Figure 5.2 An experiment to probe the drift direction of carriers under an electric field

fast recombination and small mobility of non-equilibrium carriers. The conductance, the sample's doping level, and temperature should be chosen carefully. If the optical injection turns out to be too weak to generate carrier perturbation of enough signal/noise ratio, augmentation with electrical injection can be added with capacitive, inductive, or direct coupling, with more controls such as polarity and frequency modulation, etc.

5.6.1 Highest T_c in the Excitonic Coulomb Pairing Model

The formulae for T^* and T_c of Chapter 1 are reproduced as follows:

$$k_B T^* = \frac{c_{\text{dim}} e^4 m_\mu}{32\pi^2 \epsilon_0^2 \epsilon_r^2 \hbar^2},$$

$$k_B T_c = 3.31 \frac{n_t^{2/3} \hbar^2}{m^*}, \tag{5.6}$$

where in the first formula we neglect the exponential factor of Eq. (1.132) for simplicity. Written in terms of the band mass of hole m_h (or electron m_e^b) and according to our model $m_\mu = m_h/2 = m_e^b/2$ and $m^* = 4m_h$:

$$k_B T^* = \frac{c_{\text{dim}} e^4 m_h}{64\pi^2 \epsilon_0^2 \epsilon_r^2 \hbar^2},$$

$$k_B T_c = 3.31 \frac{n_t^{2/3} \hbar^2}{4m_h}. \tag{5.7}$$

The first expression of Eq. (5.7) for T^* is an increasing function of m_h, while in the second T_c is a decreasing function of m_h. When the two are equal, a maximal T_c should be given. Equating the two, one gets the equation for m_h to achieve the maximal T_c. Then substituting the expression of m_h into either T_c or T^*, we have the expression for maximal $T_{c,\max}$

$$m_h = \left(\frac{53.0}{c_{dim}}\right)^{1/2} \frac{\pi \epsilon_0 \epsilon_r \hbar^2}{e^2} n_t^{1/3}, \tag{5.8}$$

$$k_B T_{c,\max} = \frac{(53.0 c_{dim})^{1/2}}{64} \frac{e^2}{\pi \epsilon_0 \epsilon_r} n_t^{1/3}. \tag{5.9}$$

We take an aggressive $c_{dim} = 4$ and the max T_c is very closely given by the average electrostatic Coulomb energy at an average inter-tetron distance. Notice that this condition only holds with the optimal band mass given by the above equation.

Given a typical ϵ_r of 15 and the best n_t that can be achieved in the cuprate, $\approx 1.245 \times 10^{21}$ cm^{-3} which is equivalent to inter-tetron distance of 9.3 Å, the energy is then 0.0937 eV or 1087 K. Of course this is really an overestimate. When the exponential factor is taken into account the result needs to multiply with $1/e = 1/2.71828$, which gives 400 K. The corresponding best band mass is equal to 1.286 m_e, or 1.286 times the bare mass of electron (the electron rest mass).

The current cuprates typically have hole band mass near $2m_e$. Therefore, band-engineering the material so as to have lighter quasiparticles, and combining with reduced intrinsic dielectric constant appear to be the way to higher T_c. Or one can look for new material systems where these conditions are satisfied.

5.6.2 Some Candidate SC Materials

The general recipe for finding candidate high-T_c material is summarized as follows:

(1) Check the resistivity to see if it is significantly controlled by factors like doping (composition), voltage, temperature; The resistance variation may be over 2 orders' of magnitude, with resistivity preferably between 10^{-5} and 10^{-2} Ω·cm at 200 K or the ambient.
(2) Check the optical spectra, preferably zero gap semi-metallic type in the normal state,
(3) with T variations in the DC/low-frequency optical conductivity, i.e., a pseudogap behavior.
(4) Check static or low frequency (Mhz scale) dielectric constant in the high-resistance state, preferably <40.
(5) Check the Hall coefficient to see if it is near a sign reversal, signified by small value, strong T dependence and/or magnetic-field dependence.
(6) Check band mass to see if it is near one or a few m_e.
(7) Check the existence of saddle points or VHS.

5.6.3 Transition Metal Oxides Other than Cuprates

There is no *a priori* reason as to why other transition metal oxides than the copper one should not have a high T_c. Since metallicity generally increases to the left of the periodic table, we expect reduced polarizability for oxides of transition metals to the left of copper, at least in the fourth period. This will give lower dielectric constant and if other factors match, the TM oxide is likely to give T_c of the 100 K range. We have to emphasize that many transition metal oxides are doping-controlled. The right type of doping n or p is critical to achieve a proper approach of the Fermi level to the VHS, if it does indeed exist. It is also essential to access the adequate band-structure portion to have a proper hole/electron ratio, total carrier density and band-mass.

5.6.4 The Chalcogenides

Sulfur forms many minerals with transition metals, and judging from the known band mass of chalcogenides say Cu_2S, the chalcogenides seem to have low effective mass due to increased covalence. However, the ionicity of sulfur-metal bonds comes as a disadvantage since they tend to give large static dielectric constants. Therefore, techniques such as halogen-doping may be required, as what is often applied to the iron pnictides.

5.6.5 Carbon-Based Materials

Intercalated graphite, alkali-doped C_{60}, intercalated graphene are all known to superconduct. Graphite itself is well known to have both electrons and holes (Dresselhaus and Mavroides, 1966; Slonczewski and Weiss, 1958; Soule and Mccluke, 1959), though too many of which, according to the excitonic Coulomb model, are counterproductive to superconductivity.

One major issue preventing graphite from being SC is the small band mass on the order of $0.04\ m_e$ (Soule et al., 1964; Williamson et al., 1965, 1966), which makes the formation of excitonic bound state very difficult. The small effective mass reflects the large band dispersion that is typical of the sp-block. This is exacerbated by the high conductivity of the matrix which inevitably increases the low-frequency/static dielectric constant.

However, the large sp-dispersion could be reduced by making the unit-cell larger or stretching the bond, which is likely the case of SC CaC_6 (Walters et al., 2011). And the size of the Fermi surface can be further tuned (reduced) like in the case of graphene (Bostwick et al., 2009; Geim and Novoselov, 2010), where it is reduced to the Dirac point. Of course all these controls are very sensitive to the surface states and is hence difficult to control precisely in practice.

SC of various degrees of critical temperature probably exists already in various graphite/graphene/HOPG/exfoliated-graphite species, but mostly near the surface. The surface depletion effect should make a thin region where the density and hole/electron ratio favors high T_c. The problem though is the overall diamagnetism fraction is low and reliable electrical contact to such an interfacial superconductivity is difficult to maintain, or with a critical current too low to make a reliable characterization or practical application.

With those restrictions in mind, alloying or compounding with other elements such as nitrogen/boron might be necessary to make a high-T_c SC of significant volumetric fraction. These can be further enhanced with halogenation and metallic intercalation. The latter can be achieved with doping of alkali, alkaline-earth and/or rare-earth metals.

5.6.6 Charge Density Modulation in *c*-Axis

Chapter 1 introduced the classical electrostatic basis for charge transfer between two copper oxide planes. Chapter 3 analyzed in a strong-coupling many-body BdGE and Green's function formalism the detailed microscopic picture and the resulting experimental observables such as the spectral functions and their ARPES manifestation. Chapter 4 started with the weak-coupling DFT assumption, but was, nevertheless, helped with highly sophisticated all-electron realistic and first-principles computational techniques. Chapter 4 was able to verify in high numerical precision the electron–hole segregation phenomena (and/or charge/phase segregation), which roughly match a predicted ratio of nearly 3:1 that the author predicted nearly 20 years ago. Here we mention other tests that can demonstrate the charge transfer, modulation or density waves.

First, the STM-type probe that is confined to the Å scale may see potential difference between a hole top layer and an electron top one, barring severe surface depletion effects. Carrier drifting experiments under an in-plane potential can then be utilized to further differentiate the type of hole or electron. We expect to see negative charge excitations in proper arrangements.

In single-layer cuprates, we should observe ABAB. . .-type modulation along *c*-axis under ideal assumptions. Such a modulation should show up as a CDW along the *c*-axis, asymmetry in tunneling, etc. However, we should note that inhomogeneity is a common occurrence in cuprates. Chapter 4 also has demonstrated that doping with electron or hole often depends on how close it is to the dopant ion, like the Sr in LSCO. Therefore, a complete ABABAB. . . modulation is more likely to be found in a shorter scale of several tens of Angstrom. This may pose additional challenges to the experimental technique.

For bi-layer cuprates, the modulation is likely within the cell so as to have "ababab. . ." where the lower case indicates transfer within the unit cell. Since coupling is generally weaker across a unit cell, many other forms, like "abbaabba. . ." then arise besides the energetically more stable "abababab. . ." or "abbaabab. . . ." Again, dopant randomness may overwhelm the purported order. Spin-chain and domain wall theories can be modified to study this category of density wave.

With trilayer cuprates, "aba" is the norm within the cell. "abaabaaba. . ." and "abababab a. . ." should be similar in the energetics and stability. A difference in the spectroscopic response should be discernable.

So far, we have restricted discussion to the space group of $P4/mmm$. If the space group is $I4/mmm$, there are complication due to the shifting along the direction parallel to the CuO_2 plane. One result could be an induced in-plane CDW, coupled to the *c*-axis CDW.

5.7 Conclusion

This chapter probes the mass of supercarrier in cuprates from the London penetration depth, and gives a strong evidence to the four-body pairing complex, the tetron, as the manifestation of the Cooper pair for the cuprate. A mixed pairing symmetry is inferred from an 8-point Green's function and a four-particle reduced density matrix, to reconcile the apparently incompatible symmetry results from in-plane grain boundary junctions and *c*-axis twisted junctions. An EMC upper limit T_c of 400 K is predicted for candidate material systems if their structures and SC mechanism are similar to those of the cuprates. Some potentially superconducting materials and the methods to achieve higher T_c are discussed, with a few experimental predictions made.

Chapter 6

Reflections on John Bardeen's Interest in the Role of Excitons in Superconductivity

6.1 Introduction

Dr. John Bardeen is widely known for his Nobel Prize–winning theory of superconductivity. In connection with this book, it is worth recalling his interest in the possible connection of excitons and high temperature superconductivity.

I, Dr. G. H. Miley, joined the faculty at the University of Illinois at Urbana-Champaign in 1961. I knew of Professor John Bardeen, two-time Nobel Prize winner in Physics, who left the Bell Laboratories to join the University of Illinois faculty. He had dual appointments in Physics and Electrical Engineering. I had dual appointments in Physics and Nuclear Engineering but later was transferred from Physics to Electrical Engineering because the plasma courses I was teaching were in the Electrical Engineering Department. I did not have much opportunity to interact with Professor Bardeen in my early years at the University of Illinois, but I had taken advantage of attending several of his seminars and chatting with him afterwards. Later I did get to know Professor Bardeen better, but we never actually did any joint research. Still, as I describe here,

High-T_c Superconductivity: The Excitonic Coulomb Perspective
Nie Luo and George H. Miley

ISBN 978-981-4800-57-0 (Hardcover), 978-0-429-27870-9 (eBook)
www.jennystanford.com

Figure 6.1 John Bardeen was an American physicist and electrical engineering, the only person to have won the Nobel Prize in physics twice: first in 1956 with William Shockley and Walter Brattain for the invention of the transistor; and again in 1972 with Leon N. Cooper and John Robert Schrieffer for a fundamental theory of conventional superconductivity known as the BCS theory.

my interactions with him did provide me with some insights into his interest in the role of excitons and superconductivity.

While I am sure the readers know of Professor Bardeen, I would like to briefly summarize some of his accomplishments as commonly known to the field.

6.2 Knowing Bardeen

When I arrived at the University of Illinois in 1961, one of my strong areas of interest involved plasmas and nuclear fusion. I had a joint appointment as an assistant professor in both Nuclear

Engineering and Physics. One of my colleagues in Physics who took interest in my research was David Pines. He had come to Illinois from Princeton several years before I joined the faculty. Because of his interest in excitations and solids, in discussions he brought up the connections between that physics and plasma physics. After several extended conversations with me, he ended up giving me a copy of the book he wrote on "Elementary Excitations in Solids." Indeed, up until this time, I have not thought much about how one might model phenomena in solids and the similarities in modeling plasmas. I carefully read his book, which became my first introduction to excitons. As it happens, some years later I had some informal discussions about these phenomena with John Bardeen. I got to know John Bardeen better when one of my PhD students proposed that he serve on his thesis committee. He was an international student who felt it would be very prestigious to have a Nobel Prize winner on his thesis committee. I tried to explain to the student that the members of the committee were to be experts in the area of the thesis (in this case, fusion plasma physics) but that Professor Bardeen did not work on plasmas. He countered that the Tokamak-type fusion system he was studying (specifically aspects of fusion product alpha particle plasma heating) would use superconducting magnets, which Professor Bardeen would be interested in. I responded that this connection was quite indirect, but conceded that he could ask Professor Bardeen if he would be on his committee. As described in the book about him, *True Genius* (Hoddeson and Daitch, 2002), Bardeen was very modest and unassuming. He was not inclined to say "no" to a student, so he did agree to be on the committee. As an aside, I would note that I had a disagreement with the thesis student who employed a "new algebra" that he "invented" to perform some simple algebraic operations such as taking cubic root of numbers in one chapter of the thesis. The description of the steps involved using this new algebra filled up almost two pages. I had told him that this was unnecessary and if he wanted to present this "new algebra," he should get a degree in Mathematics, not Nuclear Engineering. The student accused me of trying to suppress this "new math." He then reduced the size of the section involved but refused to remove it from the thesis. Professor Bardeen did not say much during the early part of the thesis exam. Then at one point, he spoke up loudly and said he did not feel this new algebra action should be in the body of the thesis, but should be relegated to an appendix. With Professor Bardeen saying that, the student no longer protested and agreed to move it to an appendix. I doubt that he thought Professor Bardeen would question this section when he asked him to be on the committee.

The next time I met Professor Bardeen was a few weeks after the thesis exam when we both found ourselves waiting at a lunch wagon stopped along the street at the Nuclear Engineering Laboratory where I had my office. After getting sandwiches, we went together to my office to eat lunch. At this time, I was very interested in research I was doing to study the physics of Low-Energy Nuclear Reactions (LENRs). I asked John what he thought about LENR, which was initially motivated by announcement "Cold Fusion" in 1989. John said he had not thought about this and really had no opinion. I was not surprised since he was known for not speculating about phenomena, but studying them in great detail when he thought it was appropriate. I told him that I thought there may be some connection between LENR Physics and BCS theory and possibly excitons. John replied he would like to hear more about my thoughts on this when we have more time. As it turns out, we did meet together at the lunch wagon off and on and had a number of rambling discussions of all this. Those discussions brought out his interest in studying the role of exciton in high-temperature superconductivity. In fact, we had met for lunch just weeks before he passed away.

One article that he called to my attention came from some of his work done in 1973 on "Model for an Exciton Mechanism of Superconductivity" (Allender et al., 1973). The thought process involved is described in the abstract of this article. It states:

> "The exciton mechanism of superconductivity is discussed with respect to a particular model, a thin metal layer on a semiconductor surface. In this model, the metal electrons at the Fermi surface tunnel into the semiconductor gap where they interact with virtual excitons, producing a net attractive interaction among the electrons in direct analogy with the phonon mechanism

> of superconductivity. The physical requirements for successful realization of the exciton mechanism in a metal-semiconductor system are explored in detail, and the relevant parameters are described. Estimates are made for electron tunneling and band-bending effects, and an electron-exciton coupling constant is defined and estimated. Finally, an appropriately modified integral equation for the superconducting energy gap is solved numerically to yield transition temperatures both for a pure-exciton mechanism and for the exciton and phonon mechanisms acting simultaneously."

6.3 Common Interest in Exciton

With time, I have more recently been explaining my LENR research as being caused by Bose–Einstein condensation of deuterium or hydrogen into defects created either in thin films of platinum or nickel metal or nano particles. Further explanation of this has been given by Y. Kim (Kim, 2009). Interestingly, Mohit Randeria and Edward Taylor have discussed the connection between some types of Bose–Einstein Condensation and BCS theory (Randeria and Taylor, 2014). They state in the abstract of their paper:

> "The crossover from weak coupling Bardeen-Cooper-Schrieffer (BCS) pairing to a Bose–Einstein condensate (BEC) of tightly bound pairs, as a function of the attractive interaction in Fermi systems, has long been of interest to theoretical physicists. The past decade has seen a series of remarkable experimental developments in ultra cold Fermi gases that has realized the BCS-BEC crossover in the laboratory, bringing with it fresh new insights into the very strongly interacting unitary regime in the middle of this crossover. In this review, we start with a pedagogical introduction to the crossover and then focus on recent progress in the strongly interacting regime. While our focus is on new theoretical developments, we also describe three key experiments that probe the thermodynamics, transport and spectroscopy of the unitary Fermi gas. We discuss connections between the unitary regime and other areas of physics—quark-gluon plasmas, gauge-gravity duality and high temperature superconductivity—and conclude with open questions about strongly interacting Fermi gases."

In fact, the role of excitons in a number of solid-state phenomena remains under study. For example, Einstein has discussed a connection to Quantum Hall Systems (Eisenstein, 2014). In this paper he discusses:

> "The condensation of excitons, bound electron-hole pairs in a solid, into a coherent collective electronic state was predicted over 50 years ago. Perhaps surprisingly, the phenomenon was first observed in a system consisting of two closely-spaced parallel two-dimensional electron gases in a semiconductor double quantum well. At an appropriate high magnetic field and low temperature, the bilayer electron system condenses into a state resembling a superconductor, only with the Cooper pairs replaced by excitons comprised of electrons in one layer bound to holes in the other. In spite of being charge neutral, the transport of excitons within the condensate gives rise to several spectacular electrical effects. This article describes these phenomena and examines how they inform our understanding of the unique phase of quantum electronic matter."

6.4 Exciton and Superconductivity

Rather than talking in more detail about my informal discussions with John Bardeen, I feel it would be most instructive to review a discussion along these lines written by V. L. Ginzburg, in the book *The Physics of a Lifetime: Reflections on the Problems and Personalities of 20th Century Physics* (Aksent'eva and Ginzburg, 2013). In sections of the book, Ginzburg states:

> "After the BCS theory was created (Bardeen et al., 1957), John Bardeen naturally went on developing the theory of superconductivity and published a number of papers on this subject." Ginzburg has had some insightful comments about this follow–on work, namely:
>
> "In a paper published in December, 1990 (Bardeen, 1990), John Bardeen interpreted superconductivity as an example of a macroscopic quantum phenomenon, and first of all as superfluidity. The analogy between superconductivity and superfluidity has long been understood by Bardeen, as is clearly demonstrated in his

paper. This analogy was not mentioned in the original BCS paper; I think it was an oversight."

Since 1964, W. Little and Ginzburg have discussed the problem of high-temperature superconductivity (HTSC). The state of the art at the time did not allow any exact predictions of the value of Tc: Only intuitive and qualitative arguments were presented. Despite improvements, Tc cannot yet be accurately predicted for all cases. Simply speaking, the basic idea of the so-called excitonic mechanism is the following: In the BCS expression $T_c = 1.13\theta_D \exp(-1/NV)$, the pre-exponential factor θ_D should be understood not as a Debye cut-off temperature θ_D, but as some electron (excitonic) temperature $\theta_e > \theta_D$, so that $T_c \approx \theta_e \exp(-1/NV)$. Physically, we mean the replacement of the electron-phonon interaction (or rather its enhancement) by the interaction of the conduction electrons with the "bound" electrons through their excitations (excitons). Moreover, conditions were discussed under which the excitons might "work" more efficiently, for instance, in quasi-one-dimensional and quasi-two-dimensional systems. This paper is not an appropriate place to dwell on these points (see (Ginzburg, 1989; Ginzburg and Kirzhnits, 1987)). As is typical in such cases, because the issue was neither clear-cut nor obvious, there were many skeptics and critics. John Bardeen and his colleagues (Allender et al., 1973) responded by introducing an interesting and important model of a "sandwich" with semiconductor "sides". The attempts to create HTSC had not been successful, and therefore the pessimism grew. But Bardeen remained true to his convictions and intuition. The following was his view of the HTSC problem given in 1978 (Bardeen, 1978):

"In view of the large number of experiments that have been done and the wide variety of materials tested, many including Bernd Matthias have been pessimistic about prospects for finding excitonic superconductivity. While these experiments do show that the conditions for observing it must be very exacting, they do not rule it out completely. Since the potential importance of high temperature superconductivity is so great, I feel that the search should be pursued vigorously even though the prospects for success may be small."

Concluding his note, Bardeen wrote:

> "The search for superconductivity from excitonic mechanisms has been fruitful in leading to new scientific insights and there is still hope that the dream of high temperature superconductivity will become a reality. It is a challenging problem that will intrigue material scientists for a long time to come." It is unnecessary to prove today that Bardeen's view was correct. And when the "thunder struck" and HTSC was eventually discovered in 1986–1987, Bardeen and his colleagues gave a talk on this subject at the first major conference (Bardeen et al., 1987).
>
> The mechanism of superconductivity in HTSC is not obvious and is still a subject for discussion. The discovery of HTSC and the related discussions have undoubtedly extended our insight into the theory of superconductivity. It has become particularly clear that, besides the original and the generalized (see (Carbotte, 1990)) BCS models, the Schafroth local-pair picture or other models may also be realistic. Bardeen, from (Bardeen et al., 1987), did not see any need to go beyond the BCS scheme, even with allowance for strong coupling, and not only for the electron-phonon but also the electron-exciton interaction (but possibly with a disregard of magnetic effects or, in other words, spin-wave exchange). I had been of the same opinion, although an unbiased approach is needed here, of course. The results obtained in Moscow (Mikhailovsky et al., 1991) seemed to be fairly impressive in this respect. According to this paper, the experimental data on HTSC are perfectly well explained within the generalized BCS scheme with strong coupling and with allowance for the temperature-dependent pair breaking caused by the inelastic electron-phonon scattering."

In conclusion, John Bardeen had a strong feeling that excitons played a significant role in high temperature superconductivity. He and colleagues had begun studies of this as I have reviewed here. Others continue this work. We can all be encouraged to study this field in view of Bardeen's uncanny ("True Genius") insights into condensed matter physics. One purpose of this book is to further review this issue and develop related physics.

Bibliography

Abrikosov, A. A. (1995). Possibility of reconciliation on the type of the order parameter in high-temperature superconductors, *Phys. Rev. B* **51**, pp. 11955–11957, doi:10.1103/PhysRevB.51.11955, URL https://link.aps.org/doi/10.1103/PhysRevB.51.11955.

Ahn, C., Gariglio, S., Paruch, P., Tybell, T., Antognazza, L., and Triscone, J.-M. (1999). Electrostatic modulation of superconductivity in ultrathin GdBa2Cu3O7-x films, *Science* **284**, 5417, pp. 1152–1155.

Aksent'eva, M., and Ginzburg, V. (2013). *The Physics of a Lifetime: Reflections on the Problems and Personalities of 20th Century Physics* (Springer Berlin Heidelberg), ISBN 9783662044551, URL https://books.google.com/books?id=gW78CAAAQBAJ.

Alexandrov, A. S. (1996). Bipolaron anisotropic flat bands, Hall mobility edge, and metal-semiconductor duality of overdoped high-T_c oxides, *Phys. Rev. B* **53**, pp. 2863–2869, doi:10.1103/PhysRevB.53.2863, URL https://link.aps.org/doi/10.1103/PhysRevB.53.2863.

Alexandrov, A. S., Kabanov, V. V., and Mott, N. F. (1996). Coherent *ab* and *c* transport theory of high- T_c cuprates, *Phys. Rev. Lett.* **77**, pp. 4796–4799, doi:10.1103/PhysRevLett.77.4796, URL https://link.aps.org/doi/10.1103/PhysRevLett.77.4796.

Alexandrov, A. S., and Mott, N. (1994). *High Temperature Superconductors and Other Superfluids* (CRC Press).

Ali, M., Aftabuzzaman, M., Roknuzzaman, M., Rayhan, M., Parvin, F., Ali, M., Rubel, M., and Islam, A. (2014). New superconductor (Na0.25K0.45) Ba3Bi4O12: A first-principles study, *Physica C* **506**, pp. 53–58, doi:https://doi.org/10.1016/j.physc.2014.08.010, URL http://www.sciencedirect.com/science/article/pii/S092145341400286X.

Allen, J. W., Olson, C. G., Maple, M. B., Kang, J.-S., Liu, L. Z., Park, J.-H., Anderson, R. O., Ellis, W. P., Markert, J. T., Dalichaouch, Y., and Liu, R. (1990). Resonant photoemission study of $Nd_{2-x}Ce_xCuO_{4-y}$: Nature of electronic states near the Fermi level, *Phys. Rev. Lett.* **64**, pp. 595–

598, doi:10.1103/PhysRevLett.64.595, URL https://link.aps.org/doi/10.1103/PhysRevLett.64.595.

Allen, P. B., Pickett, W. E., and Krakauer, H. (1987a). Band-theory analysis of anisotropic transport in La_2CuO_4-based superconductors, *Phys. Rev. B* **36**, pp. 3926–3929, doi:10.1103/PhysRevB.36.3926, URL https://link.aps.org/doi/10.1103/PhysRevB.36.3926.

Allen, P. B., Pickett, W. E., and Krakauer, H. (1987b). Prediction of anisotropic thermopower of La2-x MxCuO4, in *Novel Superconductivity* (Springer), pp. 489–491.

Allen, P. B., Pickett, W. E., and Krakauer, H. (1988). Anisotropic normal-state transport properties predicted and analyzed for high-Tc oxide superconductors, *Phys. Rev. B* **37**, 13, p. 7482.

Allender, D., Bray, J., and Bardeen, J. (1973). Model for an exciton mechanism of superconductivity, *Phys. Rev. B* **7**, pp. 1020–1029, doi:10.1103/PhysRevB.7.1020, URL https://link.aps.org/doi/10.1103/PhysRevB.7.1020.

Anderson, P. W. (2000). Brainwashed by Feynman? *Physics Today* **53**, 2, pp. 11–12, doi:10.1063/1.882955.

Anderson, P. W. (2013). Twenty-five years of high-temperature superconductivity—a personal review, *J. Phys. Conf. Ser.* **449**, 1, p. 012001, URL http://stacks.iop.org/1742-6596/449/i=1/a=012001.

Annett, J. F., Goldenfeld, N., and Leggett, A. J. (1996). *Experimental Constraints on the Pairing State of the Cuprate Superconductors: An Emerging Consensus*, pp. 375–461, doi:10.1142/9789814261289_0006, URL https://www.worldscientific.com/doi/abs/10.1142/9789814261289_0006.

Anukool, W., Barakat, S., Panagopoulos, C., and Cooper, J. R. (2009). Effect of hole doping on the london penetration depth in $Bi_{2.15}Sr_{1.85}CaCu_2O_{8+\delta}$ and $Bi_{2.1}Sr_{1.9}Ca_{0.85}Y_{0.15}Cu_2O_{8+\delta}$, *Phys. Rev. B* **80**, p. 024516, doi:10.1103/PhysRevB.80.024516, URL https://link.aps.org/doi/10.1103/PhysRevB.80.024516.

Bardeen, J. (1978). Excitonic superconductivity? *J. Less Common Metals* **62**, pp. 447–450, doi:https://doi.org/10.1016/0022-5088(78)90058-9, URL http://www.sciencedirect.com/science/article/pii/0022508878900589.

Bardeen, J. (1987). Theories of high tc superconductors, *MRS Proc.* **99**, p. 27, doi:10.1557/PROC-99-27.

Bardeen, J. (1990). Superconductivity and other macroscopic quantum phenomena, *Phys. Today* **43**, pp. 25–31.

Bardeen, J., Cooper, L. N., and Schrieffer, J. R. (1957). Microscopic theory of superconductivity, *Phys. Rev.* **106**, pp. 162–164, doi:10.1103/PhysRev.106.162, URL https://link.aps.org/doi/10.1103/PhysRev.106.162.

Bardeen, J., Ginsberg, D., and Salamon, M. (1987). Excitonic superconductivity in layer structures, in *Novel Superconductivity* (Springer), pp. 333–339.

Barišić, N., Chan, M. K., Li, Y., Yu, G., Zhao, X., Dressel, M., Smontara, A., and Greven, M. (2013). Universal sheet resistance and revised phase diagram of the cuprate high-temperature superconductors, *Proc. Natl. Acad. Sci.* **110**, 30, pp. 12235–12240.

Batlogg, B., Hwang, H., Takagi, H., Cava, R., Kao, H., and Kwo, J. (1994). Normal state phase diagram of (La,Sr)2CuO4 from charge and spin dynamics, *Physica C* **235-240**, pp. 130–133, doi:https://doi.org/10.1016/0921-4534(94)91330-7, URL http://www.sciencedirect.com/science/article/pii/0921453494913307.

Belchenko, Y. (1993). Surface negative ion production in ion sources, *Rev. Sci. Instrum.* **64**, 6, pp. 1385–1393.

Bennemann, K., and Ketterson, J. (2004). *The Physics of Superconductors: Vol II: Superconductivity in Nanostructures, High-T_c and Novel Superconductors, Organic Superconductors*, Physics and Astronomy Online Library (Springer Berlin Heidelberg), ISBN 9783540442325, URL https://books.google.com/books?id=_h5HIEEned0C.

Bennemann, K., and Ketterson, J. (2008). *Superconductivity: Volume 2: Novel Superconductors*, SpringerLink: Springer e-Books (Springer Berlin Heidelberg), ISBN 9783540732532, URL https://books.google.com/books?id=PguAgEQTiQwC.

Bethe, H., and Salpeter, E. (1957). *Quantum Mechanics of One- and Two-Electron Atoms* (Springer), URL https://books.google.com/books?id=1ZUuAAAAIAAJ.

Bishop, A. R., Martin, R. L., Müller, K. A., and Tešanović, Z. (1989). Superconductivity in oxides: Toward a unified picture, *Zeitschrift fur Physik B Condensed Matter* **76**, pp. 17–24, doi:10.1007/BF01323483.

Blaha, P., Schwarz, K., Madsen, G., Kvasnicka, D., and Luitz, J. (2001). *An Augmented Plane Wave Plus Local Orbitals Program for Calculating Crystal Properties*, Vienna University of Technology, Austria.

Blanco-Canosa, S., Frano, A., Schierle, E., Porras, J., Loew, T., Minola, M., Bluschke, M., Weschke, E., Keimer, B., and Le Tacon, M. (2014). Resonant x-ray scattering study of charge-density wave correlations in $YBa_2Cu_3O_{6+x}$, *Phys. Rev. B* **90**, p. 054513, doi:10.1103/PhysRevB.90.054513, URL https://link.aps.org/doi/10.1103/PhysRevB.90.054513.

Blue, M. D. (1964). Optical absorption in HgTe and HgCdTe, *Phys. Rev.* **134**, pp. A226–A234, doi:10.1103/PhysRev.134.A226, URL https://link.aps.org/doi/10.1103/PhysRev.134.A226.

Boebinger, G. S., Ando, Y., Passner, A., Kimura, T., Okuya, M., Shimoyama, J., Kishio, K., Tamasaku, K., Ichikawa, N., and Uchida, S. (1996). Insulator-to-metal crossover in the normal state of $La_{2-x}Sr_xCuO_4$ near optimum doping, *Phys. Rev. Lett.* **77**, pp. 5417–5420, doi:10.1103/PhysRevLett.77.5417, URL https://link.aps.org/doi/10.1103/PhysRevLett.77.5417.

Born, M., and Cheng, K. C. (1948). Theory of superconductivity, *Nature* **161**, 4103, p. 968.

Bostwick, A., McChesney, J., Ohta, T., Rotenberg, E., Seyller, T., and Horn, K. (2009). Experimental studies of the electronic structure of graphene, *Prog. Surf. Sci.* **84**, 11-12, pp. 380–413.

Bracker, A. S., Stinaff, E. A., Gammon, D., Ware, M. E., Tischler, J. G., Park, D., Gershoni, D., Filinov, A. V., Bonitz, M., Peeters, F., and Riva, C. (2005). Binding energies of positive and negative trions: From quantum wells to quantum dots, *Phys. Rev. B* **72**, p. 035332, doi:10.1103/PhysRevB.72.035332, URL https://link.aps.org/doi/10.1103/PhysRevB.72.035332.

Bradley, C. C., Sackett, C. A., Tollett, J. J., and Hulet, R. G. (1995). Evidence of Bose–Einstein condensation in an atomic gas with attractive interactions, *Phys. Rev. Lett.* **75**, pp. 1687–1690, doi:10.1103/PhysRevLett.75.1687, URL https://link.aps.org/doi/10.1103/PhysRevLett.75.1687.

Brinkman, A., and Hilgenkamp, H. (2005). Electron-hole coupling in high-T_c cuprate superconductors, *Physica C* **422**, 3-4, pp. 71–75.

Carbotte, J. P. (1990). Properties of boson-exchange superconductors, *Rev. Mod. Phys.* **62**, pp. 1027–1157, doi:10.1103/RevModPhys.62.1027, URL https://link.aps.org/doi/10.1103/RevModPhys.62.1027.

Carlson, E. W., Kivelson, S. A., Orgad, D., and Emery, V. J. (2004). *Concepts in High Temperature Superconductivity* (Springer Berlin Heidelberg, Berlin, Heidelberg), ISBN 978-3-642-18914-2, pp. 275–451, doi:10.1007/978-3-642-18914-2_6, URL https://doi.org/10.1007/978-3-642-18914-2_6.

Chan, M., Veit, M., Dorow, C., Ge, Y., Li, Y., Tabis, W., Tang, Y., Zhao, X., Barišić, N., and Greven, M. (2014). In-plane magnetoresistance obeys Kohler's rule in the pseudogap phase of cuprate superconductors, *Phys. Rev. Lett.* **113**, 17, p. 177005.

Chang, J., Blackburn, E., Ivashko, O., Holmes, A., Christensen, N. B., Hücker, M., Liang, R., Bonn, D., Hardy, W., Rütt, U. et al. (2016). Magnetic field controlled charge density wave coupling in underdoped YBa 2 Cu 3 O 6+ x, *Nat. Commun.* **7**, p. 11494.

Chang, L. L., and Ploog, K. (2012). *Molecular Beam Epitaxy and Heterostructures*, Vol. 87 (Springer Science & Business Media).

Chen, C., Birgeneau, R., Gabbe, D., Jenssen, H., Kastner, M., Picone, P., Preyer, N., and Thio, T. (1989). Frequency dependence of the conductivity and dielectric constant of La2CuO4+y near the insulator-metal transition, *Physica C* **162–164**, pp. 1031–1032, doi:https://doi.org/10.1016/0921-4534(89)90578-9, URL http://www.sciencedirect.com/science/article/pii/0921453489905789.

Chen, C. T., Sette, F., Ma, Y., Hybertsen, M. S., Stechel, E. B., Foulkes, W. M. C., Schulter, M., Cheong, S.-W., Cooper, A. S., Rupp, L. W., Batlogg, B., Soo, Y. L., Ming, Z. H., Krol, A., and Kao, Y. H. (1991a). Electronic states in $La_{2-x}Sr_xCuO_{4+\delta}$ probed by soft-x-ray absorption, *Phys. Rev. Lett.* **66**, pp. 104–107, doi:10.1103/PhysRevLett.66.104, URL https://link.aps.org/doi/10.1103/PhysRevLett.66.104.

Chen, C. Y., Birgeneau, R. J., Kastner, M. A., Preyer, N. W., and Thio, T. (1991b). Frequency and magnetic-field dependence of the dielectric constant and conductivity of La_2CuO_{4+y}, *Phys. Rev. B* **43**, pp. 392–401, doi:10.1103/PhysRevB.43.392, URL https://link.aps.org/doi/10.1103/PhysRevB.43.392.

Chen, Q., Stajic, J., Tan, S., and Levin, K. (2005). Bcs-bec crossover: From high temperature superconductors to ultracold superfluids, *Physics Reports* **412**, 1, pp. 1–88, doi:https://doi.org/10.1016/j.physrep.2005.02.005, URL http://www.sciencedirect.com/science/article/pii/S0370157305001067.

Ching, W. Y., Xu, Y., Zhao, G.-L., Wong, K. W., and Zandiehnadem, F. (1987). Electronic structure and excitonic-enhanced superconducting mechanism in $YBa_2Cu_3O_{7-\delta}$, *Phys. Rev. Lett.* **59**, pp. 1333–1336, doi:10.1103/PhysRevLett.59.1333, URL https://link.aps.org/doi/10.1103/PhysRevLett.59.1333.

Chuang, Y.-D., Gromko, A. D., Dessau, D. S., Aiura, Y., Yamaguchi, Y., Oka, K., Arko, A. J., Joyce, J., Eisaki, H., Uchida, S. I., Nakamura, K., and Ando, Y. (1999). Reexamination of the electronic structure of $Bi_2Sr_2CaCu_2O_{8+\delta}$ and $Bi_2Sr_2Cu_1O_{6+\delta}$: Electronlike portions of the Fermi surface and depletion of spectral weight near $\overline{m}$, *Phys. Rev. Lett.* **83**, pp. 3717–3720, doi:10.1103/PhysRevLett.83.3717, URL https://link.aps.org/doi/10.1103/PhysRevLett.83.3717.

Collins, R. T., Schlesinger, Z., Holtzberg, F., Chaudhari, P., and Feild, C. (1989). Reflectivity and conductivity of $YBa_2Cu_3O_7$, *Phys. Rev. B* **39**, pp. 6571–6574, doi:10.1103/PhysRevB.39.6571, URL https://link.aps.org/doi/10.1103/PhysRevB.39.6571.

Conder, K. (2016). A second life of the matthias's rules, *Superconductor Science and Technology* **29**, 8, p. 080502, URL http://stacks.iop.org/0953-2048/29/i=8/a=080502.

Cooper, J. R., Alavi, B., Zhou, L.-W., Beyermann, W. P., and Grüner, G. (1987). Thermoelectric power of some high-T_c oxides, *Phys. Rev. B* **35**, pp. 8794–8796, doi:10.1103/PhysRevB.35.8794, URL https://link.aps.org/doi/10.1103/PhysRevB.35.8794.

Cooper, L. N. (1956). Bound electron pairs in a degenerate Fermi gas, *Phys. Rev.* **104**, pp. 1189–1190, doi:10.1103/PhysRev.104.1189, URL https://link.aps.org/doi/10.1103/PhysRev.104.1189.

Cottenier, S. (2004). Density functional theory and the family of (L) APW-methods: A step-by-step introduction, *KU Leuven, Belgium: Instituut voor Kern-en Stralingsfysica* .

Croft, T. P., Lester, C., Senn, M. S., Bombardi, A., and Hayden, S. M. (2014). Charge density wave fluctuations in $La_{2-x}Sr_xCuO_4$ and their competition with superconductivity, *Phys. Rev. B* **89**, p. 224513, doi:10.1103/PhysRevB.89.224513, URL https://link.aps.org/doi/10.1103/PhysRevB.89.224513.

Crommie, M., Briceno, G., and Zettl, A. (1989). Transport and tunneling in Bi2Sr2CaCu2O8-x, *Physica C* **162-164**, pp. 1397–1400, doi: https://doi.org/10.1016/0921-4534(89)90749-1, URL http://www.sciencedirect.com/science/article/pii/0921453489907491.

Dacal, L. C. O., Ferreira, R., Bastard, G., and Brum, J. A. (2002). Binding energy of charged excitons bound to interface defects of semiconductor quantum wells, *Phys. Rev. B* **65**, p. 115325, doi:10.1103/PhysRevB.65.115325, URL https://link.aps.org/doi/10.1103/PhysRevB.65.115325.

Doiron-Leyraud, N., Badoux, S., René de Cotret, S., Lepault, S., Leboeuf, D., Laliberté, F., Hassinger, E., Ramshaw, B. J., Bonn, D. A., Hardy, W. N., Liang, R., Park, J.-H. ., Vignolles, D., Vignolle, B., Taillefer, L., and Proust, C. (2015). Evidence for a small hole pocket in the Fermi surface of underdoped $YBa_2Cu_3O_y$, *Nat. Commun.* **6**, 6034, doi:10.1038/ncomms7034.

Doiron-Leyraud, N., Proust, C., LeBoeuf, D., Levallois, J., Bonnemaison, J.-B., Liang, R., Bonn, D., Hardy, W., and Taillefer, L. (2007). Quantum oscillations and the Fermi surface in an underdoped high-T_c superconductor, *Nature* **447**, 7144, p. 565.

Dreizler, R. M., and da Providência, J. (2013). *Density Functional Methods in Physics*, Vol. 123 (Springer Science & Business Media).

Dresselhaus, M. S., and Mavroides, J. G. (1966). Energy band parameter determination in graphite, *Carbon* **3**, 4, pp. 465–473.

Drozdov, A., Eremets, M., Troyan, I., Ksenofontov, V., and Shylin, S. (2015). Conventional superconductivity at 203 kelvin at high pressures in the sulfur hydride system, *Nature* **525**, 7567, p. 73.

Eagles, D. (1989). Concentrations and mobilities of holes and electrons in a crystal of a 90k oxide superconductor from analysis of ab-plane resistivity and Hall data, *Solid State Commun.* **69**, 3, pp. 229–234, doi:https://doi.org/10.1016/0038-1098(89)90840-5, URL http://www.sciencedirect.com/science/article/pii/0038109889908405.

Eagles, D., and Savvides, N. (1989). A two-band model applied to resistivity data on a superconducting ceramic specimen of YBa2Cu3O7-x, *Physica C* **158**, 1, pp. 258–264, doi:https://doi.org/10.1016/0921-4534(89)90327-4, URL http://www.sciencedirect.com/science/article/pii/0921453489903274.

Eagles, D. M. (1969). Possible pairing without superconductivity at low carrier concentrations in bulk and thin-film superconducting semiconductors, *Phys. Rev.* **186**, pp. 456–463, doi:10.1103/PhysRev.186.456, URL https://link.aps.org/doi/10.1103/PhysRev.186.456.

Edwards, J. P., Gerber, U., Schubert, C., Trejo, M. A., and Weber, A. (2017). The Yukawa potential: Ground state energy and critical screening, *Progress of Theoretical and Experimental Physics* **2017**, 8, p. 083A01, doi:10.1093/ptep/ptx107, URL http://dx.doi.org/10.1093/ptep/ptx107.

Efetov, K. B. (1991). Electron-hole pairing and anomalous properties of layered high-T_c compounds, *Phys. Rev. B* **43**, pp. 5538–5553, doi:10.1103/PhysRevB.43.5538, URL https://link.aps.org/doi/10.1103/PhysRevB.43.5538.

Eisenstein, J. (2014). Exciton condensation in bilayer quantum Hall systems, *Annu. Rev. Condensed Matter Phys.* **5**, 1, pp. 159–181, doi:10.1146/annurev-conmatphys-031113-133832, URL https://doi.org/10.1146/annurev-conmatphys-031113-133832.

Eisenstein, J., and MacDonald, A. (2004). Bose–Einstein condensation of excitons in bilayer electron systems, *Nature* **432**, 7018, p. 691.

Ekino, T., Gabovich, A. M., Li, M. S., Pekala, M., Szymczak, H., and Voitenko, A. I. (2011). The phase diagram for coexisting d-wave superconductivity and charge-density waves: Cuprates and beyond, *J. Phys. Condensed Matter* **23**, 38, p. 385701, URL http://stacks.iop.org/0953-8984/23/i=38/a=385701.

Emery, N., Hérold, C., d'Astuto, M., Garcia, V., Bellin, C., Marêché, J. F., Lagrange, P., and Loupias, G. (2005). Superconductivity of bulk CaC_6, *Phys. Rev. Lett.* **95**, p. 087003, doi:10.1103/PhysRevLett.95.087003, URL https://link.aps.org/doi/10.1103/PhysRevLett.95.087003.

Emery, V., and Kivelson, S. (1995). Importance of phase fluctuations in superconductors with small superfluid density, *Nature* **374**, 6521, pp. 434–437.

Emery, V. J., Kivelson, S. A., and Tranquada, J. M. (1999). Stripe phases in high-temperature superconductors, *Proc. Natl. Acad. Sci.* **96**, 16, pp. 8814–8817, doi:10.1073/pnas.96.16.8814, URL http://www.pnas.org/content/96/16/8814.

Emery, V. J., and Reiter, G. (1988). Mechanism for high-temperature superconductivity, *Phys. Rev. B* **38**, pp. 4547–4556, doi:10.1103/PhysRevB.38.4547, URL https://link.aps.org/doi/10.1103/PhysRevB.38.4547.

Engel, E., and Dreizler, R. (2011). *Density Functional Theory: An Advanced Course*, Theoretical and Mathematical Physics (Springer Berlin Heidelberg), ISBN 9783642140891, URL https://books.google.com/books?id=o9byjwEACAAJ.

Ernzerhof, M., Perdew, J., and Burke, K. (1996). Density functionals: Where do they come from, why do they work? *Density Functional Theory I*, pp. 1–30.

Finkelstein, G., Umansky, V., Bar-Joseph, I., Ciulin, V., Haacke, S., Ganière, J.-D., and Deveaud, B. (1998). Charged exciton dynamics in GaAs quantum wells, *Phys. Rev. B* **58**, pp. 12637–12640, doi:10.1103/PhysRevB.58.12637, URL https://link.aps.org/doi/10.1103/PhysRevB.58.12637.

Forro, L., Carr, G. L., Williams, G. P., Mandrus, D., and Mihaly, L. (1990). Far-infrared transmission study of single-crystal $Bi_2Sr_2Ca_1Cu_2O_x$ superconductors, *Phys. Rev. Lett.* **65**, pp. 1941–1944, doi:10.1103/PhysRevLett.65.1941, URL https://link.aps.org/doi/10.1103/PhysRevLett.65.1941.

Fournier, P., Jiang, X., Jiang, W., Mao, S. N., Venkatesan, T., Lobb, C. J., and Greene, R. L. (1997). Thermomagnetic transport properties of $Nd_{1.85}Ce_{0.15}CuO_{4+\delta}$ films: Evidence for two types of charge carriers,

Phys. Rev. B **56**, pp. 14149–14156, doi:10.1103/PhysRevB.56.14149, URL https://link.aps.org/doi/10.1103/PhysRevB.56.14149.

Freeman, A., Yu, J., and Massidda, S. (1989). Electronic structure, Fermi liquid and excitonic superconductivity in the high T c Cu-oxides, in *Mechanisms of High Temperature Superconductivity* (Springer), pp. 99–110.

Fretwell, H. M., Kaminski, A., Mesot, J., Campuzano, J. C., Norman, M. R., Randeria, M., Sato, T., Gatt, R., Takahashi, T., and Kadowaki, K. (2000). Fermi surface of $Bi_2Sr_2CaCu_2O_8$, *Phys. Rev. Lett.* **84**, pp. 4449–4452, doi:10.1103/PhysRevLett.84.4449, URL https://link.aps.org/doi/10.1103/PhysRevLett.84.4449.

Friedberg, R., and Lee, T. (1989a). Boson-fermion model of superconductivity, *Phys. Lett. A* **138**, 8, pp. 423–427, doi:https://doi.org/10.1016/0375-9601(89)90844-X, URL http://www.sciencedirect.com/science/article/pii/037596018990844X.

Friedberg, R., and Lee, T. D. (1989b). Gap energy and long-range order in the boson-fermion model of superconductivity, *Phys. Rev. B* **40**, pp. 6745–6762, doi:10.1103/PhysRevB.40.6745, URL https://link.aps.org/doi/10.1103/PhysRevB.40.6745.

Fujimoto, S., Kohno, H., and Yamada, K. (1991). Temperature dependence of electricalresistivity in two-dimensional Fermi systems, *J. Phys. Soc. Jpn.* **60**, 8, pp. 2724–2728, doi:10.1143/JPSJ.60.2724, URL https://doi.org/10.1143/JPSJ.60.2724.

Fukuyama, H., and Hasegawa, Y. (1987). Superconductivity in organics and oxides: Similarity and dissimilarity, *Physica B+C* **148**, 1, pp. 204–211, doi:https://doi.org/10.1016/0378-4363(87)90192-6, URL http://www.sciencedirect.com/science/article/pii/0378436387901926.

Gaididei, Y. B., and Loktev, V. M. (1988). On a theory of the electronic spectrum and magnetic properties of high-tc superconductors, *phys. status solidi (b)* **147**, 1, pp. 307–319, doi:10.1002/pssb.2221470135, URL http://dx.doi.org/10.1002/pssb.2221470135.

Geim, A. K., and Novoselov, K. S. (2010). The rise of graphene, in *Nanoscience and Technology: A Collection of Reviews from Nature Journals* (World Scientific), pp. 11–19.

Gerlach, E., Grosse, P., Rautenberg, M., and Senske, W. (1976). Dynamical conductivity and plasmon excitation in Bi, *phys. status solidi (b)* **75**, 2, pp. 553–558, doi:10.1002/pssb.2220750218, URL https://onlinelibrary.wiley.com/doi/abs/10.1002/pssb.2220750218.

Ginsberg, D. M. (1992). *Physical Properties of High Temperature Superconductors III* (World Scientific), doi:10.1142/1577, URL https://www.worldscientific.com/doi/abs/10.1142/1577.

Ginzburg, V. (1989). Chapter 1: High-temperature superconductivity: Some remarks, submitted to "Progress in Low-Temperature Physics" in November 1987. (Elsevier), pp. 1–44, doi:https://doi.org/10.1016/S0079-6417(08)60040-2, URL http://www.sciencedirect.com/science/article/pii/S0079641708600402.

Ginzburg, V. L. (1970). The problem of high-temperature superconductivity. II, *Soviet Physics Uspekhi* **13**, 3, p. 335, URL http://stacks.iop.org/0038-5670/13/i=3/a=R03.

Ginzburg, V. L., and Kirzhnits, D. A. (1987). High-temperature superconductivity (a review of theoretical ideas), *Soviet Physics Uspekhi* **30**, 8, p. 671, URL http://stacks.iop.org/0038-5670/30/i=8/a=R02.

Gollnik, F., and Naito, M. (1998). Doping dependence of normal- and superconducting-state transport properties of $Nd_{2-x}Ce_xCuO_{4\pm y}$ thin films, *Phys. Rev. B* **58**, pp. 11734–11752, doi:10.1103/PhysRevB.58.11734, URL https://link.aps.org/doi/10.1103/PhysRevB.58.11734.

Gorbatsevich, A., and Kopaev, Y. (1991). Selfinduced superconducting pairing, *Physica C* **185–189**, pp. 1711–1712, doi:https://doi.org/10.1016/0921-4534(91)90982-5, URL http://www.sciencedirect.com/science/article/pii/0921453491909825.

Gorbatsevich, A. A., and Kopaev, I. V. (1990). Locally unstable crystal configurations and the electron mechanism of superconductivity, *Pisma v Zhurnal Eksperimentalnoi i Teoreticheskoi Fiziki* **51**, pp. 327–331.

Gor'kov, L. P., and Teitel'baum, G. B. (2008). The two-component physics in cuprates in the real space and in the momentum representation, in *J. Phys. Conf. Ser.*, Vol. 108 (IOP Publishing), p. 012009.

Gor'kov, L. P., and Teitel'baum, G. B. (2014). Two regimes in conductivity and the Hall coefficient of underdoped cuprates in strong magnetic fields, *J. Phys. Condensed Matter* **26**, 4, p. 042202, URL http://stacks.iop.org/0953-8984/26/i=4/a=042202.

Gor'kov, L. P., and Teitel'Baum, G. B. (2015). Two-component energy spectrum of cuprates in the pseudogap phase and its evolution with temperature and at charge ordering, *Sci. Rep.* **5**, p. 8524.

Greene, L. H. (2008). Private communication.

Grynberg, M., Le Toullec, R., and Balkanski, M. (1974). Dielectric function in HgTe between 8 and 300ø k, *Phys. Rev. B* **9**, pp. 517–526, doi:10.1103/

PhysRevB.9.517, URL https://link.aps.org/doi/10.1103/PhysRevB.9.517.

Gurauskas, E., Kavaliauskas, J., Krivaite, G., and Łileika, A. (1983). Infrared absorption spectra of Cd0.2Hg0.8Te crystals, *phys. status solidi (b)* **115**, 1, pp. 195–202, doi:10.1002/pssb.2221150122, URL https://onlinelibrary.wiley.com/doi/abs/10.1002/pssb.2221150122.

Ha, Y. S., Chung, H. M., Park, Y. W., Mun, M.-O., Lee, S., Bae, M.-K., and Lee, S.-I. (1995). Thermoelectric power of a single-phase hgba2ca2cu3oy superconductor, *J. Mater. Chem.* **5**, pp. 71–73, doi:10.1039/JM9950500071, URL http://dx.doi.org/10.1039/JM9950500071.

Haase, J., Slichter, C. P., and Williams, G. V. (2009). Evidence for two electronic components in high-temperature superconductivity from NMR, *J. Phys. Condensed Matter* **21**, 45, p. 455702.

Hagen, S. J., Lobb, C. J., Greene, R. L., Forrester, M. G., and Kang, J. H. (1990). Anomalous Hall effect in superconductors near their critical temperatures, *Phys. Rev. B* **41**, pp. 11630–11633, doi:10.1103/PhysRevB.41.11630, URL https://link.aps.org/doi/10.1103/PhysRevB.41.11630.

Halperin, B., and Rice, T. (1968). The excitonic state at the semiconductor-semimetal transition, a summary of this paper was presented at the meeting of the American Physical Society, at Toronto, Canada, June 1967. (Academic Press), pp. 115–192, doi:https://doi.org/10.1016/S0081-1947(08)60740-7, URL http://www.sciencedirect.com/science/article/pii/S0081194708607407.

Halperin, B. I., and Rice, T. M. (1968). Possible anomalies at a semimetal-semiconductor transistion, *Rev. Mod. Phys.* **40**, pp. 755–766, doi:10.1103/RevModPhys.40.755, URL https://link.aps.org/doi/10.1103/RevModPhys.40.755.

Harris, J. M., Yan, Y. F., Matl, P., Ong, N. P., Anderson, P. W., Kimura, T. and Kitazawa, K. (1995). Violation of Kohler's Rule in the normal-state magnetoresistance of $YBa_2Cu_3O_{7-\delta}$ and $La_2Sr_xCuO_4$, *Phys. Rev. Lett.* **75**, pp. 1391–1394, doi:10.1103/PhysRevLett.75.1391, URL https://link.aps.org/doi/10.1103/PhysRevLett.75.1391.

Harrison, N., and Sebastian, S. E. (2015). Magnetotransport signatures of a single nodal electron pocket constructed from Fermi arcs, *Phys. Rev. B* **92**, p. 224505, doi:10.1103/PhysRevB.92.224505, URL https://link.aps.org/doi/10.1103/PhysRevB.92.224505.

Harshman, D. R., Dow, J. D., and Fiory, A. T. (2011). Coexisting holes and electrons in high-T_c materials: Implications from normal state transport, *Philos. Mag.* **91**, 5, pp. 818–840, doi:10.1080/14786435.2010.527864, URL https://doi.org/10.1080/14786435.2010.527864.

Hebard, A., Rosseinsky, M., Haddon, R., Murphy, D., Glarum, S., Palstra, T., Ramirez, A., and Kortan, A. (1991). Superconductivity at 18 k in potassium-doped c60, *Nature* **350**, 6319, p. 600.

Hirsch, J., and Marsiglio, F. (1992a). Normal state properties of high-T_c oxides, *Physica C* **195**, 3, pp. 355–366, doi:https://doi.org/10.1016/0921-4534(92)90361-F, URL http://www.sciencedirect.com/science/article/pii/092145349290361F.

Hirsch, J. E., and Marsiglio, F. (1991). Hole superconductivity in oxides: A two-band model, *Phys. Rev. B* **43**, pp. 424–434, doi:10.1103/PhysRevB.43.424, URL https://link.aps.org/doi/10.1103/PhysRevB.43.424.

Hirsch, J. E., and Marsiglio, F. (1992b). London penetration depth in hole superconductivity, *Phys. Rev. B* **45**, pp. 4807–4818, doi:10.1103/PhysRevB.45.4807, URL https://link.aps.org/doi/10.1103/PhysRevB.45.4807.

Hoddeson, L., and Daitch, V. (eds.) (2002). *True Genius: The Life and Science of John Bardeen: The Only Winner of Two Nobel Prizes in Physics* (The National Academies Press, Washington, DC), ISBN 978-0-309-09511-2, doi:10.17226/10372, URL https://www.nap.edu/catalog/10372/true-genius-the-life-and-science-of-john-bardeen-the.

Hohenberg, P., and Kohn, W. (1964). Inhomogeneous Electron Gas, *Phys. Rev.* **136**, pp. 864–871, doi:10.1103/PhysRev.136.B864.

Holmes, A. J. T. (1996). A one-dimensional model of a negative ion source, *Plasma Sources Science and Technology* **5**, 3, p. 453, URL http://stacks.iop.org/0963-0252/5/i=3/a=014.

Honma, T., and Hor, P. H. (2015). Quantitative connection between the nanoscale electronic inhomogeneity and the pseudogap of Bi2Sr2CaCu2O8+δ superconductor, *Physica C* **509**, pp. 11–15, doi:https://doi.org/10.1016/j.physc.2014.12.003, URL http://www. sciencedirect. com/ science/article/pii/S0921453414003621.

Huard, V., Cox, R. T., Saminadayar, K., Arnoult, A., and Tatarenko, S. (2000). Bound states in optical absorption of semiconductor quantum wells containing a two-dimensional electron gas, *Phys. Rev. Lett.* **84**, pp. 187–190, doi:10.1103/PhysRevLett.84.187, URL https://link.aps.org/doi/10.1103/PhysRevLett.84.187.

Hurd, C. (1972). *The Hall Effect in Metals and Alloys*, International Congresses of Quantum Chemistry Series (Plenum Press), ISBN

9780306305306, URL https://books.google.com/books?id=DjFRAAAAMAAJ.

Hwu, S.-J., Song, S. N., Thiel, J., Poeppelmeier, K. R., Ketterson, J. B., and Freeman, A. J. (1987). High-T_c superconductivity in regions of possible compound formation: $Y_{2-x}Ba_xCuO_{4-x/2+\delta}$ and $Y_{2-x}Ba_{1+x}Cu_2O_{6-x/2+\delta}$, *Phys. Rev. B* **35**, pp. 7119–7121, doi:10.1103/PhysRevB.35.7119, URL https://link.aps.org/doi/10.1103/PhysRevB.35.7119.

Hylleraas, E. A. (1947). Electron affinity of positronium, *Phys. Rev.* **71**, pp. 491–493, doi:10.1103/PhysRev.71.491, URL https://link.aps.org/doi/10.1103/PhysRev.71.491.

Imada, M., Fujimori, A., and Tokura, Y. (1998). Metal-insulator transitions, *Rev. Mod. Phys.* **70**, pp. 1039–1263, doi:10.1103/RevModPhys.70.1039, URL https://link.aps.org/doi/10.1103/RevModPhys.70.1039.

Inkson, J. C., and Anderson, P. W. (1973). Comment on "Model for an exciton mechanism of superconductivity", *Phys. Rev. B* **8**, pp. 4429–4432, doi:10.1103/PhysRevB.8.4429, URL https://link.aps.org/doi/10.1103/PhysRevB.8.4429.

Jehl, G., Zetterer, T., Otto, H., Schützmann, J., Shulga, S., and Renk, K. (1992). Infrared reflectivity and conductivity of tl2ba2cacu2o8 crystals: Evidence for nonconducting behaviour of the tlo layers, *EPL (Europhys. Lett.)* **17**, 3, p. 255.

Jenkins, G. S., Schmadel, D. C., Sushkov, A. B., Gu, G. D., Kontani, H., and Drew, H. D. (2010). Terahertz Hall measurements on optimally doped single-crystal $Bi_2Sr_2CaCu_2O_{8+x}$, *Phys. Rev. B* **82**, p. 094518, doi:10.1103/PhysRevB.82.094518, URL https://link.aps.org/doi/10.1103/PhysRevB.82.094518.

Jérome, D., Rice, T. M., and Kohn, W. (1967). Excitonic insulator, *Phys. Rev.* **158**, pp. 462–475, doi:10.1103/PhysRev.158.462, URL https://link.aps.org/doi/10.1103/PhysRev.158.462.

Kaga, H. (1991). Slave-boson approach to the anomalous doping effects of the Hall constants of La2-xSrxCuO4, *Physica C* **172**, 5, pp. 400–406, doi:https://doi.org/10.1016/0921-4534(91)90205-D, URL http://www.sciencedirect.com/science/article/pii/092145349190205D.

Kaga, H., and Saikawa, T. (1992). Origins of the anomalous doping effects of the Hall constants of La2-xSrxCuO4 and Nd2-xCexCuO4-y, *Solid State Commun.* **82**, 3, pp. 189–193, doi:https://doi.org/10.1016/0038-1098(92)90262-8, URL http://www.sciencedirect.com/science/article/pii/0038109892902628.

Kaiser, A. (1997). Comparison of thermopower behaviour in different superconductors, *Physica C* **282-287**, pp. 1251–1252, doi:https://doi.org/10.1016/S0921-4534(97)00665-5, URL http://www.sciencedirect.com/science/article/pii/S0921453497006655, Proceedings of the International Conference on Materials and Mechanisms of Superconductivity High Temperature Superconductors V.

Kamihara, Y., Watanabe, T., Hirano, M., and Hosono, H. (2008). Iron-based layered superconductor La [O1-x F x] feas (x= 0.05- 0.12) with T c = 26 k, *J. Am. Chem. Soc.* **130**, 11, pp. 3296–3297.

Kaminski, A., Rosenkranz, S., Fretwell, H. M., Norman, M. R., Randeria, M., Campuzano, J. C., Park, J.-M., Li, Z. Z., and Raffy, H. (2006). Change of Fermi-surface topology in $Bi_2Sr_2CaCu_2O_{8+\delta}$ with doping, *Phys. Rev. B* **73**, p. 174511, doi:10.1103/PhysRevB.73.174511, URL https://link.aps.org/doi/10.1103/PhysRevB.73.174511.

Karppinen, M., Fukuoka, A., Niinistö, L., and Yamauchi, H. (1996). Determination of oxygen content and metal valences in oxide superconductors by chemical methods, *Superconductor Science and Technology* **9**, 3, p. 121.

Keren, A., Kanigel, A., and Bazalitsky, G. (2006). Evidence for two fluids in cuprate superconductors from a nuclear resonance study of (Ca x La 1-x)(Ba 1.75- x La 0.25+ x) Cu 3 O y, *Phys. Rev. B* **74**, 17, p. 172506.

Ketterson, J. B. (2000). Private communication.

Kheng, K., Cox, R. T., d' Aubigné, M. Y., Bassani, F., Saminadayar, K., and Tatarenko, S. (1993). Observation of negatively charged excitons x^- in semiconductor quantum wells, *Phys. Rev. Lett.* **71**, pp. 1752–1755, doi:10.1103/PhysRevLett.71.1752, URL https://link.aps.org/doi/10.1103/PhysRevLett.71.1752.

Kiehl, W., Duan, H., and Hermann, A. (1995). Anisotropic Seebeck coefficient in Tl2Ba2CuO6-δ single crystals, *Physica C* **253**, 3, pp. 271–278, doi:https://doi.org/10.1016/0921-4534(95)00448-3, URL http://www.sciencedirect.com/science/article/pii/0921453495004483.

Kim, H.-J., Chowdhury, P., Jo, I.-S., and Lee, S.-I. (2002). Equilibrium magnetization of $Tl_2Ba_2CaCu_2O_{8+\delta}$ single crystals, *Phys. Rev. B* **66**, p. 134508, doi:10.1103/PhysRevB.66.134508, URL https://link.aps.org/doi/10.1103/PhysRevB.66.134508.

Kim, J. H., Levin, K., and Auerbach, A. (1989). Assessment of Fermi-liquid description for the normal state of high-T_c superconductors, *Phys. Rev. B* **39**, 16, p. 11633.

Kim, Y. E. (2009). Theory of Bose–Einstein condensation mechanism for deuteron-induced nuclear reactions in micro/nano-scale metal grains and particles, *Naturwissenschaften* **96**, 7, pp. 803–811, doi: 10.1007/s00114-009-0537-6, URL https:// doi.org/ 10.1007/s00114-009-0537-6.

Kittel, C. (1996). *Introduction to Solid State Physics* (Wiley), ISBN 9780471111818, URL https://books.google.com/books?id=1X8pAQAAMAAJ.

Klemm, R. A. (2005). The phase-sensitive c-axis twist experiments on Bi2Sr2CaCu2O8+δ and their implications, *Philos. Mag.* **85**, 8, pp. 801–853, doi:10.1080/14786430412331314573.

Koch, W., and Holthausen, M. C. (2015). *A Chemist's Guide to Density Functional Theory* (John Wiley & Sons).

Kohn, W. (1967). Excitonic phases, *Phys. Rev. Lett.* **19**, pp. 439–442, doi: 10.1103/PhysRevLett.19.439, URL https://link.aps.org/doi/10.1103/PhysRevLett.19.439.

Kohn, W., and Sham, L. J. (1965). Self-consistent equations including exchange and correlation effects, *Phys. Rev.* **140**, pp. A1133–A1138, doi: 10.1103/PhysRev.140.A1133, URL https://link.aps.org/doi/10.1103/PhysRev.140.A1133.

Kokalj, A. (1999). Xcrysden–a new program for displaying crystalline structures and electron densities, *J. Mol. Graph. Model.* **17**, 3-4, pp. 176–179.

Kopnin, N. B., Ivlev, B. I., and Kalatsky, V. A. (1993). The flux-flow Hall effect in type II superconductors. An explanation of the sign reversal, *J. Low Temperature Phys.* **90**, 1, pp. 1–13, doi:10.1007/BF00682008, URL https://doi.org/10.1007/BF00682008.

Kosterlitz, J. M., and Thouless, D. J. (1973). Ordering, metastability and phase transitions in two-dimensional systems, *J. Phys. C Solid State Phys.* **6**, 7, p. 1181.

Kristoffel, N., Ord, T., and Rubin, P. (2006). Doping dependence of cuprate coherence length, supercarrier effective mass and penetration depth in a two-component scenario, *Electron Correlation in New Materials and Nanosystems* , p. 275.

Kubler, J., Hock, K. H., Sticht, J., and Williams, A. R. (1988). Density functional theory of non-collinear magnetism, *J. Phys. F Metal Phys.* **18**, 3, p. 469, URL http://stacks.iop.org/0305-4608/18/i=3/a=018.

Kubo, Y., and Manako, T. (1992). T2-dependence of inverse Hall mobility observed in overdoped Tl-cuprates, *Physica C* **197**, 3, pp. 378–384, doi: https://doi.org/10.1016/0921-4534(92)90020-D, URL http://www.sciencedirect.com/science/article/pii/092145349290020D.

Kubo, Y., and Manako, T. (1994). T2-dependence of inverse Hall mobility observed in overdoped tl-cuprates, *Physica B: Condensed Matter* **194–196**, pp. 2047–2048, doi:https://doi.org/10.1016/0921-4526(94)91523-7, URL http://www.sciencedirect.com/science/article/pii/0921452694915237.

Kuroda, Y., and Varma, C. M. (1990). Superconductive state of Cu-O metals, *Phys. Rev. B* **42**, pp. 8619–8622, doi:10.1103/PhysRevB.42.8619, URL https://link.aps.org/doi/10.1103/PhysRevB.42.8619.

Kusar, P., Demsar, J., Mihailovic, D., and Sugai, S. (2005). A systematic study of femtosecond quasiparticle relaxation processes in $La_{2-x}Sr_xCuO_4$, *Phys. Rev. B* **72**, p. 014544, doi:10.1103/PhysRevB.72.014544, URL https://link.aps.org/doi/10.1103/PhysRevB.72.014544.

Lampert, M. A. (1958). Mobile and immobile effective-mass-particle complexes in nonmetallic solids, *Phys. Rev. Lett.* **1**, pp. 450–453, doi: 10.1103/PhysRevLett.1.450, URL https://link.aps.org/doi/10.1103/PhysRevLett.1.450.

Latyshev, Y. I., Orlov, A. P., Nikitina, A. M., Monceau, P., and Klemm, R. A. (2004). *c*-Axis transport in naturally grown $Bi_2Sr_2CaCu_2O_{8+\delta}$ cross-whisker junctions, *Phys. Rev. B* **70**, p. 094517, doi:10.1103/ PhysRevB.70.094517, URL https://link.aps.org/doi/10.1103/PhysRevB.70.094517.

Laughlin, R. B. (1983). Anomalous quantum Hall effect: An incompressible quantum fluid with fractionally charged excitations, *Phys. Rev. Lett.* **50**, pp. 1395–1398, doi:10.1103/PhysRevLett.50.1395, URL https://link.aps.org/doi/10.1103/PhysRevLett.50.1395.

LeBoeuf, D., Doiron-Leyraud, N., Levallois, J., Daou, R., Bonnemaison, J.-B., Hussey, N., Balicas, L., Ramshaw, B., Liang, R., Bonn, D. et al. (2007). Electron pockets in the Fermi surface of hole-doped high-T_c superconductors, *Nature* **450**, 7169, p. 533.

Lee, P. A., Nagaosa, N., and Wen, X.-G. (2006). Doping a Mott insulator: Physics of high-temperature superconductivity, *Rev. Mod. Phys.* **78**, 1, p. 17.

Leggett, A. J. (1980). Diatomic molecules and cooper pairs, in *Modern Trends in the Theory of Condensed Matter* (Springer), pp. 13–27.

Leggett, A. (1998). Where is the energy saved in cuprate superconductivity? *J. Phys. Chem. Solids* **59**, 10, pp. 1729–1732, doi:https://doi.org/10.1016/S0022-3697(98)00091-2, URL http://www.sciencedirect.com/science/article/pii/S0022369798000912.

Leggett, A. J. (1999). Cuprate superconductivity: Dependence of T_c on the c-axis layering structure, *Phys. Rev. Lett.* **83**, pp. 392–395, doi:10.1103/PhysRevLett.83.392.

Leggett, A. J. (1999). A "midinfrared" scenario for cuprate superconductivity, *Proc. Natl. Acad. Sci.* **96**, 15, pp. 8365–8372, doi:10.1073/pnas.96.15.8365, URL http://www.pnas.org/content/96/15/8365.

Leggett, A. J. (2006). Some thoughts about two dimensionality and cuprate superconductivity, *J. Superconductivity Novel Magnetism* **19**, 3, pp. 187–192, doi:10.1007/s10948-006-0154-y, URL https://doi.org/10.1007/s10948-006-0154-y.

Li, Q., Tsay, Y. N., Suenaga, M., Klemm, R. A., Gu, G. D., and Koshizuka, N. (1999). $Bi_2Sr_2CaCu_2O_{8+\delta}$ bicrystal *c*-axis twist Josephson junctions: A new phase-sensitive test of order parameter symmetry, *Phys. Rev. Lett.* **83**, pp. 4160–4163, doi:10.1103/PhysRevLett.83.4160, URL https://link.aps.org/doi/10.1103/PhysRevLett.83.4160.

Li, Z. G., Feng, H. H., Yang, Z. Y., Hamed, A., Ting, S. T., Hor, P. H., Bhavaraju, S., DiCarlo, J. F., and Jacobson, A. J. (1996). Carrier-controlled doping efficiency in $La_2CuO_{4+\delta}$, *Phys. Rev. Lett.* **77**, pp. 5413–5416, doi:10.1103/PhysRevLett.77.5413, URL https://link.aps.org/doi/10.1103/PhysRevLett.77.5413.

Lieb, E. H. (1984a). Atomic and molecular negative ions, *Phys. Rev. Lett.* **52**, pp. 315–317, doi:10.1103/PhysRevLett.52.315, URL https://link.aps.org/doi/10.1103/PhysRevLett.52.315.

Lieb, E. H. (1984b). Bound on the maximum negative ionization of atoms and molecules, *Phys. Rev. A* **29**, pp. 3018–3028, doi:10.1103/PhysRevA.29.3018, URL https://link.aps.org/doi/10.1103/PhysRevA.29.3018.

Little, W. A. (1964). Possibility of synthesizing an organic superconductor, *Phys. Rev.* **134**, pp. A1416–A1424, doi:10.1103/PhysRev.134.A1416, URL https://link.aps.org/doi/10.1103/PhysRev.134.A1416.

Little, W. A. (1965). Superconductivity at room temperature, *Sci. Am.* **212**, pp. 21–27, doi:10.1038/scientificamerican0265-21.

Loram, J. W., Mirza, K. A., Cooper, J. R., and Liang, W. Y. (1993). Electronic specific heat of $YBa_2Cu_3O_{6+x}$ from 1.8 to 300 k, *Phys. Rev. Lett.* **71**, pp. 1740–1743, doi:10.1103/PhysRevLett.71.1740, URL https://link.aps.org/doi/10.1103/PhysRevLett.71.1740.

Lovett, D. (1977). *Semimetals & Narrow-Bandgap Semiconductors*, Applied Physics Series (Pion), ISBN 9780850860603, URL https://books.google.com/books?id=pT7xAAAAIAAJ.

Ludbrook, B., Levy, G., Nigge, P., Zonno, M., Schneider, M., Dvorak, D., Veenstra, C., Zhdanovich, S., Wong, D., Dosanjh, P., et al. (2015). Evidence for superconductivity in li-decorated monolayer graphene, *Proc. Natl. Acad. Sci.* **112**, 38, pp. 11795–11799.

Lundqvist, S., and March, N. (1983). *Theory of the Inhomogeneous Electron Gas*, Physics of Solids and Liquids (Springer US), ISBN 9780306412073, URL https://books.google.com/books?id=mqa4Q90dTz0C.

Luo, N. (2000a). An Alternative Explanation on the Two Relaxation Rates in Cuprate Superconductors, *eprint arXiv:cond-mat/0011371* .

Luo, N. (2000b). Subtle features in transport properties: Evidence for a possible coexistence of holes and electrons in cuprate superconductors, *eprint arXiv:cond-mat/0003074* .

Luo, N. (2001). Electron-like carriers in p-type superconductor La2-xSrxCuO4-δ: A joint transport, infrared and defect chemistry study, *Physica C* **350**, 1, pp. 132–138, doi:https://doi.org/10.1016/S0921-4534(00)01558-6, URL http://www.sciencedirect.com/science/article/pii/S0921453400015586.

Luo, N. (2001). Variational bounds on the ground-state energy of three electrons and one hole in two-dimension, *eprint arXiv:cond-mat/0101004*.

Luo, N. (2008). Electronlike quasiparticles in hole-doped high-T_c superconductors, Presented at the 2008 GRC Correlated Electron Systems, University of New England, Biddeford, ME.

Luo, N., and Miley, G. (2002). Kohler's rule and relaxation rates in high-T_c superconductors, *Physica C* **371**, 4, pp. 259–269, doi:https://doi.org/10.1016/S0921-4534(01)01101-7, URL http://www.sciencedirect.com/science/article/pii/S0921453401011017.

Luo, N., and Miley, G. H. (2009). An alternative theory on relaxation rates in cuprate superconductors, *J. Phys. Condensed Matter* **21**, 2, p. 025701, URL http://stacks.iop.org/0953-8984/21/i=2/a=025701.

Ma, F., and Lu, Z.-Y. (2008). Iron-based layered compound LaFeAsO is an antiferromagnetic semimetal, *Phys. Rev. B* **78**, p. 033111, doi:10.1103/PhysRevB.78.033111, URL https://link.aps.org/doi/10.1103/PhysRevB.78.033111.

Maebashi, H., and Fukuyama, H. (1998). Electrical conductivity of interacting fermions. II. Effects of normal scattering processes in the presence of umklapp scattering processes, *J. Phys. Soc. Jpn.* **67**, 1, pp. 242–251, doi:10.1143/JPSJ.67.242, URL https://doi.org/10.1143/JPSJ.67.242.

Markiewicz, R. (1990). Van Hove excitons and high-T_c superconductivity, *Physica B: Condensed Matter* **165-166**, pp. 1053–1054, doi: https://doi.org/10.1016/S0921-4526(09)80112-9, URL http://www.sciencedirect.com/science/article/pii/S0921452609801129, lT-19.

Markiewicz, R. S. (1991). Excitons at a van Hove singularity, *J. Phys. Condensed Matter* **3**, 21, p. 3859, URL http://stacks.iop.org/0953-8984/3/i=21/a=019.

Markiewicz, R. S., Buda, I. G., Mistark, P., and Bansil, A. (2015). Van Hove singularity as the driver of pseudogap physics in cuprate high-T superconductors, *ArXiv e-prints* .

Markiewicz, R. S., Sahrakorpi, S., Lindroos, M., Lin, H., and Bansil, A. (2005). One-band tight-binding model parametrization of the high-T_c cuprates including the effect of k_z dispersion, *Phys. Rev. B* **72**, p. 054519, doi:10.1103/PhysRevB.72.054519, URL https://link.aps.org/doi/10.1103/PhysRevB.72.054519.

Martin, C., Hejtmanek, J., Simon, C., Maignan, A., and Raveau, B. (1995). Systematic thermopower measurements of the thallium cuprates Tl(Ba,Sr)2Cam-1CumO2m+3-δ and Tl2Ba2Cam-1CumO2m+4+δ, *Physica C* **250**, 3, pp. 235–239, doi: https://doi.org/10.1016/0921-4534(95)00366-5, URL http://www.sciencedirect.com/science/article/pii/0921453495003665.

Martin, S., Fiory, A. T., Fleming, R. M., Schneemeyer, L. F., and Waszczak, J. V. (1990). Normal-state transport properties of $Bi_{2+x}Sr_{2-y}CuO_{6+\delta}$ crystals, *Phys. Rev. B* **41**, pp. 846–849, doi:10.1103/PhysRevB.41.846, URL https://link.aps.org/doi/10.1103/PhysRevB.41.846.

Massey, H. (1938). *Negative ions*, Cambridge Physical Tracts (Cambridge University Press), URL https://books.google.com/books?id=MUxIAQAAIAAJ.

Matthias, B. T. (1955). Empirical relation between superconductivity and the number of valence electrons per atom, *Phys. Rev.* **97**, pp. 74–76, doi:10.1103/PhysRev.97.74, URL https://link.aps.org/doi/10.1103/PhysRev.97.74.

Matthias, B. T., Geballe, T. H., and Compton, V. B. (1963). Superconductivity, *Rev. Mod. Phys.* **35**, pp. 1–22, doi:10.1103/RevModPhys.35.1, URL https://link.aps.org/doi/10.1103/RevModPhys.35.1.

McMillan, W. L. (1968). Transition temperature of strong-coupled superconductors, *Phys. Rev.* **167**, pp. 331–344, doi:10.1103/PhysRev.167.331, URL https://link.aps.org/doi/10.1103/PhysRev.167.331.

Meaden, G. (1965). *Electrical Resistance of Metals*, International Congresses of Quantum Chemistry Series (Plenum Press), URL https://books.google.com/books?id=vSl8AAAAIAAJ.

Miao, H., Lorenzana, J., Seibold, G., Peng, Y., Amorese, A., Yakhou-Harris, F., Kummer, K., Brookes, N., Konik, R., Thampy, V. et al. (2017). High-temperature charge density wave correlations in La1. 875Ba0. $125CuO_4$ without spin–charge locking, *Proc. Natl. Acad. Sci.*, p. 201708549.

Michenaud, J. P., and Issi, J. P. (1972). Electron and hole transport in bismuth, *J. Phys. C Solid State Phys.* **5**, 21, p. 3061, URL http://stacks.iop.org/0022-3719/5/i=21/a=011, p. Michenaud confirmed to the author that tick mark labels for the vertical axis of Fig. 2 should be 10^{-7}, 10^{-8} and 10^{-9} downwards. Data are from Table 2.

Mikhailovsky, A., Shulga, S., Karakozov, A., Dolgov, O., and Maksimov, E. (1991). Thermal pair-breaking in superconductors with strong electron-phonon interaction, *Solid State Commun.* **80**, 7, pp. 511–515, doi:https://doi.org/10.1016/0038-1098(91)90062-Z, URL http://www.sciencedirect.com/science/article/pii/003810989190062Z.

Mirzaei, S. I., Stricker, D., Hancock, J. N., Berthod, C., Georges, A., van Heumen, E., Chan, M. K., Zhao, X., Li, Y., Greven, M., Barišić, N. and van der Marel, D. (2013). Spectroscopic evidence for Fermi liquid-like energy and temperature dependence of the relaxation rate in the pseudogap phase of the cuprates, *Proc. Natl. Acad. Sci.* **110**, 15, pp. 5774–5778, doi:10.1073/pnas.1218846110, URL http://www.pnas.org/content/110/15/5774.

Mitsuhashi, R., Suzuki, Y., Yamanari, Y., Mitamura, H., Kambe, T., Ikeda, N., Okamoto, H., Fujiwara, A., Yamaji, M., Kawasaki, N. et al. (2010). Superconductivity in alkali-metal-doped picene, *Nature* **464**, 7285, p. 76.

Mott, N. F. (1961). The transition to the metallic state, *Philos. Mag.* **6**, 62, pp. 287–309, doi:10.1080/14786436108243318, URL https://doi.org/10.1080/14786436108243318.

Müller, K. A., and Giorgio, B. (1993). *Phase Separation in Cuprate Superconductors—Proceedings of the Workshop*, The science and culture Series—Physics (World Scientific Publishing Company), ISBN 9789814553513, URL https://books.google.com/books?id=jIbsCgAAQBAJ.

Nagamatsu, J., Nakagawa, N., Muranaka, T., Zenitani, Y., and Akimitsu, J. (2001). Superconductivity at 39 K in magnesium diboride, *nature* **410**, 6824, p. 63.

Nagaoka, T., Matsuda, Y., Obara, H., Sawa, A., Terashima, T., Chong, I., Takano, M., and Suzuki, M. (1998). Hall anomaly in the superconducting state of high-T_c cuprates: Universality in doping dependence, *Phys. Rev. Lett.* **80**, pp. 3594–3597, doi:10.1103/PhysRevLett.80.3594, URL https://link.aps.org/doi/10.1103/PhysRevLett.80.3594.

Nagata, S., and Atake, T. (1999). Survey of chalcogenide superconductors, *J. Thermal Anal. Calorimetry* **57**, 3, pp. 807–821, doi:10.1023/A:1010142225299, URL https://doi.org/10.1023/A:1010142225299.

Nakai, Y., Ishida, K., Kamihara, Y., Hirano, M., and Hosono, H. (2008). Evolution from itinerant antiferromagnet to unconventional superconductor with fluorine doping in LaFeAs(O1-xFx) revealed by 75As and 139La nuclear magnetic resonance, *J. Phys. Soc. Jpn.* **77**, 7, p. 073701, doi:10.1143/JPSJ.77.073701, URL https://doi.org/10.1143/JPSJ.77.073701.

Nakayama, K., Sato, T., Dobashi, T., Terashima, K., Souma, S., Matsui, H., Takahashi, T., Campuzano, J. C., Kudo, K., Sasaki, T., Kobayashi, N., Kondo, T., Takeuchi, T., Kadowaki, K., Kofu, M., and Hirota, K. (2006). Shadow bands in single-layered $Bi_2Sr_2CuO_{6+\delta}$ studied by angle-resolved photoemission spectroscopy, *Phys. Rev. B* **74**, p. 054505, doi:10.1103/PhysRevB.74.054505, URL https://link.aps.org/doi/10.1103/PhysRevB.74.054505.

Nguyen, N., Choisnet, J., Hervieu, M., and Raveau, B. (1981). Oxygen defect K2NiF4-type oxides: The compounds La2-xSrxCuO4-x2+δ, *J. Solid State Chem.* **39**, 1, pp. 120–127, doi:https://doi.org/10.1016/0022-4596(81)90310-8, URL http://www.sciencedirect.com/science/article/pii/0022459681903108.

Nideröst, M., Frassanito, R., Saalfrank, M., Mota, A. C., Blatter, G., Zavaritsky, V. N., Li, T. W., and Kes, P. H. (1998). Lower critical field H_{c1} and barriers for vortex entry in $Bi_2Sr_2CaCu_2O_{8+\delta}$ crystals, *Phys. Rev. Lett.* **81**, pp. 3231–3234, doi:10.1103/PhysRevLett.81.3231, URL https://link.aps.org/doi/10.1103/PhysRevLett.81.3231.

Niedermayer, C., Bernhard, C., Binninger, U., Glückler, H., Tallon, J. L., Ansaldo, E. J., and Budnick, J. I. (1993). Muon spin rotation study of the correlation between T_c and n_s/m^* in overdoped $Tl_2Ba_2CuO_{6+\delta}$, *Phys. Rev. Lett.* **71**, pp. 1764–1767, doi:10.1103/PhysRevLett.71.1764, URL https://link.aps.org/doi/10.1103/PhysRevLett.71.1764.

Nomura, Y., Sakai, S., Capone, M., and Arita, R. (2015). Unified understanding of superconductivity and Mott transition in alkali-doped fullerides from first principles, *Science advances* **1**, 7, p. e1500568.

Norman, M. R. (2010). Trend: Fermi-surface reconstruction and the origin of high-temperature superconductivity, *Physics* **3**, p. 86.

Norman, M. R., and Proust, C. (2014). Focus on fermiology of the cuprates, *N. J. Phys.* **16**, 4, p. 045004.

Noto, K., Shinzawa, S., and Muto, Y. (1976). Hall effect in intrinsic type II superconductors near the lower critical field, *Solid State Commun.* **18**, 8, pp. 1081–1084, doi:https://doi.org/10.1016/0038-1098(76)91245-X, URL http://www.sciencedirect.com/science/article/pii/003810987691245X.

Ono, S., Ando, Y., Murayama, T., Balakirev, F. F., Betts, J. B., and Boebinger, G. S. (2000). Metal-to-insulator crossover in the low-temperature normal state of $Bi_2Sr_{2-x}La_xCuO_{6+\delta}$, *Phys. Rev. Lett.* **85**, pp. 638–641, doi:10.1103/PhysRevLett.85.638, URL https://link.aps.org/doi/10.1103/PhysRevLett.85.638.

Orlando, M., Rouver, A., Rocha, J., and Cavichini, A. (2018). Correlation among the effective mass (m*), λ_{ab} and T_c of superconducting cuprates in a Casimir energy scenario, *Phys. Lett. A* **382**, 22, pp. 1486–1491, doi:https://doi.org/10.1016/j.physleta.2018.04.001, URL http://www.sciencedirect.com/science/article/pii/S0375960118303529.

Padilla, D. J. (2010). Hall effect study of iron-arsenide-based superconducting compounds under pressure, PhD thesis, San Diego State University.

Padilla, W. J., Lee, Y. S., Dumm, M., Blumberg, G., Ono, S., Segawa, K., Komiya, S., Ando, Y., and Basov, D. N. (2005). Constant effective mass across the phase diagram of high-T_c cuprates, *Phys. Rev. B* **72**, p. 060511, doi:10.1103/PhysRevB.72.060511, URL https://link.aps.org/doi/10.1103/PhysRevB.72.060511.

Parr, R., and Weitao, Y. (1994). *Density-Functional Theory of Atoms and Molecules*, International Series of Monographs on Chemistry (Oxford University Press), ISBN 9780195357738, URL https://books.google.com/books?id=mGOpScSIwU4C.

Patil, S., Medicherla, V. R. R., Ali, K., Singh, R. S., Manfrinetti, P., Wrubl, F., Dhar, S. K., and Maiti, K. (2017). Observation of pseudogap in MgB 2, *J. Phys. Condensed Matter* **29**, 46, p. 465504, URL http://stacks.iop.org/0953-8984/29/i=46/a=465504.

Pavarini, E., Dasgupta, I., Saha-Dasgupta, T., Jepsen, O., and Andersen, O. K. (2001). Band-structure trend in hole-doped cuprates and correlation with T_{cmax}, *Phys. Rev. Lett.* **87**, p. 047003, doi:10.1103/PhysRevLett.87.047003, URL https://link.aps.org/doi/10.1103/PhysRevLett.87.047003.

Peeters, F., Riva, C., and Varga, K. (2001). Trions in quantum wells, *Physica B: Condensed Matter* **300**, 1, pp. 139–155, doi:https://doi.org/10.1016/S0921-4526(01)00577-4, URL http://www.sciencedirect.com/science/article/pii/S0921452601005774, jubilee issue Volume 300.

Pelc, D., Popčević, P., Požek, M., Greven, M., and Barišić, N. (2019). Unusual behavior of cuprates explained by heterogeneous charge localization, *Science Advances* **5**, 1, doi:10.1126/sciadv.aau4538, URL https://advances.sciencemag.org/content/5/1/eaau4538.

Pickett, W. E. (1989). Electronic structure of the high-temperature oxide superconductors, *Reviews of Modern Physics* **61**, 2, p. 433.

Pines, D. (1994). dx2-y2 pairing and spin fluctuations in the cuprate superconductors: Experiment meets theory, *Physica C* **235-240**, pp. 113–121, doi:https://doi.org/10.1016/0921-4534(94)91327-7, URL http://www.sciencedirect.com/science/article/pii/0921453494913277.

Prozorov, R., Giannetta, R. W., Carrington, A., Fournier, P., Greene, R. L., Guptasarma, P., Hinks, D. G., and Banks, A. R. (2000). Measurements of the absolute value of the penetration depth in high-T_c superconductors using a low-T_c superconductive coating, *Appl. Phys. Lett.* **77**, 25, pp. 4202–4204, doi:10.1063/1.1328362, URL https://doi.org/10.1063/1.1328362.

Puźniak, R., Usami, R., and Yamauchi, H. (1996). Overdoped regime of the high-T_c superconductor $HgBa_2CuO_{4+\delta}$ and the relation between normal and superconducting carrier densities, *Phys. Rev. B* **53**, pp. 86–89, doi:10.1103/PhysRevB.53.86, URL https://link.aps.org/doi/10.1103/PhysRevB.53.86.

Puźniak, R. (1997). Temperature dependence of superconducting carrier density and effective mass in HgBa2Can- 1CunOy, *Physica C* **282**, pp. 1459–1460.

Quan, Y., and Pickett, W. E. (2016). Van Hove singularities and spectral smearing in high-temperature superconducting H_3S, *Phys. Rev. B* **93**, p. 104526, doi:10.1103/PhysRevB.93.104526, URL https://link.aps.org/doi/10.1103/PhysRevB.93.104526.

Quijada, M. A., Tanner, D. B., Kelley, R. J., Onellion, M., Berger, H., and Margaritondo, G. (1999). Anisotropy in the ab-plane optical properties of $Bi_2Sr_2CaCu_2O_8$ single-domain crystals, *Phys. Rev. B* **60**, pp. 14917–14934, doi:10.1103/PhysRevB.60.14917, URL https://link.aps.org/doi/10.1103/PhysRevB.60.14917.

Randeria, M., and Taylor, E. (2014). Crossover from Bardeen-Cooper-Schrieffer to Bose–Einstein condensation and the unitary Fermi gas, *Annu. Rev. Condensed Matter Phys.* **5**, 1, pp. 209–232, doi:10.1146/annurev-conmatphys-031113-133829, URL https://doi.org/10.1146/annurev-conmatphys-031113-133829.

Raveau, B., Michel, C., Hervieu, M., and Groult, D. (1991). *Crystal Chemistry of High-T_c Superconducting Copper Oxides*, Springer Series in Materials Science (Springer Berlin Heidelberg), ISBN 9783540515456, URL https://books.google.com/books?id=_WZ_mQEACAAJ.

Razzoli, E., Sassa, Y., Drachuck, G., Månsson, M., Keren, A., Shay, M., Berntsen, M. H., Tjernberg, O., Radovic, M., Chang, J., Pailhès, S., Momono, N., Oda, M., Ido, M., Lipscombe, O. J., Hayden, S. M., Patthey, L., Mesot, J. and Shi, M. (2010). The Fermi surface and band folding in La2-xSrxCuO4 , probed by angle-resolved photoemission, *N. J. Phys.* **12**, 12, p. 125003, URL http://stacks.iop.org/1367-2630/12/i=12/a=125003.

Ren, Z.-A., Lu, W., Yang, J., Yi, W., Shen, X.-L., Li, Z.-C., Che, G.-C., Dong, X.-L., Sun, L.-L., Zhou, F., and Zhao, Z.-X. (2008). Superconductivity at 55 K in iron-based F-doped layered quaternary compound Sm[O1-xFx]FeAs, *ArXiv e-prints* .

Romberg, H., Alexander, M., Nücker, N., Adelmann, P., and Fink, J. (1990). Electronic structure of the system $La_{2-x}Sr_xCuO_{4+\delta}$, *Phys. Rev. B* **42**, pp. 8768–8771, doi:10.1103/PhysRevB.42.8768, URL https://link.aps.org/doi/10.1103/PhysRevB.42.8768.

Romero, D., Porter, C., Tanner, D., Forro, L., Mandrus, D., Mihaly, L., Carr, G., and Williams, G. (1992a). On the phenomenology of the infrared properties of the copper-oxide superconductors, *Solid State Commun.* **82**, 3, pp. 183–187, doi:https://doi.org/10.1016/0038-1098(92)90261-7, URL http://www.sciencedirect.com/science/article/pii/0038109892902617.

Romero, D. B., Porter, C. D., Tanner, D. B., Forro, L., Mandrus, D., Mihaly, L., Carr, G. L., and Williams, G. P. (1992b). Quasiparticle damping in $Bi_2Sr_2CaCu_2O_8$ and $Bi_2Sr_2CuO_6$, *Phys. Rev. Lett.* **68**, pp. 1590–1593, doi:10.1103/PhysRevLett.68.1590, URL https://link.aps.org/doi/10.1103/PhysRevLett.68.1590.

Rourke, P. M. C., Bangura, A. F., Benseman, T. M., Matusiak, M., Cooper, J. R., Carrington, A., and Hussey, N. E. (2010). A detailed de Haas–van Alphen effect study of the overdoped cuprate Tl 2 Ba 2 CuO 6+δ, *N. J. Phys.* **12**, 10, p. 105009, URL http://stacks.iop.org/1367-2630/12/i=10/a=105009.

Sahrakorpi, S., Lindroos, M., Markiewicz, R. S., and Bansil, A. (2005). Evolution of midgap states and residual three dimensionality in $La_{2-x}Sr_xCuO_4$, *Phys. Rev. Lett.* **95**, p. 157601, doi: 10.1103/PhysRevLett.95.157601, URL https://link.aps.org/doi/10.1103/PhysRevLett.95.157601.

Samara, G. A., Hammetter, W. F., and Venturini, E. L. (1990). Temperature and frequency dependences of the dielectric properties of $YBa_2Cu_3O_{6+x}$ ($x \approxeq 0$), *Phys. Rev. B* **41**, pp. 8974–8980, doi:10.1103/PhysRevB.41.8974, URL https://link.aps.org/doi/10.1103/PhysRevB.41.8974.

Sandratskii, L. M., and Guletskii, P. G. (1986). Symmetrised method for the calculation of the band structure of noncollinear magnets, *J. Phys. F Metal Phys.* **16**, 2, p. L43, URL http://stacks.iop.org/0305-4608/16/i=2/a=002.

Sanna, S., Coneri, F., Rigoldi, A., Concas, G., and De Renzi, R. (2008). Experimental evidence of two distinct charge carriers in underdoped cuprate superconductors, *Phys. Rev. B* **77**, 22, p. 224511.

Sano, W., Koretsune, T., Tadano, T., Akashi, R., and Arita, R. (2016). Effect of van Hove singularities on high-T_c superconductivity in h_3S, *Phys. Rev. B* **93**, p. 094525, doi:10.1103/PhysRevB.93.094525, URL https://link.aps.org/doi/10.1103/PhysRevB.93.094525.

Sato, M., Shamoto, S., Onoda, M., Sera, M., Fukuda, K., Hosoya, S., Akimitsu, J., Ekino, T., and Imaeda, K. (1987). *Single Crystal Studies and Electron Tunneling of $(La_{1-x}M_x)_2CuO_{4-\delta}$ (M=Ba and Sr)* (Plenum, New York), p. 927, doi:10.1007/978-1-4613-1937-5, URL https://www.springer.com/us/book/9780306426919.

Scheller, M. K., Compton, R. N., and Cederbaum, L. S. (1995). Gas-phase multiply charged anions, *Science* **270**, pp. 1160–1166, doi:10.1126/science.270.5239.1160.

Schlesinger, Z., Collins, R. T., Holtzberg, F., Feild, C., Blanton, S. H., Welp, U., Crabtree, G. W., Fang, Y., and Liu, J. Z. (1990a). Superconducting energy gap and normal-state conductivity of a single-domain $YBa_2Cu_3O_7$ crystal, *Phys. Rev. Lett.* **65**, pp. 801–804, doi:10.1103/PhysRevLett.65.801, URL https://link.aps.org/doi/10.1103/PhysRevLett.65.801.

Schlesinger, Z., Collins, R. T., Holtzberg, F., Feild, C., Koren, G., and Gupta, A. (1990b). Infrared studies of the superconducting energy gap and normal-state dynamics of the high-T_c superconductor $YBa_2Cu_3O_7$, *Phys. Rev. B* **41**, pp. 11237–11259, doi:10.1103/PhysRevB.41.11237, URL https://link.aps.org/doi/10.1103/PhysRevB.41.11237.

Schlesinger, Z., Rotter, L., Collins, R., Holtzberg, F., Feild, C., Welp, U., Crabtree, G., Liu, J., Fang, Y., and Vandervoort, K. (1991). Infrared properties of high T_c superconductors, *Physica C* **185-189**, pp. 57–64, doi:https://doi.org/10.1016/0921-4534(91)91950-9, URL http://www.sciencedirect.com/science/article/pii/0921453491919509.

Schrieffer, J. R., Wen, X.-G., and Zhang, S.-C. (1988). Spin-bag mechanism of high-temperature superconductivity, *Phys. Rev. Lett.* **60**, pp. 944–947, doi:10.1103/PhysRevLett.60.944, URL https://link.aps.org/doi/10.1103/PhysRevLett.60.944.

Sebastian, S. E., Harrison, N., Goddard, P. A., Altarawneh, M. M., Mielke, C. H., Liang, R., Bonn, D. A., Hardy, W. N., Andersen, O. K., and Lonzarich, G. G. (2010). Compensated electron and hole pockets in an underdoped high-T_c superconductor, *Phys. Rev. B* **81**, p. 214524, doi:10.1103/PhysRevB.81.214524, URL https://link.aps.org/doi/10.1103/PhysRevB.81.214524.

Seeger, K. (2013). *Semiconductor Physics: An Introduction*, Advanced Texts in Physics (Springer Berlin Heidelberg), ISBN 9783662050255, URL https://books.google.com/books?id=LLboCAAAQBAJ.

Seng, P., Diehl, J., Klimm, S., Horn, S., Tidecks, R., Samwer, K., Hänsel, H. and Gross, R. (1995). Hall effect and magnetoresistance in $Nd_{1.85}Ce_{0.15}CuO_{4-\delta}$ films, *Phys. Rev. B* **52**, pp. 3071–3074, doi:10.1103/PhysRevB.52.3071, URL https://link.aps.org/doi/10.1103/PhysRevB.52.3071.

Sera, M., Shamoto, S., and Sato, M. (1988). Anisotropic thermoelectric powers of YBa2Cu3O7- δ and (La1- x) 2CuO4 single crystals, *Solid State Commun.* **68**, 7, pp. 649–654.

Shafer, M. W., Penney, T., and Olson, B. L. (1987). Correlation of T_c with hole concentration in $La_{2-x}Sr_xCuO_{4-\delta}$ superconductors, *Phys. Rev. B* **36**, pp. 4047–4050, doi:10.1103/PhysRevB.36.4047, URL https://link.aps.org/doi/10.1103/PhysRevB.36.4047.

Shaked, H. (1994). *Crystal Structures of the High-T_c Superconducting Copper-Oxides*, Physica: Superconductivity (Elsevier Science Publishers B.V.).

Shen, Z.-X., and Dessau, D. (1995). Electronic structure and photoemission studies of late transition-metal oxides—Mott insulators and high-temperature superconductors, *Phys. Rep.* **253**, 1, pp. 1–162, doi: https://doi.org/10.1016/0370-1573(95)80001-A, URL http://www.sciencedirect.com/science/article/pii/037015739580001A.

Shi, L., Schmadel, D., Drew, H., Tsukada, I., and Ando, Y. (2005). Infrared Hall effect in underdoped and optimally doped LSCO, *arXiv preprint cond-mat/0510794* .

Singh, J., Birkedal, D., Lyssenko, V., and Hvam, J. M. (1996). Binding energy of two-dimensional biexcitons, *Phys. Rev. B* **53**, 23, p. 15909.

Slater, J. C. (1937). Wave functions in a periodic potential, *Phys. Rev.* **51**, pp. 846–851, doi:10.1103/PhysRev.51.846, URL https://link.aps.org/doi/10.1103/PhysRev.51.846.

Slocombe, D. R., Kuznetsov, V. L., Grochala, W., Williams, R. J. P., and Edwards, P. P. (2015). Superconductivity in transition metals, *Philosophical Transactions of the Royal Society of London A: Mathematical, Physical and Engineering Sciences* **373**, 2037, doi:10.1098/rsta.2014.0476, URL http://rsta.royalsocietypublishing.org/content/373/2037/20140476.

Slonczewski, J. C., and Weiss, P. R. (1958). Band structure of graphite, *Phys. Rev.* **109**, pp. 272–279, doi:10.1103/PhysRev.109.272, URL https://link.aps.org/doi/10.1103/PhysRev.109.272.

Soule, D., and Mccluke, J. (1959). Band structure and transport properties of single-crystal graphite, *J. Phys. Chem. Solids* **8**, pp. 29–35.

Soule, D. E., McClure, J. W., and Smith, L. B. (1964). Study of the shubnikov-de haas effect. Determination of the Fermi surfaces in graphite, *Phys. Rev.* **134**, pp. A453–A470, doi:10.1103/PhysRev.134.A453, URL https://link.aps.org/doi/10.1103/PhysRev.134.A453.

Sparlin, D. M., and Marcus, J. A. (1966). Empirical Fermi-surface parameters for w and mo, *Phys. Rev.* **144**, pp. 484–494, doi:10.1103/PhysRev.144.484, URL https://link.aps.org/doi/10.1103/PhysRev.144.484.

Stébé, B., Munschy, G., Stauffer, L., Dujardin, F., and Murat, J. (1997). Excitonic trion X^- in semiconductor quantum wells, *Phys. Rev. B* **56**, pp. 12454–12461, doi:10.1103/PhysRevB.56.12454, URL https://link.aps.org/doi/10.1103/PhysRevB.56.12454.

Stojković, B. P., and Pines, D. (1996). Anomalous Hall effect in $YBa_2Cu_3O_7$, *Phys. Rev. Lett.* **76**, pp. 811–814, doi:10.1103/PhysRevLett.76.811, URL https://link.aps.org/doi/10.1103/PhysRevLett.76.811.

Stojković, B. P., and Pines, D. (1997). Theory of the longitudinal and Hall conductivities of the cuprate superconductors, *Phys. Rev. B* **55**, pp. 8576–8595, doi:10.1103/PhysRevB.55.8576, URL https://link.aps.org/doi/10.1103/PhysRevB.55.8576.

Storey, J., and Tallon, J. (2012). Two-component electron fluid in underdoped high-T_c cuprate superconductors, *EPL (Europhys. Lett.)* **98**, 1, p. 17011.

Suzuki, M. (1989). Hall coefficients and optical properties of $La_{2-x}Sr_xCuO_4$ single-crystal thin films, *Phys. Rev. B* **39**, pp. 2312–2321, doi:10.1103/PhysRevB.39.2312, URL https://link.aps.org/doi/10.1103/PhysRevB.39.2312.

Suzuki, Y., and Varga, K. (1998). *Stochastic Variational Approach to Quantum-Mechanical Few-Body Problems*, Vol. 54 (Springer Science & Business Media).

Takagi, H., Ido, T., Ishibashi, S., Uota, M., Uchida, S., and Tokura, Y. (1989). Superconductor-to-nonsuperconductor transition in $(La_{1-x}Sr_x)_2CuO_4$ as investigated by transport and magnetic measurements, *Phys. Rev. B* **40**, pp. 2254–2261, doi:10.1103/PhysRevB.40.2254, URL https://link.aps.org/doi/10.1103/PhysRevB.40.2254.

Takayama-Muromachi, E., and Rice, D. E. (1991). Oxygen deficiency in the superconducting oxides $(La_{1-x}M_x)_2CuO_4$(M=Sr, Ba), *Physica C* **177**, pp. 195–206, doi:10.1016/0921-4534(91)90318-S.

Tamasaku, K., Ito, T., Takagi, H., and Uchida, S. (1994). Interplane charge dynamics in La 2- x Sr x CuO 4, *Phys. Rev. Lett.* **72**, 19, p. 3088.

Tanigaki, K., Ebbesen, T., Saito, S., Mizuki, J., Tsai, J., Kubo, Y., and Kuroshima, S. (1991). Superconductivity at 33 K in CsxRbyC60, *Nature* **352**, 6332, pp. 222–223.

Tanner, D. B., and Timusk, T. (2013). *Optical Properties of High-Temperature Superconductors* (World Scientific), pp. 363–469, doi:10.1142/9789814439688_0005, URL https://www.worldscientific.com/doi/abs/10.1142/9789814439688_0005.

Teitel'baum, G. B. (2017). Phenomenological view at the two-component physics of cuprates, *JETP Lett.* **106**, 3, pp. 199–208.

Thilagam, A. (1997). Two-dimensional charged-exciton complexes, *Phys. Rev. B* **55**, pp. 7804–7808, doi:10.1103/PhysRevB.55.7804, URL https://link.aps.org/doi/10.1103/PhysRevB.55.7804.

Timusk, T., Herr, S. L., Kamarás, K., Porter, C. D., Tanner, D. B., Bonn, D. A., Garrett, J. D., Stager, C. V., Greedan, J. E., and Reedyk, M. (1988). Infrared studies of ab-plane oriented oxide superconductors, *Phys. Rev. B* **38**, pp. 6683–6688, doi:10.1103/PhysRevB.38.6683, URL https://link.aps.org/doi/10.1103/PhysRevB.38.6683.

Timusk, T., and Statt, B. (1999). The pseudogap in high-temperature superconductors: An experimental survey, *Rep. Prog. Phys.* **62**, pp. 61–122, doi:10.1088/0034-4885/62/1/002.

Tinkham, M. (2004). *Introduction to Superconductivity*, Dover Books on Physics Series (Dover Publications), ISBN 9780486134727, URL https://books.google.com/books?id=VpUk3NfwDIkC.

Torrance, J. B., Tokura, Y., Nazzal, A. I., Bezinge, A., Huang, T. C., and Parkin, S. S. P. (1988). Anomalous disappearance of high-T_c superconductivity at high hole concentration in metallic $La_{2-x}Sr_xCuO_4$, *Phys. Rev. Lett.* **61**, pp. 1127–1130, doi:10.1103/PhysRevLett.61.1127, URL https://link.aps.org/doi/10.1103/PhysRevLett.61.1127.

Tsuei, C. C., and Kirtley, J. R. (2000). Pairing symmetry in cuprate superconductors, *Rev. Mod. Phys.* **72**, pp. 969–1016, doi:10.1103/RevModPhys.72.969, URL https://link.aps.org/doi/10.1103/RevModPhys.72.969.

Uchida, S., Eisaki, H., and Tajima, S. (1993). Electron correlation and optical conductivity in high-T_c copper oxides–origin of the mid-infrared absorption band, *Physica B: Condensed Matter* **186**, pp. 975–980.

Uchida, S., Ido, T., Takagi, H., Arima, T., Tokura, Y., and Tajima, S. (1991). Optical spectra of $La_{2-x}Sr_xCuO_4$: Effect of carrier doping on the electronic structure of the CuO_2 plane, *Phys. Rev. B* **43**, pp. 7942–7954, doi:10.1103/PhysRevB.43.7942, URL https://link.aps.org/doi/10.1103/PhysRevB.43.7942.

Uchida, S., Takagi, H., and Tokura, Y. (1989). Doping effect on the transport and optical properties of p-type and n-type cuprate superconductors, *Physica C* **162–164**, pp. 1677–1680, doi:https://doi.org/10.1016/0921-4534(89)90877-0, URL http://www.sciencedirect.com/science/article/pii/0921453489908770.

Uemura, Y., Keren, A., Le, L., Luke, G., Wu, W., Kubo, Y., Manako, T., Shimakawa, Y., Subramanian, M., Cobb, J. et al. (1993). Magnetic-field penetration depth in TI2Ba2CuO6+ δ in the overdoped regime, *Nature* **364**, 6438, p. 605.

Uemura, Y. J., Luke, G. M., Sternlieb, B. J., Brewer, J. H., Carolan, J. F., Hardy, W. N., Kadono, R., Kempton, J. R., Kiefl, R. F., Kreitzman, S. R., Mulhern, P., Riseman, T. M., Williams, D. L., Yang, B. X., Uchida, S., Takagi, H., Gopalakrishnan, J., Sleight, A. W., Subramanian, M. A., Chien, C. L., Cieplak, M. Z., Xiao, G., Lee, V. Y., Statt, B. W., Stronach, C. E., Kossler, W. J., and Yu, X. H. (1989). Universal correlations between T_c and $\frac{n_s}{m^*}$ (carrier density over effective mass) in high-T_c cuprate superconductors, *Phys. Rev. Lett.* **62**, pp. 2317–2320, doi:10.1103/PhysRevLett.62.2317, URL https://link.aps.org/doi/10.1103/PhysRevLett.62.2317.

Valla, T., Fedorov, A. V., Johnson, P. D., Li, Q., Gu, G. D., and Koshizuka, N. (2000). Temperature dependent scattering rates at the Fermi surface of optimally doped $Bi_2Sr_2CaCu_2O_{8+\delta}$, *Phys. Rev. Lett.* **85**, pp. 828–831, doi: 10.1103/PhysRevLett.85.828, URL https://link.aps.org/doi/10.1103/PhysRevLett.85.828.

Valla, T., Fedorov, A. V., Johnson, P. D., Wells, B. O., Hulbert, S. L., Li, Q., Gu, G. D., and Koshizuka, N. (1999). Evidence for quantum critical behavior in the optimally doped cuprate Bi2Sr2CaCu2O8+δ, *Science* **285**, 5436, pp. 2110–2113, doi:10.1126/science.285.5436.2110, URL http://science.sciencemag.org/content/285/5436/2110.

Van Harlingen, D. J. (1995). Phase-sensitive tests of the symmetry of the pairing state in the high-temperature superconductors—evidence for $d_{x^2-y^2}$ symmetry, *Rev. Mod. Phys.* **67**, pp. 515–535, doi: 10.1103/RevModPhys.67.515, URL https://link.aps.org/doi/10.1103/RevModPhys.67.515.

van Veenendaal, M. A., Sawatzky, G. A., and Groen, W. A. (1994). Electronic structure of $Bi_2Sr_2Ca_{1-x}Y_xCu_2O_{8+\delta}$: Cu 2p x-ray-photoelectron spectra and occupied and unoccupied low-energy states, *Phys. Rev. B* **49**, pp. 1407–1416, doi:10.1103/PhysRevB.49.1407, URL https://link.aps.org/doi/10.1103/PhysRevB.49.1407.

Varga, K. (2008). Solution of few-body problems with the stochastic variational method II: Two-dimensional systems, *Comput. Phys. Commun.* **179**, 8, pp. 591–596, doi:https://doi.org/10.1016/j.cpc.2007.07.015, URL http://www.sciencedirect.com/science/article/pii/S0010465508001926.

Varma, C. M., Littlewood, P. B., Schmitt-Rink, S., Abrahams, E., and Ruckenstein, A. E. (1989). Phenomenology of the normal state of Cu-O high-temperature superconductors, *Phys. Rev. Lett.* **63**, pp. 1996–1999, doi:10.1103/PhysRevLett.63.1996, URL https://link.aps.org/doi/10.1103/PhysRevLett.63.1996.

Varyukhin, S., and ZAkharov, A. (1991). Anisotropic dielectric constant of La2CuO4, *Physica C* **185-189**, pp. 975–976, doi:https://doi.org/10.1016/0921-4534(91)91711-C, URL http://www.sciencedirect.com/science/article/pii/092145349191711C.

Vinattieri, A., Shah, J., Damen, T. C., Kim, D. S., Pfeiffer, L. N., Maialle, M. Z., and Sham, L. J. (1994). Exciton dynamics in GaAs quantum wells under resonant excitation, *Phys. Rev. B* **50**, pp. 10868–10879, doi: 10.1103/PhysRevB.50.10868, URL https://link.aps.org/doi/10.1103/PhysRevB.50.10868.

Viskadourakis, Z., Sunku, S., Mukherjee, S., Andersen, B., Ito, T., Sasagawa, T., and Panagopoulos, C. (2015). Ferroelectricity in underdoped La-based cuprates, *Scientific reports* **5**, p. 15268.

von Barth, U., and Hedin, L. (1972). A local exchange-correlation potential for the spin polarized case: I, **5**.

Walters, A., Howard, C., Upton, M., Dean, M., Alatas, A., Leu, B., Ellerby, M., McMorrow, D., Hill, J., Calandra, M. et al. (2011). Comparative study of the phonons in nonsuperconducting bac 6 and superconducting cac 6 using inelastic x-ray scattering, *Physical Review B* **84**, 1, p. 014511.

Wang, Z. D., Dong, J., and Ting, C. S. (1994). Unified theory of mixed state Hall effect in type-II superconductors: Scaling behavior and sign reversal, *Phys. Rev. Lett.* **72**, pp. 3875–3878, doi:10.1103/PhysRevLett.72.3875, URL https://link.aps.org/doi/10.1103/PhysRevLett.72.3875.

Wang, Z. Z., Chien, T. R., Ong, N. P., Tarascon, J. M., and Wang, E. (1991). Positive Hall coefficient observed in single-crystal $Nd_{2-x}Ce_xCuO_{4-\delta}$ at low temperatures, *Phys. Rev. B* **43**, pp. 3020–3025, doi:10.1103/PhysRevB.43.3020, URL https://link.aps.org/doi/10.1103/PhysRevB.43.3020.

Weber, M., Birrer, P., Gygax, F., Hitti, B., Lippelt, E., Maletta, H., and Schenck, A. (1991). Measurements of the London penetration depth in Bi-based high-T_c compounds, *Hyperfine Interactions* **63**, 1-4, pp. 93–101.

Weber, W., Shelankov, A., and Zotos, X. (1988). A Cu d-d excitation model for the pairing in the high-T_c cuprates, *Physica C* **153–155**, pp. 1305–1306, doi:https://doi.org/10.1016/0921-4534(88)90293-6, URL http://www.sciencedirect.com/science/article/pii/0921453488902936, Proceedings of the International Conference on High Temperature Superconductors and Materials and Mechanisms of Superconductivity Part II.

Williamson, S., Foner, S., and Dresselhaus, M. (1965). De Haas-van Alphen effect in pyrolytic and single-crystal graphite, *Physical Review* **140**, 4A, p. A1429.

Williamson, S., Foner, S., and Dresselhaus, M. (1966). De Haas-van Alphen effect in pyrolytic and single crystal graphite, *Carbon* **4**, 1, pp. 29–40.

Wulin, D., Guo, H., Chien, C.-C., and Levin, K. (2012). Two-component optical conductivity in the cuprates: A necessary consequence of preformed pairs, *Phys. Rev. B* **86**, p. 134518, doi:10.1103/PhysRevB.86.134518, URL https://link.aps.org/doi/10.1103/PhysRevB.86.134518.

Xing, D. Y., and Liu, M. (1991). Single-band model of normal-state transport properties of high-T_c copper oxides, *Phys. Rev. B* **43**, pp. 3744–3747, doi:10.1103/PhysRevB.43.3744, URL https://link.aps.org/doi/10.1103/PhysRevB.43.3744.

Xing, D. Y., and Ting, C. S. (1988). Two-band model for anisotropic Hall effect in high-T_c Y-Ba-Cu-O, *Phys. Rev. B* **38**, pp. 5134–5137, doi:10.1103/PhysRevB.38.5134, URL https://link.aps.org/doi/10.1103/PhysRevB.38.5134.

Xue, M., Cao, T., Wang, D., Wu, Y., Yang, H., Dong, X., He, J., Li, F., and Chen, G. (2012). Superconductivity above 30 k in alkali-metal-doped hydrocarbon, *Scientific reports* **2**, p. srep00389.

Yang, C. N. (1962). Concept of off-diagonal long-range order and the quantum phases of liquid He and of superconductors, *Rev. Mod. Phys.* **34**, pp. 694–704, doi:10.1103/RevModPhys.34.694, URL https://link.aps.org/doi/10.1103/RevModPhys.34.694.

Yang, C. N. (1989). η pairing and off-diagonal long-range order in a Hubbard model, *Phys. Rev. Lett.* **63**, pp. 2144–2147, doi:10.1103/PhysRevLett.63.2144, URL https://link.aps.org/doi/10.1103/PhysRevLett.63.2144.

Yang, C. N., and Zhang, S. (1990). So4 symmetry in a Hubbard model, *Modern Phys. Lett. B* **04**, 11, pp. 759–766, doi: 10.1142/ S0217984990000933, URL https://www.worldscientific.com/doi/abs/10.1142/S0217984990000933.

Yin, Z. P., Lebègue, S., Han, M. J., Neal, B. P., Savrasov, S. Y., and Pickett, W. E. (2008). Electron-hole symmetry and magnetic coupling in antiferromagnetic LaFeAsO, *Phys. Rev. Lett.* **101**, p. 047001, doi: 10.1103/PhysRevLett.101.047001, URL https://link.aps.org/doi/10.1103/PhysRevLett.101.047001.

Yoshida, T., Zhou, X. J., Nakamura, M., Kellar, S. A., Bogdanov, P. V., Lu, E. D., Lanzara, A., Hussain, Z., Ino, A., Mizokawa, T., Fujimori, A., Eisaki, H., Kim, C., Shen, Z.-X., Kakeshita, T., and Uchida, S. (2001). Electronlike Fermi surface and remnant $(\pi, 0)$ feature in overdoped $La_{1.78}Sr_{0.22}CuO_4$, *Phys. Rev. B* **63**, p. 220501, doi:10.1103/PhysRevB.63.220501, URL https://link.aps.org/doi/10.1103/PhysRevB.63.220501.

Yoshida, T., Zhou, X. J., Sasagawa, T., Yang, W. L., Bogdanov, P. V., Lanzara, A., Hussain, Z., Mizokawa, T., Fujimori, A., Eisaki, H., Shen, Z.-X., Kakeshita, T., and Uchida, S. (2003). Metallic behavior of lightly doped $La_{2-x}Sr_xCuO_4$ with a fermi surface forming an arc, *Phys. Rev. Lett.* **91**, p. 027001, doi:10.1103/PhysRevLett.91.027001, URL https://link.aps.org/doi/10.1103/PhysRevLett.91.027001.

Zhang, F. C., and Rice, T. M. (1988). Effective hamiltonian for the superconducting Cu oxides, *Phys. Rev. B* **37**, pp. 3759–3761, doi:10.1103/PhysRevB.37.3759, URL https://link.aps.org/doi/10.1103/PhysRevB.37.3759.

Zhou, J.-S., and Goodenough, J. B. (1995). Thermoelectric power in single-layer copper oxides, *Phys. Rev. B* **51**, pp. 3104–3115, doi:10.1103/PhysRevB.51.3104, URL https://link.aps.org/doi/10.1103/PhysRevB.51.3104.

Zhu, J. X. (2008). Private communication.

Ziman, J. (1960). *Electrons and Phonons: The Theory of Transport Phenomena in Solids*, International Series of Monographs on Physics (OUP Oxford), ISBN 9780198507796, URL https://books.google.com/books?id=UtEy63pjngsC.

Index